# Millennials, Spirituality and Tourism

Edited by Sandeep Kumar Walia and Aruditya Jasrotia

Routledge
Taylor & Francis Group
LONDON AND NEW YORK

First published 2022
by Routledge
2 Park Square, Milton Park, Abingdon, Oxon OX14 4RN

and by Routledge
605 Third Avenue, New York, NY 10158

*Routledge is an imprint of the Taylor & Francis Group, an informa business*



*British Library Cataloguing-in-Publication Data*
A catalogue record for this book is available from the British Library

*Library of Congress Cataloging-in-Publication Data*
Names: Walia, Sandeep, editor. | Jasrotia, Aruditya, editor.
Title: Millennials, spirituality and tourism / Edited by Sandeep Kumar Walia & Aruditya Jasrotia.
Description: Abingdon, Oxon ; New York, NY : Routledge, 2022. | Series: Routledge insights in tourism series | Includes bibliographical references and index.
Identifiers: LCCN 2021025443 (print) | LCCN 2021025444 (ebook) | ISBN 9780367772826 (hardback) | ISBN 9780367772857 (paperback) | ISBN 9781003170631 (ebook)
Subjects: LCSH: Spiritual tourism. | Generation Y--Social conditions. | Tourism--Management.
Classification: LCC G156.5.S65 M55 2022 (print) | LCC G156.5.S65 (ebook) | DDC 306.4/819--dc23/eng/20211001
LC record available at https://lccn.loc.gov/2021025443
LC ebook record available at https://lccn.loc.gov/2021025444

ISBN: 978-0-367-77282-6 (hbk)
ISBN: 978-0-367-77285-7 (pbk)
ISBN: 978-1-003-17063-1 (ebk)

DOI: 10.4324/9781003170631

Typeset in Times New Roman
by SPi Technologies India Pvt Ltd (Straive)

***Dedicated to our Parents and Family***

# Contents

# Figures

# Tables

# Contributors

**Ann Suwaree Ashton** graduated with a PhD and master's degree from the University of Queensland, Australia, and Bachelor degree in Physical Education, Kasetsart University, Thailand. She is currently working as the Associate Dean for Administration at the Graduation School of Tourism Management, National Institute of Development Administration (NIDA), Bangkok Thailand. She has experience working in the hotel industry and spent the 2008-2010 school years as a teaching assistant at the University of Queensland in Australia and universities in New Zealand. Her area of interest are; wellness tourism; spiritual tourism; spa and beauty tourism; sport for health tourism; long distance running destination development; slow food tourism; street food tourism; destination image development; international retirement tourism development; active aging wellness activity experience design, and hotel restaurant co-branding. Ann is Editor in Chief of International and Thai Tourism.

**Pedro Ricardo Coelho de Azevedo** holds a degree in History and Master in Heritage and Cultural Tourism by the University of Minho. He is currently a researcher at CETRAD (Centre for Transdisciplinary Development Studies) of University of Trás-os-Montes and Alto Douro (Portugal). His current research interests focus on the Ways of Saint James in the region of Trás-os-Montes and Alto Douro (Northern interior of Portugal).

**Gül Erkol Bayram** is associate Professor in the School of Tourism and Hotel Management, Department of Tour Guiding, at the University of Sinop where she has been a faculty member since 2016. She is also head of the tour guiding department. ERKOL BAYRAM completed her Ph.D. at Sakarya University and his undergraduate studies at Balıkesir University. She has also been a professional tour guide for ten years. His research interests lie in the area of tourism management, policy and planning, organizational behaviour, tour guiding.

**Anna Carr** is an Associate Professor and co-director of the Centre for Recreation Research at the Department of Tourism, University of Otago, New Zealand. Before academia, Anna was the owner-operator of two adventure tourism businesses and worked for the Department of Conservation.

Her current research interests focus on protected areas, cultural landscapes, and indigenous development.

**Neil Carr** is a Professor in the Department of Tourism at the University of Otago, Dunedin, New Zealand. His research focuses on understanding behaviour within tourism and leisure experiences; with a particular emphasis on children and families, sex, and animals. Having completed his PhD in 1998 at the University of Exeter, he has since gone on to publish over 100 journal articles and book chapters, as well as several books and edited books. In addition, he has supervised over 40 PhD students.

**Inge Yan Fong Chan** was born and raised in Hong Kong. She obtained her Bachelor from Translation and Interpretation Studies at Chang Jung Christian University, Tainan, Taiwan R. O. C. Currently, she is a Master student of Tourism Management at Budapest Metropolitan University in Hungary and working on payments and receivables in Avis Budget Group. Her thesis topic focuses on sustainable tourism of festivals.

**Kevser Çınar** holds PhD in Tourism Management from Necmettin Erbakan University, Turkey. She is Assistant Professor at the Faculty of Tourism in the same university. She is certified in curriculum development program by the Institute of Educational Sciences. She is also the vice president of Euro-Asia Tourism Studies Association. She has also been working as an EU Projects expert for 14 years. She has won several European Projects grants including TOURBAN, TOURIBOOST, STARTUPS, SCOUT and she has organized several international tourism events such as HACKATOUR Konya, Science Centers as Tourism Assets and the 5th EATSA International Tourism Conference. She was also a TEDx speaker at TEDxDenHelder. Her research interests include consumer behaviour, innovation and branding.

**Shaheema Hameed** is UGC NET JRF and works as Assistant Professor (Management), Banasthali Vidyapith, India. She is MBA, PhD majoring in Finance and Human Resources. Her doctoral research focused on the managerial competencies of the Generation Z. She has presented and published her research work in various national and international forums. Her areas of specialization are Generation Z, Human Resource Management, Organizational and Financial Behavior and Business Communication. She has developed course content on student engagement under the aegis of the National Resource Centre, Ministry of Human Resource Development, India. She has authored book chapters and research papers in Emerald and Springer on Generation Z. She is also an accredited trainer of English to speakers of other languages.

**Diego Hannon Ovies** is presently working as a consultant analyst at Nenenki, a consulting firm that assists public and private entities in designing sustainable tourist products in indigenous and rural areas. He has a bachelor's degree in Tourism and a master's degree in Local Development and Territorial Studies by the University of Guadalajara. He is still conducting independent research in association with the Department of Geography of the University of Guadalajara with the aim of building a robust curriculum to apply for a PhD. His area of research interest lies at the intersection between cultural tourism and endogenous development in rural and indigenous communities. He has authored 2 book chapters and 2 research papers in scientific journals

**Ankur Joshi** is Assistant Professor at FMS-WISDOM, Banasthali Vidyapith. He is Fellow of MDI-Gurugram in Public Policy and Governance. He has contributed as co-coordinator of National Resource Center (Management) at Banasthali for developing online refresher course in management under ARPIT, MHRD project. He is a researcher at OMRISE research group set up in collaboration with Banasthali with Buurtzorg and Praan Group, The Netherlands. He has authored research papers on the theme of Gurukul education, social entrepreneurship, sustainable development and indigenous research. He is also Visiting Faculty of Hinduism and Spirituality at Vrije University, The Netherlands

**Aruditya Jasrotia** is presently working as an Assistant Professor in Amity Institute of Travel and Tourism at Amity University, Greater Noida, India. He completed his Ph. D. from the Department of Tourism and Travel Management from Central University of Jammu, Jammu and Kashmir, India. His research interests are smart tourism, smart tourism destinations, tourism planning, community-based tourism, technology in tourism and sustainable tourism. He has published 04 research papers and 03 book chapters in various reputed journals and books.

**Lina Munirah Kamarudin** is currently working as a Senior Lecturer at School of Tourism, Hospitality and Event Management (STHEM), Universiti Utara Malaysia (UUM), Kedah, Malaysia. She also holds a position of an Advisor for Tourism Management Society in UUM. Her research focuses on tourism management studies related to tourist motivation, tourism system, religious tourism and Muslim tourism. Lina has written and presented various papers locally and internationally in regards to spiritual and religious tourism. As millennial is one of the important groups in sustaining future tourism development, her current research is to study the millennial's preferences to travel at religious attribute destinations

**Sıla Karacaoğlu** is Assistant Professor in the Bilecik s,eyh Edebali University Faculty of Applied Sciences, department of Tourism Guiding. She received her PhD degree from the Mersin University in Tourism Management in 2017. In her doctoral dissertation, she investigates "Community Perceptions, Attitudes and Support for Community Based Tourism: The Case of Misi Village". Her research interests are tourism marketing, sustainable tourism, cultural heritage and tourist guiding.

**Melanie Kay Smith** is an Associate Professor, Researcher and Consultant whose work focuses on urban planning, cultural tourism, wellness tourism experiences and the relationship between tourism and wellbeing. She is

Programme Leader for BSc and MSc Tourism Management at Budapest Metropolitan University in Hungary. She has lectured in the UK, Hungary, Estonia, Germany, Austria and Switzerland as well as being an invited Keynote speaker in many countries worldwide. She was Chair of ATLAS (Association for Tourism and Leisure Education) for seven years and has undertaken consultancy work for UNWTO and ETC as well as regional and national projects on cultural and health tourism. She is the author or editor of 15 books as well as more than 80 journal articles or book chapters. Her most recent research focuses on wellness tourism, spa experience creation, the impacts of 'over tourism' in cultural cities and the changing nature of cultural tourism.

**Robert Kiss** is an Assistant Professor of the Department of International Tourism and Hospitality of I-Shou University, Kaohsiung, Taiwan R. O. C. He was the former Head of the School of Tourism, Leisure and Hospitality of the Budapest Metropolitan University, Budapest, Hungary. He obtained his PhD from Tourism Geography and prior to that he graduated from Glion Institute of Higher Education and he also holds three further Masters degrees in the fields of education, teaching and business. His international research and publications concentrate on tourism management, including leisure, recreation and sport tourism, tourism geography, niche and SIT tourism, consumer behaviour and generational issues. Within his international teaching activities, he was a visiting lecturer at Andrzej Frycz Modrzewski Krakow University and had teaching experiences in six other European universities in different countries. He started his professional career as a Manager of the DMO Baranya County, Hungary, representing and coordinating that destination's tourism industry, including decision makers, stakeholders, entrepreneurs, before turning towards academia.

**Reni Polus** is a PhD candidate in the Department of Tourism at the University of Otago, Dunedin, New Zealand. Her research interest is focused on spirituality within tourism and leisure experiences particularly in volunteering, pilgrimage, and heritage tourism.

**Wai Ching Poon** is an Associate Professor of Economics and the Director of Graduate Research Programs at the School of Business, Monash University Malaysia. She is an Editor for Cogent Economics and Finance, and serves in Editorial Boards for Corporate Governance: An International Review, and Management and Organization Review. Her key research areas are Business and Financial Economics, and Sustainable Development. She has published actively in high-quality mainstream and multidisciplinary journals, such as Energy Economics, Journal of Contemporary Accounting and Economics, Pacific-Basin Finance Journal, International Review of Finance, Scientometrics, International Business Review, Journal of Cross-Cultural Psychology, Personality and Individual Differences, etc. She is the top 25% economist in Malaysia, ranked by RePEc. She has secured 18 external research fundings from 2011-2020. Of these, she is the Principal Investigator for 5 of the 18 external grants from Malaysia, UK and Japan. She has been honoured numerous scholarly teaching, research, and reviewer awards.

**Zahra Rahmani** is a PhD student at the Department of Tourism, University of Otago, New Zealand. She is working on virtual retreats and psychological well-being as her PhD thesis. Her research interests include wellness retreats, positive tourism, tourist experience, well-being, and happiness.

**Raymond Rastegar** joined the University of Queensland in 2010 after working on several projects on tourism development, community empowerment and environmental conservation in developing countries. He completed his PhD in the field of tourism management at the University of Queensland. Raymond's research focuses on justice and sustainability in tourism, hospitality and events. His areas of interest include the impact of tourism, tourism policy and planning, and tourism sustainability. Recent research projects have investigated the impacts of tourism at local level and delivered new insights into the tourism phenomenon to advocate a more just and sustainable tourism future. He has developed and taught different postgraduate and undergraduate courses both in Australian universities and overseas. In 2019, Raymond was recognised as a Fellow of the Higher Education Academy (UK) for teaching and supporting learning in higher education. Raymond is also a member of the CAUTHE 2021 scientific committee. He has published in top academic journals including Annals of Tourism Research and Journal of Sustainable Tourism and participated in many conferences.

**Johnnel Smith** is a PhD Scholar at Griffith University, Australia, and the University of Otago, New Zealand. She has been a Lecturer in Tourism, Management and Marketing with Universities in the Caribbean and Pacific regions for over 13 years. She lectured at the University of Otago, University of the West Indies (UWI), the University of the Commonwealth Caribbean and the Caribbean Maritime University. She was also a Research Advisor and Consultant for the U.W.I Centre for Tourism Research and Policy Development and the University of Technology- School of Hospitality and Tourism Management in Jamaica. Miss Smith holds a Master's and a Bachelor's degree (honors) in Tourism and Management from the University of the West Indies. Her scope of research include: Tourism and COVID-19, Sustainable Development Goals, Indigenous Tourism, Gender and Luxury Management. She is also a three- time recipient of the Caribbean Tourism Organization Foundation- Graduate Scholarship Award. In addition to her achievements in academia, she has also worked extensively in the consumer goods and tourism industries with several international brands.

**Sandeep Kumar Walia** is presently working as Associate Professor and Head of Department in School of Hotel Management and Tourism at Lovely Professional University, Jalandhar, Punjab, India. He is an alumnus of Himachal Pradesh University Shimla and IGNOU, New Delhi. Sandeep is associated as Visiting Faculty to School of Tourism, Hospitality and Event Management, Universiti Utara Malaysia Kedah (A QS ranked University),

Malaysia and Himalayan Institutes of Alternative Learning (An upcoming Mountain University) in Leh, Laddakh as Curriculum Development / Board of Studies Expert and Visiting Faculty. His areas of interest are community-based tourism, tourism marketing and cultural tourism. He has worked as Program Director/ Paper Coordinator for the subject Aviation Industry – Ticketing and Frontier Formalities and developed E-content under MHRD-Sponsored UGC Project titled EPG Pathshala for the online learning of Post Graduate Level. He has also acted as a faculty coordinator for Peer to Peer: Facebook Global Digital Challenge Spring 2017. Sandeep Kumar Walia is Founder Editor in Chief of Tourism and Hospitality Research Review (International Journal of LPU), Managing Editor & Editor of 'Tourism Dimensions' & 'Tourism Spectrum' (International Refereed Journals). He has authored 9 books with various International Publishers and 18 Research papers in various International Journals. He has recently published *The Routledge Handbook of Community-Based Tourism Management* . His forthcoming book titles include *Social Debates in Tourism with Good fellow publishers, Millennials and Tourism: A New Fangled Trend for the Tourism Industry*. Sandeep Kumar Walia has also Guest Edited the Special Issues of International *Journal of Religious Tourism and Pilgrimage* and *Journal of Hospitality, Leisure, Sport & Tourism Education.* He is currently Guest Editing Special Issues of *Foresight* and *International Journal of Spa and Wellness.*

**Hasibe Yazıt** is presently working at Sinop University in School of Tourism and Hotel Management. She has graduated from Kocaeli University undergraduate education in tourism in 2009 and has done her master program about tourism management from Sakarya University. She has completed her PhD from Çanakkale University doctorate program about tourism business. She has worked as a lecturer in Tekirdağ Namık Kemal University in tourism and hotel management.

**Seda Yetimoğlu** is a Research Fellow of the Faculty of Tourism at Necmettin Erbakan University in Konya, Turkey. She completed her Bachelor degree from the Department of Tourist Guiding, College of Tourism and Hotel Management at Erciyes University. After her Bachelor degree, she worked as a Freelance Tour Guide in the tourism industry for some time. She received her Master degree from the Department of Tourism Management at Necmettin Erbakan University. She is currently a PhD candidate. Her research interests include tourism marketing, experiential marketing, tour guiding, tourist and consumer behaviour and cultural and heritage tourism development and gastronomy tourism.

**Zohreh (Zara) Zarezadeh** has completed her second PhD in Technology and Tourism Behavior at the University of Queensland recently. Her research interests include the role of technology in tourism marketing and tourism behavior. Her recent projects focus on tourism online information search behaviour, social media marketing and personalization. Zara has teaching experience in both overseas universities and the University of Queensland

# Acknowledgements

To acknowledge, write and express gratitude seems to be our toughest task. The book *Millennials, Spirituality and Tourism* would have remained a dream without the support of all these people to whom we would like to express and extend our gratitude and appreciation in this page. We would like to extend our sincere gratitude to all our academic colleagues around the globe who have contributed the masterpiece of their unmatched authentic research in this book. We are thankful to all of them for their continued, sincere, committed, time-bound and generous support. They incorporated the suggested comments and modifications with full zeal and vigor within the specified timelines.

We treat this section as to place our sincere gratitude to Faye Leerink, Commissioning Editor for her consistent help and guidance to complete this handbook. We are also thankful to Nonita Saha, Editorial Assistant of Taylor & Francis, for her kind support. We would also like to express our word of appreciation for all other members of Taylor & Francis editorial, production and marketing teams for successful execution of the book.

We do not have any words to express our debt of gratitude to our parents and family members who have always been there for us.

*Sandeep Kumar Walia & Aruditya*

# Preface

The Millennial generation is discarding the philosophies and concept of organized travel which provides a rule-book set in stone, as the concept of having an individualized spiritual experience allows them a sense of control. Furthermore, in academics, the Millennial generation has recently been receiving a lot of attention and exposure. Many of the web searches, research publications, blogs, online articles and so on are focusing on the characteristics, travel behaviour, and travel patterns of Millennials. There is an ongoing emphasis on understanding the ambitions, goals, aspirations, beliefs, actions, hobbies, health and lifestyle of millennials. Studies showcase that in the 21st century Millennials are in search of mental peace and an escape from their daily lives through spirituality and spiritual travel and tourism practices. Through spiritual tourism, the Millennials get the opportunity to educate, reflect and exercise spirituality in a way that is not usually available in day-to-day life. The spiritual tourism is an amalgamation of spirituality and tourism. It is a gateway for the young tourists to attain brilliance, insights and enlightenment. Spirituality provides knowledge by providing opportunity to acquire new information while performing activities under the concept of spiritual tourism. It can be said that spiritual tourism is the kind of tourism which offers persistence and purpose in a tourist's life.

To this end, tourism has emerged as a noteworthy activity which has contributed towards the growth of the spiritual marketplace. This is because of the contemporary revolutions in the lifestyle of Millennials and the ways in which they are seeking peace and state of happiness in life. Spiritual tourism refers to travel for finding meaning and purpose in your life. It uplifts your mental, physical and emotional dynamisms. Moreover, it develops, sustains and enhances the synchronization between one's body and spirit. Spiritual tourism is not linked with a particular religion. It is not the same as religious tourism. Spiritual tourism aims to connect your body, mind and soul. On the other hand, religious tourism aims to search for blessings from God. Therefore, spiritual tourism is a wide-ranging perspective in comparison to religious tourism. Religious tourism is a subset of spiritual tourism. In a nutshell, spiritual tourism connects your body, mind and soul. Undoubtedly, spiritual tourism is going to be an emergent field and spirituality will become a key driver for tourism.

The young tourists, also referred to as Millennials, are prophesied to be the travellers of the forthcoming age. Modern tourism research talks intensively of young people, also known as Generation Y or Millennials. Millennials are a significant section of the international tourism and travel industry. Modern Millennials are inclined towards unexpected solo trips relating to interaction with nature, intercultural exchange, meditation, wellness, spiritual healing, active engagement and immersion in the local practices and lifestyle. The significance of faith, spirituality, belief and religion in human life will always remain the key reason behind the inclination of modern travellers towards spirituality and allied practices. This book will be one of a kind as there is dearth of books which address the growing inclination of Millennials towards spirituality. This book will help to understand Millennials and how the new generation is embracing new ways of finding peace and solace in this fast-moving world.

To reflect upon the various aspects of spirituality and Millennials, this edited collection addresses broad areas concerning spiritual tourism as a pathway to connect with body, mind and soul, inclinations of youngsters towards spirituality, holistic tourism and the future of spiritual tourism. The contributions cover a variety of themes: spirituality and tourism: defining the linkages from literature; theorizing the tourism terminology in spirituality among Millennials; globalization and the predicament of spirituality in tourism among Millennials; the anthropological view of spiritual tourism in the Millennial world; travel motivations and experiences among Millennials; linking spirituality to wellness; health and medical travel prospects among Millennials; the impact of spiritual tourism on work–life balance and the quality of life of Millennials; tourist behaviour and travel adoption among Millennials; spiritual tourism: a pathway for learning sustainability among Millennials; determinants of spiritual tourism rejuvenation among Millennials, the role of digitization in spiritual tourism adoption, marketing and management of spiritual tourism destinations with special emphasis on Millennials; issues, threats and prospects of spiritual tourism in the emerging era; analysis of policies and practices for spiritual tourism development; the strategic approach to spiritual destination branding among Millennials and spiritual tourism: finding space in the future. In order to quality contributions, experts were invited from diverse backgrounds, to contribute to the existing body of knowledge on linkages between spiritual tourism and Millennials. Hence, in the present book an effort has been made to gather the knowledge and expertise from all over the world and to consolidate it in the form of this edited collection.

The purpose of this edited book is to bring together high-quality research that focuses on new knowledge and understanding on the concept of spirituality, spiritual tourism and allied terminologies which has the power to influence the lives of Millennials. To contribute in this respect, this edited book aims to offer an extensive collection of papers regarding the Millennials and spirituality and the emergence of spiritual tourism in the global tourism industry. Divided into four parts, the chapters provide new perspectives and approaches towards spiritual tourism which are based on qualitative, quantitative and mixed methodologies. The four parts cover theoretical and practical aspects on spiritual tourism and key lessons through various destination cases from all over the globe. As an edited book, it will add on new research and knowledge base with high-quality contributions from researchers and practitioners. This book will provide an exhaustive source of information and seeks to inspire leaders, academicians, researchers and students at the graduate and pos-graduate level so as to equip them with knowledge about spirituality and spiritual tourism. This edited book will support the existing body of knowledge about spirituality and Millennials, not only in tourism but also in other allied disciplines.

Sandeep Kumar Walia
Aruditya Jasrotia

# Part I

# Spiritual tourism

## Pathway to connect body, mind, and soul

The book opens with Part I, which introduces the reader to the concepts of spirituality and tourism. It contains four chapters, which illustrate the conceptual and theoretical base of spirituality and its associated linkages from tourism perspectives. It further elaborates on the different conceptual models and theoretical backgrounds of spiritual tourism and the influence of spirituality on the Millennial generation. This part also aims to theorize the tourism terminology in spirituality among Millennials, the predicament of spirituality in tourism among Millennials and an anthropological view of spiritual tourism in the Millennial world. The objective of this part is to provide readers with insights into the minds of Millennials and how this new generation is embracing new ways of finding peace and solace in this fast-moving world.

The first chapter, titled "Spirituality and Tourism: Defining the linkages from literature", written by Kevser Çınar and Seda Yetimoglu, provides an updated literature review of this important topic, presenting spirituality from the point of view of the tourism industry and individuals involved in many activities. Another focus of this research is to clarify the notion that spirituality in tourism is not just composed of religious tourism, journey or faith. Although they are used interchangeably, it is aimed to raise awareness that each of them is different at certain points.

Chapter 2, "The Quest For Exploring Self: Generation Z and Spiritual Tourism In India", written by Shaheema Hameed and Ankur Joshi, discussed that, along with the spiritual element which the Millennial craves, spiritual tourism in India is a niche part in the tourism industry. This chapter stressed that in the post-COVID world, tourism will be the one of the most affected sectors; however, this will be only in the short term and everything is expected to be normal in due course.

Similarly, Chapter 3, "Influence of Spirituality on Generation Y and Z Working in the Tourism & Hospitality Industry" by Hasibe Yazıt, draws on the perspectives of earlier conceptual papers to establish theoretical and practical gaps that need to be filled about this issue. This chapter aims to identify the influence of spirituality on Generations Y and Z working in the

DOI: 10.4324/9781003170631-1

tourism and hospitality industry. Today's human being seeks something beyond materiality in his quest to give meaning to his life, because, despite the use of the most advanced and modern technologies, the sense of monotony and discontent with life in his daily life leads him to find a way out of the crisis in which he finds himself.

Moving ahead, Chapter 4, "Millennials and spiritual journeys: the ways of Saint James in Northern interior Portugal as a contemporary pilgrimage", written by Pedro Azevedo, adopts the designation of tourist-pilgrim because it unites the concepts of tourist and pilgrim. The objective of this study is to demonstrate the importance of the Ways as a pilgrimage route and that it is increasingly visited by tourists and pilgrims, especially young people, namely Millennials.

# 1 Spirituality and tourism

## Defining the linkages from literature

*Seda Yetimoğlu and Kevser Çınar*

### 1.1 Introduction

Discovering and achieving the meaning of life is one of the main goals of human creation. Whenever a person feels a void inside, their main problem is the absence of spirituality. Factors such as historical events, technological innovations, global socio-cultural changes, economic changes, terrorism, natural disasters, and epidemics that have ravaged the world over the past year have caused people to experience both moral and material pressures. These events have not only have changed the daily state of affairs; but they also have disrupted the rhythms of life, a sense of loss regarding both time and place, corruption of culture and social alienation, the collapse of lifestyles, increasing risk, personal insecurity and ambiguous life changes and a trend toward increased skepticism (Beck, 1992). Therefore, people want to move away from the factors that are having a negative impact on their spirituality. This may prove to be a solution to meet the psychological and social challenges arising from contemporary lifestyles (Marjani & Pourjam, 2016:101).

At this point, tourism emerges as the most important sector to meet the needs of these people. The tourism industry provides the opportunity to meet the need for both spiritual relaxation and an opportunity for people to rest physically. The phenomenon of tourism has always been seen as a process whereby people renew themselves, as well as relaxing, educating themselves and becoming more tolerant (Ryan, 1997). According to Seaton and Bennett (1996), tourism motivation is more about mental than physical refreshment. Fosarelli (2002:207) explains the concept of spirituality in tourism as "paying attention to people's emotional and spiritual lives in addition to physical care and looking for new ways to make themselves healthier".

Research studies suggest that travel reinforces one of the most fundamental and effective components in the development of one's spiritual intelligence which helps it to comprehend spirituality in his own life. Lloyd (1998) states that the increase in spirituality motivated by travel has been a significant driver in the development of modern-age tourism. For people seeking calmness in life, travel – considered in its broader perspective – has been an

DOI: 10.4324/9781003170631-2

important event with spirituality coming to the forefront (Heelas & Woodhead, 2005). These people are generally the tourists who seek out religious or spiritual settings for the purpose of fulfilling their desire to travel, either in whole or in part, and to have some form of religious or spiritual experience.

The sense of meaning and inner integrity of these journeys, embedded identity and ritual experience can lead to the attainment of spiritual benefits and spiritual satisfaction. It can be argued that religious and spiritually motivated travel to a sacred place, or one with considerable spiritual significance, is always important; individuals have undertaken missions for thousands of years to worship the Gods or fulfill personal religious responsibilities. The opportunity that contemplative journeys provide to discover inner solace, purpose, clarity, or redemption has propelled travellers to launch transformative voyages throughout the ages. This travel not only helps the pursuit of the knowledge, connection and understanding of other cultures and religions, but also goes much deeper, gain philosophical experience and enhance the pursuit of spirituality.

This research provides an updated literature review of this important topic, presenting spirituality from the point of view of the tourism industry and individuals involved in many activities. Another focus of the current research is to clarify that the issue of spirituality in tourism is not just composed of religious tourism, journey or faith. Although they are used interchangeably, its intention is to raise awareness that each of them is different at certain points. From another point of view there is as yet undiscovered spiritual value hidden at many points on the earth. Although there are substantial tourist flows to these destinations, there are few studies that will bring these values to the forefront and bring them to the literature. Therefore, a literature review of the concepts related to the subject has been included in as far as it deals with them in terms of general concepts of spiritualism and in the position of the tourism industry.

## 1.2 The Conceptual framework of spirituality

The word 'spirit' in English is a word that derives from the Latin word "spiritus", meaning 'soul (psyche), courage, breath of life' (Principe, 1983) and vigor, strength', which is often confused with different concepts or sometimes used interchangeably. In many sources, the word 'spirit' generally refers to a person's soul. Two different opinions have been made about spirituality. The first is the connotation that human existence exists in the universe and the desire to find its purpose. The other is faith in God's domination that controls the entire universe (Hunsberger & Jackson, 2005). Marcoen (1994:527) has shown a similar approach to this and he defended spirituality as a relationship with something greater than internality, arguing that it was associated with either a religious dimension or an unknown experience (see Figure 1.1).

Although there are very different definitions made about spirituality, each of them emphasized different points of the concept. The most common and early definition of spirituality is an individual's experience of inner peace or

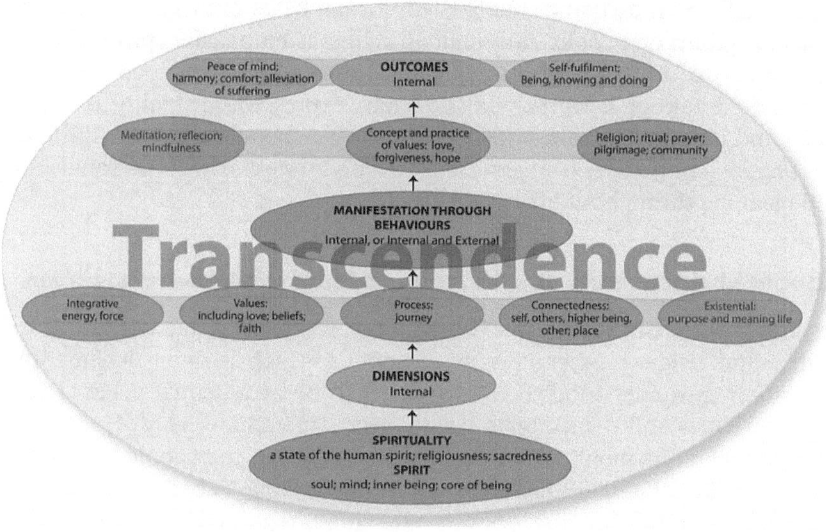


*Figure 1.1* Conceptual map of spirituality.

changing life behaviour. A spiritual follower seeks out experiences that make them blissful and peaceful, at the end of the duration, an encounter with 'the other' that is greater than one's individual self. The term 'the other' doesn't delineate with religious terms. It can be something natural or supernatural, objective or subjective. Spiritual coincidence is not something deliberate. On the contrary, this brings meanings to people' lives. This meaning is deep, detailed, abstract, powerful, unidentifiable and sincere (Kasim, 2016:118).

It has been remarked that the experience of being involved or in contact with an "other" that transcends one's individual sense of self gives one's life a deeper meaning than the intellectual level. Things should be seen in a distinctive light. The feelings of being overpowered by a powerful and enjoyable experience. According to Moal-Ulvoas and Taylor (2014), spirituality has a meaning with life, and define it as "individual construction of the meaning of one's life, the dimensions of which are the inner self, alterity (other humans and nature) and the sacred". Spirituality actually includes the acceptance of a feeling, sense or belief and idea which is something greater than ourselves. The term 'spirituality' includes discovering some universal subjects, such as love, wisdom, truth, compassion, altruism and life after death, and with the knowledge which some people, such as the saints and the holy ones, obtained and showed greater and higher levels of progress than the ordinary people. Spirituality is a broad concept with many units in terms of many perspectives.

O'Brien (2003) defined spirituality as an exact attendance in life, stated for awareness and attention to metaphysics, a peerless and connective principle in the universe. Although spirituality is confused with formal faith or religious systems, it is actually a universal phenomenon and not limited to the others. Religion is used and ritualized for formal belief events and fulfilled by a certain group of people. Although all religions emphasise spiritualism as being a part of faith, you can be spiritual without being religious (Delgado, 2005:158). Laukhof & Werner (1998:62) have expressed spirituality as "a personal and individual value system about the way people approach life" and this phenomenon varies from person to person by seeking their personal aims and meaning throughout life.

## 1.3 The characteristics of spirituality and how it differs from religion

Various views are available over the nature of any relationship between spirituality and religion. Spirituality is something which is often debated and commonly misunderstood. It is considered to be of an equal value as religious activities and events, or to use the terms interchangeably (Gorsuch & Walker, 2006; Piedmont, 2001), but the other part of the people believe the reverse. Hill et al. (2000) note that, according to some people, spirituality is included within religion; for the others, religion is regarded as a dimension of spirituality (Nolan & Crawford, 1997). In many research studies the two terms (spirituality and religion) are separated into explicit concepts. Hill and Pargament (2003) state that the distinctness between the concepts is that 'religion' is the total of the beliefs and implementations related to religious tradition. Koenig (2008:11) defines 'religion' as:

> A system of beliefs and observed by a community, supported by rituals that acknowledge, worship, communicate with, or approach the Sacred, the Divine, God (in Western cultures), or Ultimate Truth, Reality, or Nirvana (in Eastern cultures).

On the other hand, spirituality is the focus of the meaning (Pulchalski et al., 2009:887) and is defined as:

> The aspect of humanity that refers to the way individuals seek and express meaning and purpose and the way they experience their connectedness to the moment, to self, to others, to nature, and to the significant or sacred.

Spirituality is an 'innate human characteristic' and the human spirit is 'human consciousness, a constitutive (basic) dimension' of being human.

> Spirituality, whether or not it is linked to a belief in God, struggles with the mystery of the deep questions around the meaning of human life. The human spirit remains open to the search for authentic truth within the varied experiences of life, establishes normative values, exercises reasoned judgment, and involves the entire human being on the path to self-determination.
>
> (Perrin, 2007:20–21)

The term spirituality is embodied in a broader sense, and may be nihilist, materialist, humanist, or religious. The starting point of spirituality can be an effort to seek absolute truth and self-focused meaning, power, success, money, pleasure and life. Gray, Rukumnuaykit, Kittisuksathit, and Thongthai (2008) state that compassion, morality, forgiveness, hope and relations to a higher power and ancestors to the holy dead are regarded to spirituality.

Spirituality is related to the soul and inner life of man. Being spiritual includes one's personal beliefs, behaviours and search for the meaning of life. The spirituality view of each person shows an alteration throughout their life, according to self-realization, personal efforts and experiences with little connection to religion or accompanying belief in a God (Weathers, McCarthy, & Coffey, 2015). The basis of spirituality lies in the discovery of the power that one inherits in oneself to fight against all the challenges of life. Religion means sharing the same belief system and performing worship in an organized manner. The beliefs of the people are often passed on from members to converts, and are based on either a formally documented sect or founded cultural practices (Cragun, Hammer, & Nielsen, 2015; Zwingmann, Klein, & Bussing, 2011). One of the ways to approach the uncontrolled environment is religion. People's purpose in life is to experience the deep power that cannot be controlled (De, 2011). As an absolute power, God is given different names in each language: Allah, God, Yahweh, Zeus, Brahman. From this point of view, religion is considered to be an organized approach of human behaviour to supernatural power and covers many beliefs, worships and teachings. Giving God different names in each religion means establishing a strong bond with him and allows us to benefit from God's mercy. This shows that religion contributes to understanding the meaning of life and expressing community values.

Many researchers have long struggled to distinguish between psyche (the psychological soul) and pneuma (the religious spirit). Although the spiritual approach is seen as related to religion (Cragun et al., 2015:8), it is actually separated from religiosity in some aspects and includes a broader concept. A certain organized group in religiosity or those who live that culture are in search of a certain number of doctrines and teachings. The group at issue generally believes in a certain way of life within the boundaries of decency. However, spirituality makes religiosity tangible. Because religion is open to all signs from God but it also explores the meaning and significance of everything. This is a personal search due to the larger purpose of one's own existence. As human beings who adopt spiritualism also give up the drive to control the lifestyle and beliefs of others, they approach everything with more love and respect, and establish a favourable relationship with their body, self, other people, nature, and the world (Weathers et al., 2015). They also tend to feel that all of the images of the Gods symbolizing religions and of God existing forever are outward expressions of one's own inner feelings (Kasim, 2016: 119).

Spirituality is an important criterion for a patient or a person under pressure to attain good for himself and it deals with one's integrity

(Winslow & Wehtje Winslow, 2007: 63), and it is essentially experienced and expressed. The spiritual depth differs from person to person. This situation varies according to the person's ethnicity, culture, and also the environment in which they live. This situation can be different even within the same culture. For instance, the spirituality of a Christian differs from the perspective of a Tibetan priest, an Australian native, a Buddhist or an Atheist in terms of spiritualism and the way in which they experience it (Lepherd, 2015: 566). Many people confuse spirituality with religion and therefore bring pre-existing thoughts about the influence of religion to discussions about spirituality. In these studies the term 'religion' means both spirituality (the concept of transcendence and meaning of life) and religiousness (indicating faith and devotion for God) (Mandhouj, Perroud, Hasler, Younes, & Huguelet, 2016: 861). While all religions emphasize spirituality as a significant part of faith, it is possible to be "spiritual" without being a part of an organized religious community. Accepting the difference between spirituality and religion can be a great way to start understanding what spirituality means to different people. Despite the differences between both terms, they give motive and structure to people' lives with sacred and meaningful rituals (Campbell & Miles, 2008a) or meditation (Lewis & Seponski, 2012).

## 1.4 Types of spirituality

From a spiritual angle, the main purpose of human life is to grow. This often means to becoming less self-serving and materialistic and more respectful to the assumed spiritual space. The precise features of spirituality that supply this purpose and meaning is different for every individual. These differences seem to be related to various people's personalities. These differences in personality may affect the types of spiritual or religious interests which an individual pursues. It can be seen that there are many types of spirituality on which people sometimes base their beliefs, and many different reasons for people to practice spirituality.

**Authoritarian spirituality** is a particularly strong form of spirituality based on the needs for definitions and rules. It has a hierarchical structure of things, set of rules and some restrictions. This spirituality form is especially associated with precise religious applications. The ones who adopt the authoritarian spirituality type may develop a form of fundamentalist religion. The followers of authoritarian spirituality prefer to adopt the rules of their religion and have a good relationship with God (Altemeyer, 1996; Stern, 2003).

**Mystical spirituality** is founded around the desire to go beyond the material world, beyond senses, ego and even time and it concentrates on individuals' connections, a feeling of integration, and the intuitional part of the soul. According to Keirsey (1998), mystical spiritualists have a tendency to feel individual experiences of the transcendent rather than authority. Every experience goes beyond the material and physical world, and everything may be brought into one greater unity.

**Intellectual spirituality** is based on consisting knowledge and understanding of spirituality through analyzing history, spiritual theories and writings. Intellectual spirituality may be found in the study of theology, religious beliefs and history.

**Service spirituality** is the most common form of spiritual expression to serve to others. People experience spiritual peace when they serve others. This is why helping other people without expecting anything in return. To do something for someone without having something is a most usual method for people to contact with their own spiritual selves.

**Social spirituality** is a common form of spirituality in many religious faiths. It is predominantly established on serving others as a form of spiritual expression. Attendance in a religious or spiritual group is a social activity or event that attract people.

**Devotional/religious spirituality** is the remembrance of deities and involves rituals or mantras. Devotions are defined as "love, loyalty or enthusiasm for a person, activity or cause". Devotional spirituality focuses on forming a relationship of trust with God and the love of God leads to the surrender of the soul to the Divine. The ultimate scope of devotion is to attain a permanent proximity to God. Devotions and rituals, if done in the proper manner, might lead to spiritual gains. Devotion takes different shapes in all religions as a way of feeling the existence of God. The semah of Sufis, the use of Catholics' prayer beads, the chanting of both Protestants and Buddhists, meditating of Orthodox congregation on icons, gathering of Hindus to be blessed in temples, Jews wrapping themselves in a prayer shawl and the pilgrimage of Muslims are all substantial processes by which religious believers aim to get to know God better.

## 1.5 The emergence of spiritual tourism

Human beings and societies have always been engaged with a spiritualist aspect of life since the beginnings of human civilization. Spiritualism is not a new phenomenon, but it now has a new content (Ambroz & Ovsenik, 2011), although it is seen for different purposes such as healing and atonement within tourism. It is based on a history as old as religion and constitutes the oldest type of tourism (Hung, 2013; Rinschede, 1992). Visiting religiously holy places such as Mecca, Vatican, Bethlehem (Jerusalem), and India for centuries shows that spiritual purposes were also dominant in those times. Spiritual flows and journeys associated with tourism supply a detailed assessment of the primary issues and concepts related to the intersection of tourism and religion.

Spiritualism is a universal experience, and it is unique to every individual. The spiritual approach in tourism refers to an activity for people to escape voluntarily from daily life at their own request rather than a real reason for travel (Ardakani & Solemimanabadi, 2019). Spiritual tourism is a journey to a holy place or shrine or temple that allows a person to believe, as well as a quest to find their own essence. This type of tourism, in contrast to the principle of 'tourism behaviour is temporary and short-term', aims to provide the tourist with experiences that can be valid not only during his or her journey, but also throughout his or her life. Because spiritual tourism is based on travel that allows the inner world and self to be discovered rather than acting with religious beliefs (Vargheese & Vargheese, 2012). According to the results of the research on spirituality and tourism conducted by Willson, McIntosh, and Aahra (2013):

- A journey has a spiritual content if it is accompanied by intuition, perception, peace, acceptance and compassion.
- The tangible experiences of the journey should be transcendent and energetic in a way that even seeing children and ordinary people can have significant meanings.
- Experiencing evolving moments and spiritual connections on the journey (understanding transcendence and connection)

Travelling has always been an escape from everyday life, offering individuals the chance to take a break from routine and having new and unknown experiences. Myths, arts, festivals and religions make what is lost in the routine cycle accessible and perceptible (Herntrei & Pechlaner, 2011). At the heart of spiritual tourism, the concept of "the more difficult the journey, the greater the reward obtained" which expresses the conventional concept of a desire for a very comprehensive change from the comfort of daily life lies (Phukan, Rahman, & ve Devdutt, 2012). For this reason, spiritual tourism is, at its core, a drive to find the inner self by experiencing difficult and ordinary life (see Figure 1.2). This desire to find oneself is

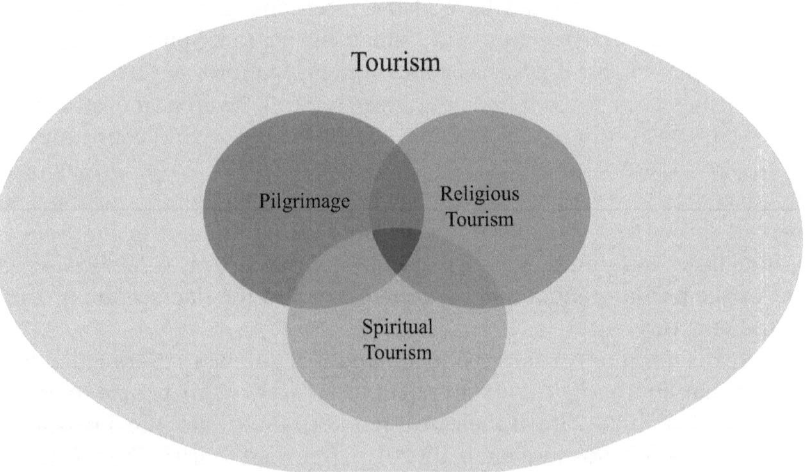


*Figure 1.2* Spiritual tourism in relation to religious tourism and pilgrimage traditions.

addressed from not only a religious point of view, but also a spiritual point of view. A spiritualism is not always characterized by religion; it is also closely related to nature or trips to the countryside, recreation, culture and health.

Spiritual tourism, which includes all these areas, gives tourists the opportunity to shape their own lives by staying alone with themselves. Contrary to what is known incorrectly, spiritualism is not always associated with religion, and trips to nature are also closely related to recreational activities and health. In the research conducted by Heidari, Yazdani, Saghafi, and Jalilvand (2017) religious tourism is defined as the visit of holy places, to attendance in religious ceremonies and/or events and take part in the pilgrimage in the form of visit and to fulfill religious duties. As religious site visits or pilgrimages are generally realized by organized groups or family members (Heidari, Yazdani, Saghafi, & Jalilvand, 2018:748), spiritual travels are preferred to participate alone or a couple of people together. Because of the increasing problems and crises around the world, people have sought to be alone in order to comfort and purify themselves in terms of spirituality and travel to spiritual and cultural sites has increased spectacularly in recent years (Phukan, 2014; Phukan et al., 2012). A visit to a spiritual site (especially to second and third world countries) – and one which mostly attracts foreign visitors (Schnell, 2012) – is often an aesthetic experience for the person who gain emotional, physical, intellectual or spiritual levels. Lehto, Brown, Chen, and Morrison (2006) note that spirituality motivates the tourists to visit certain places, but different practices have also become a motivation for travelling, as individual and unique experiences are central to contemporary spirituality.

Spiritual tourism has a concept beyond religious tourism. The majority of the researchers believe that spiritual tourism cannot be categorized under religious tourism. This is because spiritualism is like an umbrella term that can cover different types of tourism itself, resulting in the creation of a sense of excellence, elevation and sacred values. However, the scientific literature on religious and pilgrimage tourism concentrated on the general and outer elements of tourism experiences (Willson et al., 2013). Religious tourism and pilgrimage often focuses on actions and beliefs performed in holy places and one of the most important features of these people is their loyalty to these cultural sites they visit (Figure 1.3). The divisions among pilgrimage, religious tourism and spiritual tourism are actually hard to distinguish as different types of travellers are stimulated by the wish to visit sites and experience moments during their journey that will bring meaning to their lives (Collins-Kreiner, 2018). As some spiritual tourists and seekers may begin their journeys with similar religious beliefs, spiritual tourism can also include non-religious travellers and seekers based upon their experiential intentions (Griffin & Raj, 2017). Xin and Yang (2017) state that spiritual tourism consists of a wider concept of a traveller's wish for change and relief from their routine lives. So, various individuals experience similar activities in different ways.

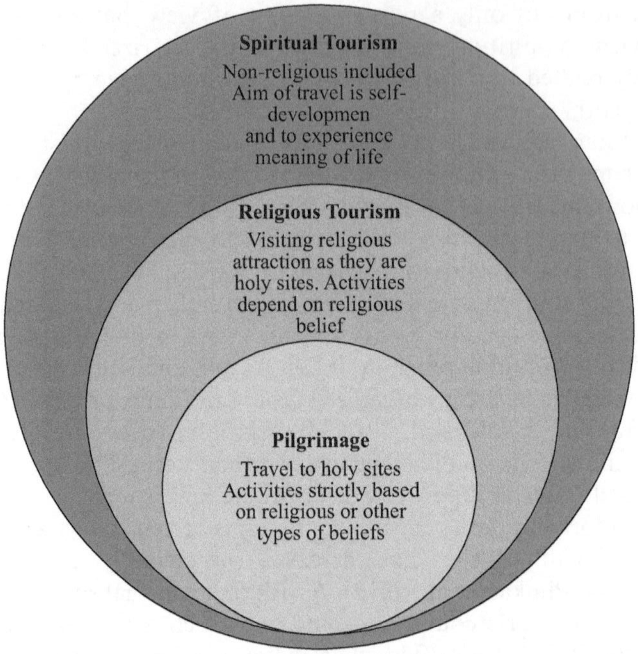


*Figure 1.3* The differences among spiritual tourism, religious tourism and pilgrimage.

In order for a place to be spiritual, it must have unusual features rather than being surrounded by only religious factors. This situation requires the destination to have its own characteristics and aura. These centres of attraction are places of magical gardens, cliffs, interesting and unusual structures, contrary to what is seen everywhere. Although it is not enough that only sacred places and temples come to mind to experience spiritual tourism activities, purging therapies such as yoga and meditation, gaining fitness with the aim of medical treatment are also among the examples of spiritual tourism. Özdemir, Acar & Dinçer (2018:595) explain that when considering the locations where spiritual tourism is witnessed. These include several iconic sights: Nepal, famous for its Hindu and Buddhism temples; India, the heart of Hinduism; the holy land of the Islamic world Mecca; and Lourdes, which is important for Christianity. The Bahai Gardens in France, along with Jerusalem and Machu Picchu, are also renowned.

There are differences between the Western and Eastern worlds about the approach to spiritual life. In the West, this approach is a phenomenon that is implemented by the spiritual guidance of religions. By contrast, in the East, many practices, such as yoga, meditation, aimed at health and wellness, not only with religion, have spread to many regions of the world. There are various faiths and religions around the world. As the number of destinations with a diversity of religious beliefs and natural wealth increases, so do known spiritual areas. According to these religions, especially with regard to the lands where religions originated, are home to sacred sites. Among these, Mecca, Jerusalem, Uluru (Australia), Nepal and India are the best-known spiritual places. Collins-Kreiner (2010), Norman (2011), Timothy and Olsen (2006), Timothy and Conover (2006) state that a large majority of the trips to these destinations are now recognized as an important aspect of spiritual tourism and the prominence of the places have been gaining much more value.

India is now regarded as one of the most preferred spiritual destinations. Sacred sites, such as Haridwar and Rishikesh in northern India, in particular, retain this importance. The Indian state of Uttarakhand is another major spiritual destination, as it promotes spiritual tourism through both its religious diversity and its historical heritage (Phukan, 2014). India shows significant potential for the development of spiritual tourism as mentioned. According to Banerjee (2013), the spiritual features unique to India are:

- Ashrams (Reclusion/Yoga houses): These are usually built beautiful or quieter natural surroundings. They are visited to deepen their understanding the natural world. Ashram residents often take up activities, such as yoga, music or meditation.
- Guru (Indian sage/Clergy): The role of the guru changes according to different religious practices. However, it is equally respected. Gurus are to be found in ashrams, but there is no need to reside in a family to follow a particular guru.
- Yoga: This is a way of life that offers guidelines on behaviours and beliefs, and *asanas* (poses) offer one small aspect of these guidelines. Yoga is done by people of many different religions and beliefs, although they originate from farewells (ancient Hindu scripts) and Hindu religion.
- Visiting temples: The movement of spiritualism seen in Japan is known as the 'Spiritual World', similar to India. Tenkawa Benzaitensha in Japan is a famous pilgrimage site of the spiritual world, but it is an ancient shrine found in Central Japan in the Yamato mountains. Since it became an important shrine in the 1980s, it has become the centre of the Japanese spiritualism movement, equipped with mystical forces (Shimazono, 1999).

Tenkawa Benzaitensha in Japan is a famous pilgrimage site of the spiritual world, but it is an ancient shrine found in Central Japan in the Yamato mountains. Since becoming an important shrine in the 1980s, it has grown to be the centre of the Japanese spiritualism movement equipped with mystical powers (Shimazono, 1999). Among the spiritual destinations of Japan, its temples, which are intimate with nature, are quite famous. The most prominent are Todai-ji, Golden Pavillion Zen, Kiyamizu, Kinkkuji, Ginkaku-ji, Nanzen-ji temples. Due to the presence of Buddhism, Shintoism and Christianity in Japan, it is common to encounter a large number of spiritual places (Özdemir, Acar & Dinçer, 2018:599).

Apart from being a tourist centre, Nepal also has both religious and spiritual characteristics. It has a spiritual energy with its unique nature and hidden aura, and is considered an important centre of pilgrimage and worship by people of the Buddhist and Hindu faiths. Through the use of esoteric elements such as yoga, meditation, astrology and ayurveda, Nepal is one of the outstanding spiritual destinations on earth. More than half of the tourists visiting Nepal travel there to strengthen their spirituality and find their own identity. The effect of the high mountains around Nepal on tourists is that they are alone with God and themselves, allowing them to make their own internal reckoning. Another location in the world, Peru – one of the most popular spiritual tourism destinations in South America – attracts a lot of tourists around the world because of its values. Peruvian people especially travel to the coastal areas, where traditional medical practices are performed. This destination grows the spiritual tourism market and combines the spiritual understanding of the people with the search for healing (Joralemon & Douglas, 1993). The busiest spiritual cities in Peru are the cities of Cusco, Iquitos and Machu Picchu. With the sites at Stonehenge as well as the Avebury Circle, England is seen as the centre of the ancient priest tradition, and priests and pagans worshipped at daybreak in Stonehenge as a symbol of respect for the Sun.

From the point of view of Islam, spiritual tourism can include visiting shrines, tombs, sacred places and holy people. Timothy and Iverson (2006) divide these practices in Islam into three: Hajj/Umrah, journey of knowledge, and visit. Hajj and umrah are a sacred worship performed by Muslims by visiting Mecca to fulfill their duties of pilgrimage financially and physically and Muslims consider it an Islamic spiritual tourism experience. In terms of Turkey, spiritual tourism is not widely remembered, but spiritualism is trending with a number of activities. Spiritual tourism movements in Turkey are mainly the experience of both cultural and religious values. These are reflected in the Seb-i Arus ceremonies, demonstrations by whirling dervishes and sufism ceremonies, where Maulawi life is still shown to be vibrant. In addition, the visit of tombs, tombs and mosques is another spiritual experience in Turkey. Turkey not only provides resources to spiritual tourism from a religious point of view, but also stands out for its natural beauty. With the camping activities that have been increasingly popular in recent years, individuals reinvent themselves by being alone with themselves in nature and are free of what can be called physical and mental rebirth. Yoga, chakra therapies, meditation, ayurveda, theta healing, and going on a silent retreat. Spiritual, physical and mental purification practices, such as spending a week without ever speaking and other relaxation techniques (whose origins extend to the Far East and Asian geography) and wellness practices, whose attraction has increased in recent years, are also different dimensions of spiritual gains. Turkey is home to many civilizations, as it is a location where many religions intersect. It not only allows domestic and foreign visitors to visit places such as mosques and shrines, but also offers the opportunity to experience places that are significant to their religion.

## 1.6 Conclusion

Spiritualism has always existed since the birth of civilization, but this phenomenon has been difficult to notice. Different practices began to emerge with the spiritual experience, which has self-discovery at its core, and the subsequent taking place of this fact in the tourism sector. While evaluating the literature in this field, there are many areas closely related to spiritual tourism. Religion, health and culture are the three main important phenomena gathered under the spiritual umbrella. Religion is considered the most important of these. People have been trying to find their own self by devoting themselves to their religion and worshipping, and feeling closer to God. While seeking the divine from a holy places (churches, mosques etc.) their feelings of happiness, inner peace and appreciation for the life multiply. Religion and spirituality are neither the same thing nor completely separate from one another.

In modern life, people experience spiritual tourism by engaging in different pursuits not only with the phenomenon of religion, but also in health and culture. In addition, in spiritual tourism, people are searching for inner self in places where they can spend time intertwined with nature alone and retreat. In spirituality, one may ponder, "Where do I find meaning in life, and what are my true values?" Meanwhile, in religion, common questions might be "What is the truth?" and "What is right and wrong?" While religion aims to build one's character and shapes one's beliefs, attitudes, and actions by giving importance to the adherence of rules, spirituality concentrates more on each person's individual soul and the quality of the message that is being offered than the differences in the details of the original story.

Spiritual journeys accompany the search for the meaning of life for travellers who want to move away from modern living conditions and bring together cultural discoveries about themselves and the world. Travellers can connect more deeply with nature on such trips, assimilating local culture, belief systems and traditions, and transfer their experience and spiritual awakening back into their own world. Thus, they find happiness, serenity and peace. These experiences obtained with spiritual tourism should be considered as a "journey" rather than reaching the final destination. Spiritual tourism can bring important values to societies with guidance that takes into account the principles of responsible and sustainable development. it is very clear that the best tool for a peaceful future is spiritual trips that increase tolerance for differences.

In order to realize of spiritual tourism, the main purpose of tourism planning is to provide confidence that there are elements necessary for tourists to experience spiritual holidays. Many countries that provide income from spiritual tourism have also made spiritual tourism a market within their own countries by creating this necessary trust environment. Every country or region with religious, cultural values and health opportunities is a potential destination for spiritual tourism. It is seen that destinations that effectively promote cultural and belief values and create a sense of trust in tourists are also spiritually valued.

## References


Ambroz, M., & Ovsenik, R. (2011). Tourist origin and spiritual motives, *Management*, 16(2), 71–86.

Ardakani, K. S., & Solemimanabadi, F. (2019). A methodology study of the concept of spirituality in tourism according to fuzzy logic, *International Journal of Tourism and Spirituality*, 3(2), 85–105.

Beck, U. (1992). *Risk society: Towards a new modernity*. London: Sage.

Campbell, L., & Miles, M. S. (2008a). Implementing parenting programs for custodial grandparents. In B. Hayslip Jr. & P. Kaminski (Eds.), *Parenting the custodial grandparent: Implications for clinical practice* (pp. 115–131). New York: Springer.

Campbell, L., & Miles, M. S. (2008b). Implementing parenting programs for custodial grandparents. In B. Hayslip Jr. & P. Kaminski (Eds.), *Parenting the custodial grandparent: Implications for clinical practice* (pp.115–131). New York: Springer.

Collins-Kreiner, N. (2010). The geography of pilgrimage and tourism: Transformations and implications for applied geography, *Applied Geography*, 30, 153–164.

Collins-Kreiner, N. (2018). Pilgrimage-Tourism: Common themes in different religions, *International Journal of Religious Tourism and Pilgrimage*, 6(1), 8–17.

Cragun, R. T., Hammer, J. H., & Nielsen, M. (2015). The NonReligious–NonSpiritual scale (NRNSS): Measuring everyone from atheists to Zionists, *Science Religion and Culture*, 2, 36–53.

Dam, A., Khanna, P., Kumar, N., & Banerjee, A. (2013, November). Measure of Hope & Spirituality in End Stage Renal Disease. In *Psycho-Oncology* (Vol. 22, pp. 176–176). Hoboken, NJ: Wiley-Blackwell.

De Blot, P. (2011). *Handbook of Spirituality and Business*. London: Palgrave Macmillan.

Delgado, C. (2005). A discussion of the concept of spirituality, *Nursing Science Quarterly*, 18(2), 157–162.

Fosarelli, P. (2002). Fearfully wonderfully made: The interconnectedness of body-mind-spirit, *Journal of Religion and Health*, 41(3), 207–229.

Gorsuch, R. L., & Walker, D. (2006). Measurement and research design in studying spiritual development. In E. C. Roehlkepartain, P. E. King, L. M. Wagener, P. L. Benson (Eds.), *The handbook of spiritual development in childhood and adolescence* (pp. 207–229). Thousand Oaks, CA: Sage Publications.

Gray, R. S., Rukumnuaykit, P., Kittisuksathit, S., & Thongthai, V. (2008). Inner happiness among Thai elderly, *Journal of Cross-Cultural Gerontology*, 23, 211–224.

Griffin K. A., & Raj, R. (2017). The importance of religious tourism and pilgrimage: Reflecting on definitions, motives and data, *International Journal of Religious Tourism and Pilgrimage*, 5(3), 2–9.

Heelas, P., & Woodhead, L. (2005). *The spiritual revolution: Why religion is giving way to spirituality*. Malden, MA: Blackwell Publishing.

Heidari, A., Yazdani, H. R., Saghafi, F., & Jalilvand, M. R. (2017). Developing strategic relationships for religious tourism businesses: A systematic literature review, *EuroMed Journal of Management*, 2(1), 77–98.

Heidari, A., Yazdani, H. R., Saghafi, F., & Jalilvand, M. R. (2018). The perspective of religious and spiritual tourism research: A systematic mapping study, *Journal of Islamic Marketing*, 9(4), 748–798.

Herntrei, M., & Pechlaner. H. (2011). Spiritual tourism-The church as a partner in tourism? In R. Conrady, M. Buck (Eds.), *Trends and issues in global tourism 2010* (pp. 201–218). Berlin, Heidelberg: Springer Verlag.

Hill, P. C., & Pargament, K. I. (2003). Advances in the conceptualization and measurement of religion and spirituality, *American Psychologist*, 58, 64–74.

Hill, P. C., Pargament, K. I., Hood, R. W., McCullough, J. P., Swyers, D. B., & Zinnbauer, B. J. (2000). Conceptualizing religion and spirituality: Points of commonality, points of departure, *Journal for the Theory of Social Behaviour*, 30, 51–77.

Hung, T., (2013). *Nature as impression for Dao: A theory of spiritual tourism development- In Da Nang – Viet Nam.* Unpublished Doctorate Dissertation, University of Minnesota, Minnesota.

Hunsberger, B., & Jackson, L. M. (2005). Religion, meaning and prejudice, *Journal of Social Issues*, 61(4), 807–826.

Joralemon, D., & Douglas, S. (1993). *Sorcery and Shamanism: Curanderos and clients in Northern Peru.* Salt Lake City: University of Utah Press

Kasim, A. (2016). The spirituality of green travel and tourism, *International Journal of Travel and Tourism*, 1(1), 116–124.

Keirsey, D. (1998). *Please understand me II: Temperament, character, intelligence*, Del Mar, CA: Prometheus Nemesis Book Company.

Koenig, H. G. (2008). *Medicine, religion, and health: Where science and spirituality meet.* West Conshohocken: Templeton Foundation Press.

Laukhof, G., & Werner, H. (1998). Spirituality: The missing link, *Journal of Neuroscience Nursing*, 30(1), 60–67.

Lehto, X. Y., Brown, S., Chen, Y., & Morrison, A. M. (2006). Yoga tourism as a niche within the wellness tourism market, *Tourism Recreation Research*, 31(1), 25–35.

Lepherd, L. (2015). Spirituality: Everyone has it, but what is it? *International Journal of Nursing Practice*, 21, 566–574.

Lewis, C. D., & Seponski, M. D. (2012). Transcendent sacrifice and spirituality: Cambodian grandparents raising orphaned grandchildren, *Journal of Intergenerational Relationships*, 10(4), 355–369.

Lloyd, D.W. (1998). *Battlefield tourism.* New York: Berg

Mandhouj, O., Perroud, N., Hasler, R., Younes, N., & Huguelet, P. (2016). Characteristics of spirituality and religion among suicide attempters, *The Journal of Nervous and Mental Disease*, 204(11), 861–867.

Marcoen, A. (1994). Spirituality and personal well-being in old age, *Ageing and Society*, 14(4), 521–536.

Marjani, T., & Pourjam, A. (2016). Entrepreneurship as a tool presenting authentic experiences through spiritual tourism development, *International Journal of Tourism and Spirituality*, 1(1), 100–115.

Moal-Ulvoas, G., & Taylor, V. (2014). The spiritual benefits of travel for senior adults. *Journal of Consumer Behaviour*, 13(6), 453–462.

Nolan, P., & Crawford, P. (1997). Towards a rhetoric of spirituality in mental health care, *Journal of Advanced Nursing*, 26, 289–294.

Norman, A. (2011). *Spiritual tourism: Travel and religious practice in Western society.* London: Continuum.

O'Brien, M. E. (2003). *Spirituality in nursing: Standing on holy ground.* Sudbury, MA: Jones and Bartlett.

Özdemir Altınay M., Acar A., & Dinçer İstanbullu, F. (2018). Dünden bugüne spiritüel turizm, *Uluslararası Sosyal Araştırmalar Dergisi*, 11(58), 594–604.

Perrin, D. (2007). *Studying Christian spirituality.* London: Routledge.

Phukan, H. (2014). A study on tourism logistics in the spiritual sites of Haridwar and Rishikesh in Uttarakhand, *International Journal of Emerging Technology and Advanced Engineering*, 4(9), 165–171.

Phukan, H., Rahman, Z., & ve Devdutt, P. (2012). Emergence of Spiritual Tourism in India. *International Journal of Marketing and Technology*, 2(4), 300–317.

Piedmont, R. L. (2001). Spiritual transcendence and the scientific study of spirituality, *Journal of Rehabilitation*, 67, 4–14.

Principe, W. H. (1983). Toward defining spirituality, *Studies in Religion/Sciences Religieuses*, 12(2), 127–141.

Rinschede, G. (1992). Forms of religious tourism, *Annals of Tourism Research*, 19(1), 51–67

Ryan, C. (1997). *The tourist experience: A new introduction*. London: Cassell.

Schnell, T. (2012). Spirituality with and without religion-differential relationship with personality, *Archive for the Psychology of Religion*, 34, 33–61.

Seaton, A. V., & Bennett, M. M. (1996). *The marketing of tourism products: concepts, issues and cases*. London: International Thomson Business Press.

Shimazono, S. (1999). New age movement or new spirituality movements and culture? *Social Compass*, 46(2), 121–133.

Stern, J. (2003). *Terror in the name of God*. New York: HarperCollins.

Timothy, D., & Olsen, D. (2006). *Tourism, religion and spiritual journeys*. New York: Routledge.

Timothy, D. J., & Iverson, T. (2006). Tourism and Islam: Considerations of culture and duty. In *Tourism, religion and spiritual journeys* (pp. 202–221). London: Routledge.

Timothy, D. J., & Conover, P. J. (2006). Nature religion, self-spirituality and new age tourism. In D. Timothy & D. Olsen (Eds.), *Tourism, religion & spiritual journeys* (pp.139–155). New York: Routledge.

Vargheese, A. J. S., & Varghees, P. S. (2012). Dimensions of spiritual tourism in Tuiticorin district of Tamil Nadu in India – A critical analysis. *Business Intelligence Journal*, 5(2), 245–251.

Weathers, E., McCarthy, G., & Coffey, A. (2015). Concept analysis of spirituality: An evolutionary approach. *Nursing Forum*, 51, 79–96.

Willson, G., McIntosh, A., & Aahra, A. (2013). Tourism and spirituality: A phenomenological analysis, *Annals of Tourism Research*, 42, 150–168,

Winslow, G. R., & Wehtje Winslow, B. J. (2007). Ethical boundaries of spiritual care, *Medical journal of Australia*, 186, 63–66.

Xin, S., & Yang, Z. (2017). *A Measure of Tourism Experience for Spiritual Tourists: Scale Development and Validation, 5th Annual International Conference on Tourism & Hospitality Research Proceedings (THOR 2017)*, pp. 26–31.

Zwingmann, C., Klein, C., & Bussing, A. (2011). Measuring religiosity/spirituality: Theoretical differentiations and categorization of instruments, *Religions*, 2, 345–357.

# 2 The quest for exploring self

## Millennials and spiritual tourism in India

*Shaheema Hameed and Ankur Joshi*

### 2.1 Introduction

India, the birthplace of religions like Hinduism, Buddhism, and Sikhism and home to other major world religions such as Islam and Christianity (Majumdar, 2018), also has the distinction of possessing the world's youngest population (Agarwal & Vaghela, 2018). The reason for connecting these two threads is that, contrary to popular beliefs and researches on the Millennials that critique their inability to focus on lack of self-exploration, there are many emerging trends that require thorough research. Most research on the Millennials of India, however, has been conducted in order to study the demographic with regard to the economic domains of marketing, human resources and trends and behavior in the workplace. The technologically savvy attitude, 24*7 connectivity, diversity tolerance and the entrepreneurial mindset of the 'digital natives' are exponential value additions in the corporate scenario (Hameed & Mathur, 2016). However, studies also show that Millennials possesses negative association with this lifestyle of constantly 'being on the go'. Alexithymia, Nomophobia, stress, depression, a sense of detachment and strife within oneself (The Economist, 2019) are among the phenomena that are unique to this demographic. Thus, the concept of travelling to find peace or a sense of contentment and solace is a trend that is increasingly being observed among Millennials. The youth of India have started to recognize the importance of returning to their roots in order to establish purpose and achieve a heightened sense of spirituality and connection to self (Padmanabhan, 2019). This renewed interest of Millennials in spirituality brings opportunities to enhance tourism in India. In the post-COVID world, tourism will be the one of the most affected sectors; however, this is likely to be a short-term phenomenon with everything expected to return to normal in due course. Modern times have seen the prevalence of materialism attitudes and a limitless consumption. In contrast, with research and awareness to the concept of minimalism and an increasing in spiritual awareness at the expense of materialism, the need for spiritual vacations and travels require detailed investigation and analysis. Millennials show a rising trend towards 'off-the-beaten-path' tourism (Ketter, 2019) and favor destinations that have not yet been promoted as tourist destinations. Along

DOI: 10.4324/9781003170631-3

with the spiritual element that the Millennial craves, spiritual tourism in India is a niche area in the tourism industry. This chapter is divided into six sections. Following this introduction, we discuss the literature to broaden readers' understanding of Millennials and Tourism. We then deal with spiritual experiences in India and Millennials, spirituality and tourism. We then discuss the primary research to date, and its contribution towards this study. We further discuss the outcomes and findings of the study and conclude with the future scope for research in this domain.

## 2.2 Millennials and tourism

The generational theory states that each generation of people is distinguished by a particular set of traits, beliefs, values and experiences. The term generation can be applied to a demographic extending over 20–25 years before the demarcation of the next generation. A generation is so considered because of the similar unified personality that distinguishes a particular cohort from the other age brackets. According to Pendergast (2010), there are four main elements brought together by generational theory:

> repeating trends based on generational type, recognition that exposure to a wide range of factors during formative adolescent years determines fundamental values and belief systems, recognition of the presence of life-cycle stages which include generational-type characteristics evident at each stage, and a relevant generation defined by birth years containing unique attributes at any given stage.

The Millennial generation (a terms attached to those born between 1980 and 1995) has been a topic of academic research and interest across many disciplines of social sciences, business and management. In recent years, there has been extensive research relating to various tangents of the Millennial demographic. Researchers, however, have yet to find a uniform operational definition for the Millennial generation. Cavagnaro et al. (2018) proposes that the age range of the Millennial should be considered to be those born between 1980 and the mid-1990s. While most scholars mark the beginning of the Millennial age as 1980, there could be some overlapping between the younger millennial generation and the older Generation Z, whose birth years have been set as early as 1995.

From the tourism perspective, Millennials are a generation that is bringing about a complete redefinition of the meaning of tourism and global travel. Given the ease of travel, decreased hotel and visa costs, travel is opening up avenues for Millennials to experience and make the most of what global tourism has to offer. The United Nations, for example, states that Millennial tourists generate more than $180 billion in annual tourism revenues (UNWTO, 2012). This stems from the fact that Millennials, in contrast to the previous generations, choose to spend their money on experiences rather than material luxuries. Millennials also make up around 20% of all international tourists due to their earning potential and spending capability. Research indicates that even though the global millennial share common motivations for travel, such as exploration, relaxation and personal challenges, visible differences in spiritual travel intention is seen among Millennials of different geographical locations. For example, Millennials from the Middle East, South East Asia and Northern Europe are less likely to travel for social reasons (Richards and Morrill, 2020).

Spiritual tourism has many times been equated with religious tourism. Many examples of this can be seen of the annual Hajj pilgrimage by Moslems, and the circumambulation of Mount Kailash by Hindus, Buddhists and Jains. However, the concept of spiritual tourism transcends any religion. A spiritual tourist would seek not just a religious experience; rather, the spectrum of spiritual travel and tourism would give the seeker's life a new meaning or even new direction (Melchers, 2006).

Spiritual Tourism offers avenues to individuals who seek a sense of self and travels to connect with their inner self. Despite the two terms being used interchangeably (Smith et al., 2010), spiritual tourism is an experiential phenomenon that goes beyond religion. While religious tourism is either dictated by the norms or mandated by religions, spiritual tourism is an entirely voluntary and individualistic act (Norman, 2011), which addresses aspects of restoring the connection among mind, body and spirit (Smith et al., 2010). This need not be restricted to any specific religion. Thus, the motive and the purpose of travel is what define a person's travel as either religious or spiritual. India, which prides itself on the concepts of secularism, has a diverse mix of peoples who follow all major world religions. However, there is no denying that moral and social norms have been weakened; this is why Millennials experience the need to search for a sense of self and identity. This void is filled by spiritual tourism and is one that cannot be overlooked in today's stressful times (Conrah, 2006). Another notable feature of spiritual tourism is the motive behind undertaking such a journey; the motives are not restricted and are, at best, difficult to label and quantify. Each person's motive for a specific spiritual journey is personal and spurred on by individual experiences. Religious and spiritual tourism tend to be interchangeably used as spiritual tourists are most likely being involved in the actual visiting of religious sites and imbibing activities that could connote a religious significance (Smith et al., 2010) such as meditation and chanting. Hamilton (2000) refers to this modern-day spirituality as "pick-and-mix religion". To distil the concepts used for this research, spiritual tourism is analyzed here as a healthy mix of travel, practices and experiences that is highly individualistic and personal. While spiritual travel is a phenomenon that is explained by individuals according to their own perceptions, it must be borne in mind that spiritual tourism is also experimental in nature. Many tourists who seek spiritual experiences through tourism do so by a process of trial and error. To this effect, 'word of mouth' marketing of so-called 'spiritual places' and references from peer groups play a major role. Another interesting fact to note is that spiritual tourists do not necessarily adhere to conversion. Rather, the journey is a means to find oneself and connect with the soul.

India is ranked among the world's top spiritual destinations, since it is the cradle of four major world religions: Hinduism, Buddhism, Sikhism and Jainism. People travel to India in search of spiritual experiences since the country houses some of the most sacred places of these religions. This is for the desired spiritual experience that drives travellers to these places (Fedorova, 2016). In addition, India offers affordable boarding and travel within the country, making India an attractive spiritual tourist destination. Varanasi, Bodhgaya, Dharamsala and Goa are the top choices for spiritual travellers.

### *2.2.1 Spiritual experiences in India*

Spiritual tourism enjoys a long history in India. The 'Tirthyatra', or pilgrimage, was indicative of a person's renunciation of the worldly matters and his quest for exploring the truth. The journey in itself was considered a spiritual one, where the individual's chances of coming back to his homestead after the experience were remote (Kamra, 1997; Gupta et al., 2002). Spiritual destinations have an innate power to attract devotees. This could be associated to one-off events or even a combination of miraculous happenings, sacred geography and how difficult it is to access these places (Phukan, Rahman & Devdutt 2012). While these could be distinguishing points in religious tourism, the Millennials are often drawn to spiritual sites such as national shrines, war graves and sporting venues. In the West, India has long been considered the land of spirituality and religion. Since four of the major world religions originated in India, it is natural to see pilgrimage and spiritual centres along the length and breadth of India. In addition to Buddhism, Hinduism, Jainism and Sikhism, other world religions, such as Islam and Christianity, have also put down deep roots in the country. Spiritual and retreat centers or spiritual circuits can be seen as a result of this (Medhekar & Haq 2012). A common denominator throughout the religious centres is the openness towards a spiritual experience and the welcoming of domestic and international tourists to embrace the offerings of the venue. If carefully observed, most spiritual centres in India, comprising shrines, sacred spots and temples, are scattered along the major riverbeds or in the hills and mountains of India. This makes an attractive option for seekers of peace and tranquility with which Spirituality is associated.

India has retained its position as an attractive spiritual destination. The cultural, civilizational and historical importance given to spiritual tourism are the factors that contribute to its refreshment. The mesmerizing fusion of contemporary highways with their ancient trade routes, the splendid architecture of old citadels and temples with New Age architecture, and the cultural and religious amalgamation, coupled with favourable climatic conditions, facilitating ease of travel and gamut of delicacies makes this country an ideal place for tourism. Although the proper branding of these factors is still absent, some attempts have been made (Geary 2008). When this helps the tourist to explore the higher-order purpose of existence and identify their spiritual calling, it is a unique value addition (Sharpley & Sundaram 2005).

### *2.2.2 Millennials, spirituality and tourism*

Before the last decade, spirituality was considered a matter confined to old age people and retired people. The young generation, in general, kept away from all such experiences. But, a number of factors changed the scenario and youth also began a deeper interest in spirituality. The factors include: the timely intervention by some spiritual leaders to reach out to youth; escalating problems in professional and personal spaces not being solved by contemporary science; and successful people talking openly about their spiritual inclinations (Lerner et al. 2006). These were amongst a few factors that brought spirituality into discussions of Millennials. The tourism aspect was the value addition and added to the "coolness" of this new experience. The higher penetration of digitalization, travel and communication ease supported by mobile applications, enhanced information of spiritual places, and corporate initiating wellness programmes were add-ons. The trend is spreading its roots amongst the Millennials. However, before we conclude anything, in-depth research is required to which this chapter attempts to make a start.

## 2.3 Primary research and its contribution

The authors decided to adopt a qualitative research approach to delve deeper into the experiences of the people who generally undertake spiritual tours. Furthermore, it was also essential to connect with respondents regarding their perception and views about spirituality. One step required before reaching out to respondents required a thorough understanding of spirituality and how academically these ideas are to be studied, discussed, and presented.

In the study, exploratory discussions were done with the experts of the domain of spirituality and indigenous knowledge traditions. Later, people who regularly go for spiritual travel were contacted for their inputs. This helped the authors to get insights into the phenomenon. Based on the literature review and these discussions, interview questions were framed and people who opt for spiritual travel were identified. The analysis of data collected from them was done using content analysis (Drisko & Maschi, 2016; Erlingsson & Brysiewicz, 2017). The questions, responses and analysis are thus detailed. The findings and discussions are shared in the subsequent section. Based on the extant literature reviewed and preliminary discussions with the experts, the following six central research questions were identified for this study:

R1 *What, according to you is spirituality all about?*
R2 *How does your understanding of spirituality help you in exploring self?*
R3 *Whenever you travel on such journeys, do you feel it helps to enhance your personality or develop your human qualities?*
R4 *How would you like to express the purpose of spiritual travel?*
R5 *Please share your views on the benefits / learning of spiritual travel*
R6 *How can the tourism sector re-design its processes to meet the needs of spiritual tourists?*

To gain a detailed understanding on the research questions, four in-depth interviews were conducted on the following respondents from different domains. Below we share the interviews to take the readers through their interesting experiences. The analyses of their responses is presented in the next sub-section.

**Question 1** *What, according to you, is spirituality all about?*

*Respondent 1* I feel that spirituality is a desire to meet God, desire to become one with him who created all.

*Respondent 2* Spirituality is a way of living. It is about how one perceives, believes and act based on this orientation.

*Respondent 3* Personally, I feel looking for wisdom to understand the purpose of life is spirituality; to be more responsible towards myself and others; doing something meaningful; making world a better place.

*Respondent 4* Spirituality is about achieving the ultimate goal of being one with Bhagwan (God) by performing duties.

***Interpretation and key takeaways***: The higher-order purpose of existence (Sharma, 2007); a way of living; wisdom to understand the purpose of life; responsibility and doing something meaningful; remembering God and performing duties.

**Question 2** *How does your understanding of spirituality help you in exploring self?*

*Respondent 1* Times when I tend to forget, makes me aware of that nothing is in my control.

*Respondent 2* Spiritual mindedness helps in inquiring about the life, the purpose, the goal and helps in identifying how to proceed further for it.

*Respondent 3* Helps in introspection; better version of myself; take charge of my life; be my own friend and philosopher/guide; convert value to action/behaviour.

*Respondent 4* Being active on a spiritual forefront helps finding one's true potential, having belief in your own capabilities. It builds strength of character.

***Interpretation and key takeaways***: Self-awareness, self-inquiry, knowing self, introspection, self-improvement, strength of character, self-belief.

**Question 3** *Whenever you travel on such journeys, do you feel it helps to enhance your personality or develop your human qualities?*

*Respondent 1* Yes, it helps me to break my routine and fills me with the amazement when I see around.

*Respondent 2* While traveling and looking around, it gives the sense of feeling that everything is divine and created by god. Just as there are passengers in a travel journey, similarly every being is a passenger in the journey of life and whatever one goes through in life will pass-by like the passing station.

*Respondent 3* Gives time to focus on self; calms one down from daily distraction; helps me to be more mindful – of my emotions, my actions and what I want to do.

*Respondent 4* Travelling does help in enhancing personality, be it spiritual travel or of any other kind. Spiritual travel, in particular, helps in understanding one's cultural aspects better and make one strongly attached to his/her cultural roots. It also helps in understanding how spirituality or spiritual inclination connects people from different backgrounds.

***Interpretation and key takeaways***: Time for self-reflection, seeing the creation of divine, analogous to the journey of life, opportunity to focus on self, mindful of own actions, understanding cultures, feeling the connect with each other.

**Question 4** *How would you like to express the purpose of spiritual travel?*

*Respondent 1* It's a journey within, if travelling to a place helps in strengthening the resolve to search within then one should embark on it. Going to a holy place to ask for some personal favors and gifts from god is not worth calling spiritual travel, its mere business travel.

*Respondent 2* The purpose of spiritual travel to reach the final destination of Moskha. However, this is long journey and not easy one [sic].

*Respondent 3* Any travel away from work is spiritual travel to me :) it may/may not be a religious one; Reading books; travel itself makes me more spiritual; learning new perspectives; seeing new colors (as in new cultures, language, food… etc.); helps in slowing down life

*Respondent 4* Understand culture and purpose of living

***Interpretation and key takeaways***: Self-exploration, to help reach destination of life, learning new perspectives, purpose of living, understanding culture

**Question 5** *Please share your views on the benefits/learning of spiritual travel*

*Respondent 1* It is useful if someone is able to find some light within, some clarity on actions he is doing or if it kindles some interest in knowing more about spirituality (God).

*Respondent 2* Helps a person by drawing parallels between travel and spiritual journey.

*Respondent 3* Similar points as shared in above answer.

*Respondent 4* Similar points as shared in above answer.

***Interpretation and key takeaways***: Finding the light and energy within, clarity on purpose of life

**Question 6** *How can the tourism sector re-design its processes to meet the needs of spiritual tourists?*

*Respondent 1* Needs are basic when it comes to any tourist. Clean stay, hygienic food and ease of commute at affordable prices. Any design which can facilitate spiritual tourists to visit shrines in an orderly fashion without too long or overcrowded lines could encourage more people to revisit.

*Respondent 2* Elaborating the importance of various places in our country.

*Respondent 3* How to understand –who is a spiritual tourist? Sometimes, one has a spiritual experience on a travel unplanned for spiritual purposes. Tourism, I believe, has one purpose – one wants to explore and experience. Explore new places, new people, new experience. Connect with one self/ connect with nature.

*Respondent 4* Improving facilities for stay and other requirements of travelers; Ensuring no illegal, illegitimate and non-pious activities like gambling, alcohol intake takes place in revered places of spiritual travel; Promotion of travel packages for all age groups to such places; renovation and maintenance of the sites; Avoiding interference in the spiritual practices of ancient temples.

***Interpretation and key takeaways***: Crowd management, clean and hygienic, spiritual importance, creating opportunities for new exploration and experiences, maintenance, maintaining the Saatvik and pious character of the places, avoiding interference.

## 2.4 Findings

The findings of the study reveal the inclination of Millennials towards experiential travelling and spiritual tourism. Respondents are highly favourable to the concept of going back to one's roots and thereby embarking on a journey of self-discovery that leads to inner peace and tranquility. Given the opportunity, the Millennials emphasizes the importance of a planned and systematic interval of spiritual travel in order to achieve a higher sense of being. Respondents believe that such travels are more likely to equip them with a better understanding and competency in their workplace. From the research conducted, the following terms gain meaning:

*Definition of spirituality:* The spirituality meant exploration towards a higher-order purpose of existence (Sharma, 2007) or the wisdom to understand the purpose of life with responsibility and doing something meaningful. Spirituality is something more than a particular religious belief. The research shows the perception of spirituality is varied according to the individual. Spirituality is, rather, a highly individualistic and personal perception of finding oneself through a series of varied activities or experiences. The understanding of an individual's spirituality is highly rooted in the cultural, religious or non-religious, social and personal traditions and psychological inclinations.

*Role of spirituality in exploring self:* It helps in higher self-awareness, self-inquiry, knowing self, doing self-introspection, take a step towards self-improvement, building a strength of character and self-belief. Various models and theories have been advanced by researchers on this concept. In the form of philosophical ideas such as Nyaya, Vaisheshika, Yoga and Vedanta,

India has developed different principles of philosophy and spiritualism and has given the world the largest number of systems of spiritualism. Thus, the Millennial generation gains by travelling for spiritual purposes as there is a solid foundation of spiritual principles to adhere or look up to.

*Role of travel:* It provides time for self-reflection by focusing on self. By seeing the creation of divine it makes the person calm. Further, it appears analogous to the journey of life and thus makes the person relax and contented. It makes a person mindful of own actions, helps in understanding different cultures and feeling the connection with each other.

*Purpose of spiritual travel and benefits:* It helps in self-exploration or finding the light and energy within, guiding the traveller towards the destination of life, learning new perspectives, and identification and clarity about of purpose of life. Spiritualism exhorts confidence in the universe's collective truth and imparts understanding of God's oneness with the world as a whole. Spiritualism should therefore be viewed as a metaphysical approach to fact, and religion should be treated as a realistic approach to the same objective.

*Improvements needed in the tourism sector to enhance spiritual tourism:* Better crowd management, and a clean and hygienic environment are the basic expectations of a traveller and the same also goes for the spiritual traveller. Secondly, the spiritual importance is highlighted with maintaining the 'saatvik' and pious character of the places. Further, avoiding interference with the spiritual practices can help connect with the ancient roots by creating opportunities for new exploration and experiences can help.

## 2.5 Discussion and future research

Spiritual tourism leads to multiple benefits in terms of social, psychological, educational and economic aspects. Discussing from the social perspective, we observe that government budgets on the maintenance of law and order is increasing enormously in almost every country.[1] The number of lawsuits that are being filed shows a similar steep rise.[2] This extrinsic approach to improve the social well-being of society has less chance of improvement. However, an intrinsic approach has the potential of bringing about long-lasting change. A person who thinks about exploring the right purpose of life, wants to identify a higher-order purpose of life and connects it with a responsibility to play a significant and positive role in both society and the workplace. Such persons are an asset to their nation and the whole world. The study also found that such people, through their reflective behaviour, always aim for self-improvement; thus, it shows that society can became a just one and also one with lower levels of crime by promoting spiritual tourism.

From the psychological aspect, it is very evident that many young people experience severe problems of distress and depression, emanating from feelings of lack of purpose and a discontent with the materialism of modern life. The cognitive load has been increasing exponentially given the dynamic changes observed in the business environment and social environment. In the past, the components of life had fewer dimensions to be taken care off and more resources in terms of social capital and family support to handle it. The absence of information and digitally empowering tools ensured that person remains humble. A study revealed that the Internet makes one perceive an inflated estimate of the self-intelligence (Fisher, Goddu, & Keil 2015). The most troublesome repercussion of this phenomenon is that real-life problems require a reflection-based solution that comes through experience, which, in the case of digital empowerment, is absent. A youth facing issues like depressions might be highly informed about overcoming from depression, but would still be in a fix. What could be the plausible solutions? It has been found from the current study that spiritual tourism helps in achieving clarity around the purpose of life and provides opportunity to learn new perspectives. The clarity on purpose of life has high chances of reducing depression and other psychological distress (Phillips 1980; Lester & Badro 1992; Robak & Griffin 2000). Thus, it can help to improve the mental health of society.

If the tourism sector focuses on the basic requirements of a traveller such as cleanliness and hygiene, creates the right narrative about the place and opportunities to create a unique experience for the traveler, there is huge potential to enhance economic activities in the entire region. The economic benefits of the tourism sector are huge; however, India, which has unarguably the best potential, is very far from realizing it (it still being ranked 7th). The tourism sector contributes about 10% to global GDP (WTTC, 2020). Thus, here is an opportunity to tap into. The potential of spiritual tourism is highest for India given that the country is the birthplace of religions that are based on seeking or search for truth (as mentioned in the introductory remarks to this chapter).

Finally, the attraction of youth towards seeking out the truth has surged dramatically after realizing that science also has limitations. Another factor contributing to this urge is the rise in availability of a plurality of ideas on social media and new media outlets, which educate youth beyond the confined boundaries of schools, colleges, and universities. Today's youth is no longer dependent on the textbooks or reference books for their knowledge; nor do they acquire all their understanding from their teachers. In the present day, they can connect with any source of knowledge from any part of the world. This opportunity was not so easily available before the digital revolution. The gatekeepers in the education domain now no longer have that control and youth have numerous options to 'seek' and 'explore' the truth.

## 2.6 Conclusion

Spiritual tourism can creative positive outcomes not only for individuals, but for the whole of society. The research highlighted the advantages of the Millennials, focusing more on indigenous ways to attain spiritual tranquility and peace through spiritual tourism. This study throws light on the experiences of Millennials who have attained, or are in the process of realizing, the benefits of returning to their roots through spiritual travel and tourism. The society is witnessing drastic changes because of new forms of work, global dynamics and an unprecedented increase in cognitive load that an individual is subjected to in contemporary scenario, in both the personal and the professional world. This research is the first of its kind to give an academic study of the different facets and implications of spiritual tourism of Millennials in India. Currently, there is very sparse literature on this domain of study of the Millennials. The authors can foresee an enhanced interest of the researchers in this domain. Various aspects, such as push and pull factors contributing to spiritual tourism, economics and spiritual tourism, the psychological well-being of the society and spiritual tourism, the role of government in promoting spiritual tourism, social enterprises and spiritual tourism and so on could be studied in order to produce newer insights. The tourism and hospitality industry of India would develop a better understanding of the mindset, preferences and tastes of the Millennials when it comes to spirituality and tourism. All of these would have a very positive impact on our society.

## Notes

1 https://www.business-standard.com/article/economy-policy/price-of-law-and-order-expenses-on-police-eat-into-other-budget-117061400055_1.html
https://www.washingtonpost.com/business/2020/06/04/us-spends-twice-much-law-order-it-does-social-welfare-data-show/
https://www.instituteforgovernment.org.uk/publication/public-services-spending-round/law-order

2 https://www.bloombergquint.com/law-and-policy/indias-pending-court-cases-on-the-rise-in-charts

## References


Agarwal, H., & Vaghela, P. (2018). *Work Values of Generation Z: Bridging the gap to the next generation. National Conference on Innovative Business Management Practices.*

Cavagnaro, E., Staffieri, S., & Postma, A. (2018), "Understanding millennials' tourism experience: values and meaning to travel as a key for identifying target clusters for youth (sustainable) tourism", *Journal of Tourism Futures*, 4(1), pp. 31–42.

Conrah, D. (2006). *The impact of spirituality on mental health: A review of the literature*. United Kingdom: Mental Health Foundation.

Drisko, J. W., & Maschi, T. (2016). *Content analysis*. Pocket Guides to Social Work. Oxford and New York: Oxford University Press.

Erlingsson, C., & Brysiewicz, P. (2017). A hands-on guide to doing content analysis. *African Journal of Emergency Medicine*, 7(3), 93–99.

Fedorova, A. (2016). *New Product Development in Spiritual Tourism*. Finland: Savonia University of Applied Sciences.

Fisher, M., Goddu, M., & Keil, F. (2015). Searching for explanations: How the Internet inflates estimates of internal knowledge. *Journal of Experimental Psychology: General*, 144(3), 674–687 DOI: 10.1037/xge0000070.

Geary, D. (2008). Destination enlightenment: Branding Buddhism and spiritual tourism in Bodhgaya, Bihar. *Anthropology Today*, 24(3), 11–14.

Gupta S, Lal K, & Bhattacharyya M. (2002). *Cultural Tourism in India.* New Delhi: Indraprastha Museum of Art and Technology.

Hameed, S., & Mathur, M. (2016). *A study on Behavioral competencies of Generation Z. International Conference of Management and Information Systems*, (pp. 63–71). Bangkok.

Hamilton, M. 2000. An analysis of the Festival for Mind-Body-Spirit, London. In S. Sutcliffe and B. Bowman (Eds). *Beyond new age: Exploring alternative spirituality.* Edinburgh: Edinburgh University Press, 188–200.

Kamra K. 1997. *Tourism: Theory, Planning and Practice.* New Delhi: Indus Publishing Company.

Ketter, E. (2019). Millennial travel: tourism micro-trends of European Generation Y. *Journal of Tourism Futures*, doi: 10.1108/JTF-10-2019-0106.

Lerner, R. M., Alberts, A. E., Anderson, P. M., & Dowling, E. M. (2006). On making humans human: Spirituality and the promotion of positive youth development. In E. C. Roehlkepartain, P. E. King, L. Wagener, & P. L. Benson (Eds.),*The handbook of spiritual development in childhood and adolescence.* Sage Publications, 60–72.

Lester, D., & Badro, S. (1992). Depression, suicidal preoccupation and purpose in life in a subclinical population. *Personality and Individual Differences*, 13(1), 75–76.

Majumdar, S. (2018, June). *Pew Research Centre.* May 2020, Retrieved from www.pewresearch.org.

Medhekar, A., & Haq, F. (2012). Development of spiritual tourism circuits: The case of India. *GSTF Journal on Business Review (GBR)*, 2(2).

Melchers, Christoph (2006) Spiritual Tourism: Motives, Formats and Marketing. Paper.

Norman, A. 2011. *Spiritual Tourism: Travel and Religious Practice in Western Society.* London, New York: Continuum.

Padmanabhan, V. (2019, April). *Religion and the Indian Millennial: What data shows.* May 2020, Retrieved from www.livemint.com.

Pendergast, D. (2010). Getting to know the Y generation. *Tourism and generation Y*, 1, 1–15.

Phillips, W. M. (1980), Purpose in life, depression, and locus of control. *Journal of Clinical Psychology*, 36, 661–667.

Phukan, H., Rahman, Z., & Devdutt, P. (2012). Emergence of spiritual tourism in India. *International Journal of Marketing and Technology*, 2(4), 300.

Richards, G., & Morrill, W. (2020). Motivações dos viajantes millennials globais. *Revista Brasileira de Pesquisa em Turismo*, 14(1). doi: 10.7784/rbtur.v14i1.1883

Robak, R. W., & Griffin, P. W. (2000). Purpose in life: What is its relationship to happiness, depression, and grieving? *North American Journal of Psychology*, 2(1), 113–119.

Sharma, S. (2007). *Management in the New Age: Western windows; Eastern doors.* Delhi: New Age International.

Sharpley, R., & Sundaram, P. (2005). Tourism: A sacred journey? The case of ashram tourism, *India. International journal of tourism research*, 7(3), 161–171.

Smith, M., MacLeod, N., & Hart Robertson, M. (2010). *Key Concepts in Tourist Studies.* London and Thousand Oaks, CA: Sage

The Economist. (2019, February). *Generation Z is stressed, depressed and exam obsessed.* Retrieved May 2020, Retrieved from www.theeconomist.com.

UNWTO. (2012) UNWTO and WYSE travel confederation launch global declaration to promote youth travel. [Online] Retrieved from: http://media.unwto.org/en/press-release/2012-09-25/unwto-and-wyse-travel-confederation-launch-global-declaration-promote-youth [Accessed 31 March 2019].

WTTC. (2020). *Economic Impacts Report.* Retrieved from https://wttc.org/Research/Economic-Impact.

# 3 The influence of spirituality on generations Y and Z working in the tourism & hospitality industry

*Hasibe Yazıt*

## 3.1 Introduction

Discovering the meaning of life and living it from day to day is one of the main goals of human creation. When an individual has a feeling of emptiness, the main problem is an absence of spirituality. Today's human being seeks something beyond materiality in their quest to give meaning to life. Despite the use of the most advanced and modern technologies, the predominant sense of monotony and discontent with life leads him to seek a way out of their crisis. Research studies in the areas of travel and traveling reinforce one of the most fundamental and most influential components of the development of a person's spiritual intelligence. Ultimately, his spiritual intelligence improves, which helps him to realize spirituality in his life. On the other hand, when we travel through the life and history of the science and moral elders, we witness that many distinguished personalities of human societies have traveled in order to attain moral virtues, reach the great humanity and discover the sources of truth. These travels tested these people, making them more resilient and purifying their soul through wandering in nature. Therefore, in spiritual books, morals and poetry both mention the positive effects of traveling on the glorification of the spirit and morality of the human being have been mentioned abundantly. Traveling alters the traveller's view of the world by seeing nature and historical artifacts inherited from the past. (Bansal & Eiselt, 2004). This change is accompanied by a commitment to values and a sense of obedience to God, who created nature, which is the meaning of tourism with spirituality. This type of tourism and travel creates a kind of deep friendship between tourists in their efforts to get to know both about the nations and cultures of the past. On the other hand, as people become acquainted with their nations and cultures, the ground is formed for a feeling of freedom from violence and a desire for peace.

Today, international tourism accounts for 7% of world trade and covers around 9% of worldwide professions, and, despite all of the problems, international tourism is developing continuously. There is a strong relationship between spirituality and tourism, which has many branches, and one of these branches is the spiritual understanding of tourism in travel. The purpose of spiritual tourism is not religious tourism. Most people support the thesis that

DOI: 10.4324/9781003170631-4

all kinds of tourism have valuable consequences in terms of meaning. All phenomena have an external face and a meaning, and tourism is a face that is familiar to all. However, this meaning and internal aspect has been paid less attention. By emphasizing the scientific and religious concepts in the field of tourism, we can identify the meaning and concept of spiritual tourism. Accordingly, we should not ignore the issue that the tourist wants to bring his soul to peace. Some researchers use this combination of tourism and spirituality as a shorthand for sustainable tourism (Beerli & Martin, 2004).

Contemporary societies have turned to spirituality as a solution to the agonies and troubles of contemporary life, in which religious tourism and visiting travel currently play an important role (Vargheese & Varghees, 2012). However, many religious and sacred regions are located in areas that are difficult to reach, and it is understood that in recent years many visitors, especially foreign tourists, have abandoned these areas, although air transport has contributed to the reduction of such problems and the development of cultural relations and customs to some extent. The freeing up of transport markets around the world means that airline companies have more freedom in choosing their flights. This is a good opportunity for them to increase their services. The process of change in the media is so rapid that if the local media cannot keep up, a pause in tourism is inevitable. Nothing has been more effective and more flexible in terms of the development of tourism than communication and mass media, and tourism has been the sector that has benefited the most from the communication revolution, with reviews on websites such as TripAdvisor and Booking.com (Phukan, Rahman, & Devdutt, 2012).

## 3.2 Generations Y and Z in tourism

Explaining the possible effects of Generation Y, which will have the majority of its spending power as of 2020, and Generation Z, which is expected to change the dynamics of the tourism sector in 2030, on the tourism sector, and their suggestions for the sector to touch these generations were listed. It is stated that the generations in the marketing sector are categorized as X, Y and Z generations and that the tourism sector, which cannot understand the behaviour, characteristics and expectations of Generation Y, cannot touch the Z generation born into technology, and the expectations and approaches of the Y and Z generations are listed as follows:

In the areas of technology and knowledge, the dominant Generation Y places the greatest significance on reliability and transparency (Etlican, 2012: 6). Generation Z, which has been born into a technology-rich atmosphere (Ayhün, 2013: 102), is likely to be a generation that is constantly questioning and researching. Generation Y prefers to go on vacation at least twice a year. This will also apply to Generation Z. Social media, websites, search engines, friend's suggestions and forum websites will be the leading factors in affecting the decisions for both generations. For Generations Y and Z, a vacation is more of a social activity than an opportunity for relaxation. Therefore, it will be important for the tourism sector and agencies to offer entertainment opportunities to Generations Y and Z (Behrstock-Sherratt & Coggshall, 2010). Most members of Generation Y state that they will not buy a service of which their friends do not approve. While Generation Y will listen to the suggestions of the people around them, they will also affect the purchasing behaviour of other individuals. The same will be true for Generation Z. More than nine-in-ten Millennials (93% of those who turn ages 23 to 38 this year) own smartphones, compared with 90% of Gen Xers (those ages 39 to 54 this year), 68% of Baby Boomers (ages 55 to 73) and 40% of the Silent Generation (74 to 91), according to a new analysis of a Pew Research Center survey of U.S. adults conducted in early 2019.Accordingly, hotels and agencies will have to pay increased attention to their place in the online media and the comments which they make. Generations Y and Z prefer to communicate by Facebook, WhatsApp and Instagram rather than face-to-face. Therefore, targeted messages and campaigns will enable them to communicate with these generations more easily. Versatile, simple, clear, easy to understand, honest, transparent, sincere and utilitarian approaches that make you feel free and included, are likely to appeal to both Generations Y and Z. According to the generation theory, people are classified into groups according to their age, groups born in the same time period, living in the same conditions, sharing the same responsibilities and common personalities. According to this theory, there are similarities between the members of the same generation, while equally there are significant differences between the expectations, needs, desires and goals of different generations. The issue of exactly which years demarcate the Y and Z generations is still the subject of some debate, however. Although some researchers have pointed out different time intervals (Reeves & Oh, 2008: 296), as discussed in this study, many researchers define those born between 1980 and 1999 as Generation Y, and those born either in or after 2000 as Generation Z (Kyles, 2005: 54). Generation Z is said to consists of educated individuals who are more egocentric than other generations, who have high self-confidence and expectations, are realistic, are distrustful of authority, and place a great deal of importance on their own personal freedom. Generation Y, by contrast, is defined as attaching importance to brands, entertainment and friendship. It is also passionate about technology, and has a preference for informal communication channels. In addition, the members of this generation are people who have advanced thinking and absorb knowledge quickly. They can adapt easily to change, have the ability to multitask but are easily bored, they enjoy living comfortably, and are impatient, unfaithful and not easily satisfied. Generation Y, which includes those from 20 to 40 years of age, is an important sector for the tourism market. Generation Z (the so-called Crystal generation), on the other hand, consists of individuals who are traditionalists, place a high value on family, have a high level of self-control and responsibility, and have creative and multiple decision-making skills. Among the most important characteristics of the members of Generation Z, who make up the young people of the technology age, are their membership to more than one social network, their tendency to communicate and relationships at a distance, and their rapid adaptation to change. Generation Z members have high global expectations, and are fond of their comfort. They focus on quality and low cost rather than brand, enjoy online shopping, follow campaigns and search for last-minute deals, thereby creating a different consumer profile. While some researchers emphasize the differences between the two generations on the characteristics of the members of Generations Y and Z, Furstenberg states that it is difficult to define the boundaries between the two generations and that overgeneralizations should be avoided. Studies reveal that the characteristics of generations have an effect on the choice of destination. One of these studies was conducted by Barton et al. (2012) on Generation Y. The results of the research show that Generation Y has become increasingly sensitive to the price and that social and environmental factors are effective in their touristic product purchasing decisions. Research has suggested that those factors that have the greatest effect on the destination selection of Generation Y are price, entertainment and cultural opportunities. In the study where they compared the destination image perceptions of the tourists in the X, Y and Z generations, they listed those factors that affect the choice of holiday destination for Generation Y participants as landscape/natural environment, entertainment facilities, historical and cultural attractions. It is emphasized that Generation Y makes the greatest use of the Internet while making travel decisions, more than half of the participants go on vacation at least once a year and that their main aim is the pursuit of enjoyment. In a study conducted by Veríssimo and Costa in Portugal, the expectations of the Y generation members staying in hostels were examined. As a result of the research, it was determined that Y generation members want to meet people, participate in different activities, have local experiences, feel comfortable/at home and get their money's worth. In a study where Migacz and Petrick examined the travel motivations of Generation Y and the perceived benefits of travel, Generation Y was considered as young and advanced age groups and grouped according to their income levels. As a result of the research in which quantitative and qualitative methods were used together, it was determined that Generation Y is not homogeneous, and that there are different perceptions and motives between the young and the old age groups. In research conducted by Cavagnaro in the Netherlands, the value orientations and tourism experiences of Generation Y were examined, and it was emphasized that this group was not homogeneous. In the research conducted by Monaco in Italy, it was examined whether the Y and Z generations use new technologies and especially web pages in vacation destination selection. As a result of the research, it was determined that Generation Y uses the internet together with traditional tools (phone, magazine, etc.) in their information-gathering and purchasing decisions, and almost all of Generation Z prefer websites and social media. In the study conducted by Styvén and Foster in Sweden, England and India, the tendency of Generations Y and Z to share their travel experiences on social media was examined and it was found that both generations tend to share their travel experiences on social media (Williams & Page, 2011: 2–8).

Determining the characteristics of the Y and Z generations, which constitute the largest target audience in the tourism market, is important in terms of understanding and interpreting the differences between generations. It is seen that both Y and Z generations tend to make the most use of the internet and social media as their source of information. However, while Y generation members mostly use webpages as their source of information, Generation Z members prefer to use social media. The findings obtained here are in line with the results of previous studies. This difference between the information sources used in terms of destination selection is important in terms of the marketing strategies which should be adopted. Accordingly, in marketing activities for both generations, social media marketing and mobile marketing, websites, online brochures and catalogues, etc. digital marketing tools should be used effectively. However, Generation Z, which monitors the lives of others instantly through social networks and conveys their experiences in the same way, gives importance to the experiences of other people. For this reason, using videos, corporate social networks, blogs and forums (Influencer Marketing), virtual or augmented reality applications to reach Generation Z will be much more effective in terms of destination marketing. The findings obtained from the study reveal that the two most important travel motivations of the members of Generations Y and Z are similar. Spiritual/physical relaxation and fun are the most important motivation factors for both generations. For Generation Y, escaping from the routine of daily life, having an adventure, being in new/different places, excitement, getting away from the crowd, trying new foods and drinks are other motivational factors. The motivation elements of Generation Z are being in new/different places, having an adventure, learning new things, getting away from the crowd, trying new foods and drink, and intellectual enrichment and excitement. These findings show that the first three factors that have the greatest impact on the destination selection of the Y and Z generations are the same: price, local food, beautiful scenery and natural attractions. This shows that members of Generations Y and Z consist of individuals who are sensitive to price, have a great desire to try different flavours and attach importance to the beauty of the natural world. The findings obtained are similar to the results of some previous studies. Other factors to which Generation Y attach importance in their choice of destination are: the safety of the destination, obtaining good value for money, different cultural attractions, all-inclusive facilities, nightlife and entertainment. Generation Z, by contrast, attaches greater importance to an unspoilt and uncontaminated environment, historical and archaeological attractions, easy transportation, good value for money, cheap shopping opportunities, different cultural attractions and suitability for family travel. The research findings reveal that although there are similarities in terms of factors affecting the destination selection of Generations Y and Z, the importance level of these factors has changed. Among the factors that affect the choice of destination, the average score of Generation Z is higher than Generation Y regarding local food, an unspoilt and uncontaminated environment, historical and archaeological attractions, the popularity of the destination, being accessible to their family, an exotic atmosphere, cheap shopping opportunities and adventure activities. When the findings are evaluated, it is revealed that Generation Z and Generation Y has an adventurous nature, sensitive to nature, interested in different historical and cultural attractions, curious about foreign/distant countries. In addition, it can be said that Generation Z, who loves shopping and is influenced by popular culture, behaves more consciously and selectively than Generation Y, wants to secure good value for their money, and has higher expectations. The research findings contain some important clues for destination marketers. Accordingly, the research shows that both generations are sensitive to price, and an emphasis should be placed on affordable prices in terms of transportation, accommodation and shopping opportunities in marketing communication. Similarly, both Y and Z generations attach importance to the motivations of relaxation and entertainment and find local elements and natural attractions important. Therefore, tourism professionals who want to attract Generation Y and Z tourists to their destination should highlight hedonistic motivations, such as relaxation, pleasure and entertainment, and also emphasize local culture, the natural environment, different adventure activities, recreation and entertainment opportunities in their marketing messages. In addition, it is thought that messages that emphasize emotion and experience rather than information in marketing communication, especially for Generation Z, will be more useful. When the findings obtained from the study are evaluated in general, the similarities as well as the differences between Y and Z generations in terms of travel motivation and destination selection draw attention. It is thought that this result is due to the fact that 80.6% of the sample is between the ages of 14 and 25. This situation, which constitutes the most important constraint of the research data, also prevents any generalizations being drawn from the data. However, it is important to know the differences between the two generations in terms of the generation marketing of the tourism sector for Generations Y and Z. At this point, it is thought that the research findings will provide preliminary information to inform future academic studies and practitioners about the basic characteristics, motivation factors, and expectations of Y and Z generations and inform the development of new touristic products. Finally, in future studies, it is predicted that comparing the members of Generations Y and Z in the young and advanced age groups will be more useful in determining the differences between the two generations.

## 3.3 Spirituality

Any expression of divine meaning is absent. The spirit that exists in the inner world of man is completely absent, that is, he is unknown (Rinschede, 1992). There are those who think that spirituality is limited only to this believer. The usage is completely correct, but if the meaning in the substance is not perceived, there will be a deficiency. In natural processes, a substance other than the "meaning in matter" exists, and the meaning it contains and the meaning attributed to it vary according to the type of matter. The meaning in the matter mentioned here is not absent; rather, it is the subject of complete science. "Meaning in matter" is the term given to the values that can be explained in the context of the meaning, strength, energy, function, cause and effect, emotion and impulse of matter. For example, the meaning in the atom, the cell, the DNA, the dark matter, etc. is either definite or open to examination. For example, cells reproduce, tissues, organisms and the body are formed, there is a dynamism, there is a spontaneous generation. This is natural and real. The universe started with the big bang, it was developing, it became hydrogen, helium, then stars, galaxies, etc. Scientific explanations can be expressed in this way. In fact, it can be said that organic structures and cells came into being from inorganic matter, living species came into being. All these are structures that can occupy a material place and can be observed. However, there is interaction, rule, balance, measure and bond within and between each other. One can even talk about another issue that puzzles scientists; the divine meaning intervenes in the formation of matter by affecting the meaning in matter. For example, if you explain exactly how the solar system settled in the safest part of the Milky Way and this natural process developed, this can be regarded as the intervention of the divine meaning to the meaning in the matter. Whatever the balance, measure, law, goal, purpose given to the natural functioning of the meaning in matter from the divine meaning, the result becomes manifest.

We can observe what is happening and speculate over how it happened. We renew our ideas with what we learn at each step forward. This is an increase in consciousness, just like the expansion of the universe; DNA, cells, atoms, whatever is developed, matter and meaning develop, diversify, always an advanced version emerges, the remaining ones are transformed, there is no loss, there is transformation and consciousness develops due to all this development.

Let's express the types of meaning for human. The soul of man is of the kind of meaning in matter, it is clear, it can be examined, criticized and so on. On the other hand, there is a given soul in man and this place is absent. Spirit is divine essence, related to divine spirituality. In order to make the soul understandable, I call it a bond that harmonizes with the divine meaning in the gifted person. This issue is confused or deliberately done because of the difference of opinion between different beliefs. The reason is due to the lack of meaning in the matter. Those who describe it as something that has no spirituality are making a mistake. Sometimes they can even lead to giving meaningless examples. One cannot know the explanation of the spirituality of the unknown. It is necessary to be able to talk about the soul while including human beings: is this possible? When we confuse the subject of the soul with the soul, we make an important mistake. In the material world of man, his spirituality is related to his soul. Fields such as psychology and psychiatry (based on "psyche") examine this.

## 3.4 Influence of spirituality on members of the Y and Z generations who work in the field of tourism and hospitality

Countries produce various policies within the scope of faith tourism for an average of 300 million people travelling for faith tourism in the world and an estimated annual income of $20 billion. In parallel with the developments in this sector, the competition between countries that want to get a share from this cake is increasing on a daily basis. Developing countries have started to attach great importance to the concepts of development. Tourism is the second-fastest-growing sector in the world, after oil trade. With the development of world tourism, tourists, while people of different religions seek new countries and new regions, also explore and explore the different types of tourism in that region. For this purpose, religiously important mosques, churches, monuments that are important in terms of art history, museums, houses, buildings that are the first or interesting examples of their kind due to their architectural features, historical artifacts are evaluated in this context and presented to the service of tourists. Within the scope of faith tourism developing across the world, it has become very important to protect existing tourism values, to bring them into tourism and to introduce them. In the case of religious tourism, tourists visit faith and sacred places based on two main factors as a result of their religious duties and curiosity. Naturally, belief is not a commodity that is bought, sold and traded. However, it is also a fact that people travel in order to be "pilgrims", to see holy places and to reach spiritual satisfaction as required by their beliefs

Benares in Brahmanism is just one of the sites that is known and associated by the whole world. Other universally recognised sites include Mecca and Medina in Islam; Jerusalem, Rome and Ephesus in Christianity; In Judaism, Jerusalem is again the most visited sacred place (Meltem, Acar, & Dinçer, 2018). With regard to the religion of Islam, the revenues from pilgrims travelling to pray in the city of Mecca during the Eid al-Adha are second only to Saudi Arabia's earnings from oil. Israel is a leading destination in faith tourism. It is followed by Italy, England, Ireland, Egypt and Greece. One-third of the groups participating in faith tourism travel in groups and they stay for an average of between 8 and 14 days, with individual spending rates of between $1,500 and $3,000.

Tourism, which is an increasingly important sector in the world, affects their psychological and social needs, either closely or remotely. As a result of the values it obtained, tourism is generally social in nature. Because of this, tourism, which is a versatile and complex event that deals with business, sociology, politics, economics and similar branches, has been conceptually evaluated using very different definitions. When the general definition of tourism is made in line with the common aspects of these definitions, there is a definition of vacation, recreation, entertainment, culture, travel etc. outside the central area where touristic consumers are to be found constantly. Travel and temporary accommodation movements made in order to meet these needs are given the label of tourism.

The perception, in its most general sense, is that individuals have knowledge about their inner and outer worlds. Perception is when individuals obtain information from their environment and select, comprehend, organize and interpret this information. In other words, perception is the task of selecting, organizing and interpreting the meanings of the stimuli elicited by the sensory organs of individuals. The differences in the effects of tourism also affect the tourism perception and these effects also change the perceptions of tourism. The perception of the social, cultural, economic and environmental effects of tourism creates the total perception of tourism.

Since its formation, the world has been in a process of constant change. Changes in various fields such as technology, art, culture, politics and economy are felt across all areas of life and have an impact on individuals' perceptions, expectations, attitudes, behaviors and priorities. The interactions in this process of change cause the separation of the generations. It is known that these separations between generations are experienced or will be experienced in subjects such as beliefs related to work, participation in work and work, teamwork, career development, and gender differences. Although each generation differs from the one that preceded it, the rate of these differences changes as a result of the interaction of generations and periodic effects. Some of the differences that exist between individuals regarding the environment in which they live are perceived as meaningful, while others are disregarded. Likewise, while some differences present critical challenges or opportunities for businesses that want to employ them, or other individuals with whom they interact and communicate, they may be seen as irrelevant to other business representatives or individuals. Both across the world as well as in Turkey itself, these generations are better qualified, more technologically savvy and more diverse than any previous demographic.

The expectations, behaviours, and job attitudes of individuals belonging to each generation are affected by their important life events, cultures and even perceptions formed through their five sensory organs.

Compared to Generation Y, Generation Z is more conscious about getting for value. While family ties and friendship ties are stronger in Generation Y, individualism and the desire to be alone is higher in members of Generation Z, who like to spend their time playing virtual environments and online games. However, because they are younger, the members of Generation Z are under constant family supervision. Generation Y is always optimistic, and they have an extensive dream world. Generation Z has a realistic structure and a practical, utilitarian structure rather than a dream. Generation Y is cause-oriented and questioning. Generation Z focuses on results rather than causes and deals with the returns of events. Generation Y grew up in a world rapidly changing by technology, whereas Generation Z was born into technology. While individuals need messaging with mobile devices for communication among themselves, individuals in Generation Z can communicate with symbols and symbols without the need to write messages.

Groups formed by individuals who were born during a certain period of time, lived during the same periods, whose value judgements, lifestyles and perceptions of the world are similar, and who have many common characteristics shaped under the influence of social, cultural and economic events during the period in which they live are defined as generations. It can be mentioned that individuals belonging to the same generation have significant similarities in their attitudes, communication styles, preferences, decision-making processes and purchasing decisions. In this sense, all sectors need to know and define the generations correctly in order to better understand the consumer profiles they address. Along with the changing age, the consumer profile also changes, and the effects of the Y and Z generations can be felt today. In this sense, tourism is like other every sector. It is thought that the sector should have information about the generations that make up the target audience in order to produce products and services that are compatible with the changing consumer profile and to satisfy the demand correctly. Whether there are differences is the question considered by this study. In the study, the tourism preferences of Generations Y and Z, which will shape the tourism sector in the present and future, and its decisions were determined by a questionnaire applied on the selected sample. The study considered the tourism perceptions of the Y and Z generations, their expectations from the accommodation businesses, the reasons for choosing the accommodation business, the reasons for participating in tourism activities, and the frequency of going on vacation. It was questioned whether or not there is a significant difference between the two generations, based on their gender, education level and profession.

Generation Y, who is exhausting the marketing world with creating strategies in terms of consumption habits and behaviours, is not without tourism. It is very important for this generation to live their lives comfortably. For this generation, technology is the symbol of many things in their lives; technology is one of the issues where Generation Y is superior to Generation X. Technology is a lifestyle for Generation Y, who is intertwined with technology, and the rate of interest in these communication tools is much higher than for their predecessor Generation X. Generation Y can easily complete their daily work by using their technology skills creatively. In addition, 89.6% of Generation Y tourists went on vacation twice a year, while the use of the internet at the decision-making stage is also confirmed by statistical data which shows that they also made their decisions by taking the advice of tour operators and acquaintances. This stands out as the biggest disadvantage of Generation Y, but it also makes it the most loyal follower of the advertising world. Because this generation has a consumer spirit.

The information gathering and evaluation behaviour of this generation before the purchasing decision is shaped in particular way. A traveller who is seeking out information about a location to which he has not travelled previously will draw on information from various accounts,, whether commercial or not, and will supplement this with information he receives from media channels about the region he wishes to visit. Therefore, the place of hotels in the media is an increasingly important one. Interpretation portals are the first condition to reach Generation Y, who wants to be able to seek out a comparison of the most appropriate places. This is why the position in comment portals is very important for Millennials.

When discussing this generation, there is little point in discussing the Internet coming into their lives. This generation is different, since for them the Internet is a natural necessity. They are born into a world of multiple screens, where access to information is literally "child's play". For them, accessing information is as easy as the blink of an eye. Moreover, I cannot say that they are content with access to information. They are also very adept at processing information and bringing it back to life. Well, when this is the case, fast thinking and decision-making mechanisms are also at a very advanced level. Some experts call it Generation Z; others call it the mobile generation. Frankly, I'm always in favour of labelling it the "Introduction to Digital 101" generation. Because technology is developing so fast we have not yet been able to internalize the digital transformation, brand new concepts enter our lives. It is a fact that Generation Z is a generation that brings together all existing generations with the digital world. Sometimes we have trouble understanding them. Because they're independent, they're free. They care about individualization rather than socialization. This approach paves the way for us to change many habits in our lives so that we can understand them. Of course, the most important of these is in the area of business life. Some companies were proactive and began to look for ways to deal with this generation at an early stage: courses given directly to managers on this subject, seminars on the introduction of the new generation, hundreds of articles written... Finding a way to agree with them and trying to create added value from the innovations this dynamic generation will bring us a step closer. What about others? Unfortunately for them, I cannot write positive and promising sentences. Because there is an inevitable fact: All businesses that do not structure their business according to the rules and habits of the new generation are doomed to lose. And, unfortunately, businesses operating in the tourism sector will be the first to lose. I would like to have a more optimistic view of the tourism representatives of our country, but unfortunately it is impossible to say that there is any active work on this issue. We read; we follow. Almost every day, the work carried out at the management level for tourism is reflected in the press. We set up development agencies, we shoot promotional commercials (work we have carried out for 15 years), we organize press trips to our historical locations, we bring new applications for China to visit our country more... We do it all(!) But never as dynamic, variable and vast as tourism elsewhere. We do not take the ideas of the young generation for a sector. But we never bring together young people who will support the realization of digital transformation. And we never support projects suitable for the travel habits of the new generation. The generation that we call Generation Z has already started to travel individually. Some of the differences that exist between individuals regarding the environment in which they live are perceived as meaningful, while others are regarded as meaningless. Likewise, while some differences present critical challenges or opportunities for businesses that want to employ them or other individuals with whom they interact and communicate, they may be seen as irrelevant to other business representatives or individuals.

It will be useful in solving conflicts, problems and troubles that may arise.

One should never ignore the idea that a generation that absorbs so much in a digital form could be a part of the 'co-creation' movement. This generation loves to be a part of their own travel experiences, meaning that they never focus on one type of travel. It looks for areas of experience; it wants experience points (Benckendorff, Moscardo, & Pendergast, 2010). It also wants to create its own experience. However, unfortunately, we do not yet have any meaningful work on this subject. As yet, we promote our country only with good photos and good writing. Travel programmes focusing on experiences and feelings, technological experience centres, mobile applications where tourists visiting our country can create personalized travel programmes. There is more than one example under this topic. Currently, if we were talking to the Turkish tourist authorities about their implementation we might reasonably ask them: did you not think it would be more beautiful?

With regard to hotel businesses, of course, one of our main bleeding wounds is hotels. I'm not talking about chain hotels because these are transforming in one way or another, internalizing practices that meet the needs of the new generation. The main problem is mainly experienced in prestige hotels. In these types of businesses, it would be the wrong approach to cite only hotel management as an example of this issue. Unfortunately in all sectors, such decisions are based on the logic of 'maximum income, minimum expense'. At the end of the day, the only criterion considered is room occupancy. However, if these hotels have balanced their "branding" strategy with their sales strategy, the number of loyal visitors we call repeat guests will also increase. This will bring along sustainable success. Of course, making this possible requires following trends and being able to provide the right answers to the needs of the next generation. It is hard to believe, but there are still operators who describe digital marketing as an "expense" in today's world. What a pity! Do you not take into account the power of social media to affect the whole world from a single point? It's sad, very sad! Urgent measures must be taken to avoid these scenarios and to become one of the favourite destinations of the new generation. Otherwise, no other group will visit our country other than the "retired European tourists between the ages of 50 and 75" that we longed for. What we can do is always more than what we do. Despite all the negativities, the world is changing, to brighter tomorrows where we can embrace change.

## 3.5 Conclusion

Discovering the meaning of life and achieving it is one of the main goals of human creation. Research studies into travel and travelling reinforce one of the most fundamental and most influential components of the development of a person's spiritual intelligence, and ultimately his spiritual intelligence improves, which helps him to realize spirituality in his life. On the other hand, when we travel through the life and history of the science and moral elders, we witness that many distinguished personalities of human societies travel many times in order to attain moral virtues, become increasingly enlightened and discover the sources of truth. These travels made these people more resilient and purified their soul through wandering in nature. Therefore, in spiritual books, as well as self-help and poetry books, the positive effects of travelling on the glorification of the spirit and morality of the human being have been mentioned abundantly. Travelling changes the traveller's view of the world by seeing nature and historical artifacts inherited from the past. Tourism professionals who want to attract Generation Y and Z tourists to their destinations should highlight hedonistic motivations, such as relaxation, pleasure and entertainment, and also emphasize local culture, natural environment, different adventure activities, recreation and entertainment opportunities in their marketing messages. In addition, it is thought that messages that emphasize emotion and experience rather than information in marketing communication, especially for Generation Z, will be more useful. Tourism, which is an increasingly important sector in the world, affects their psychological and social needs, either closely or remotely. As a result of the values it obtained, tourism is generally becoming increasingly social in character. Because of this feature, tourism, which is a versatile and complex event that deals with business, sociology, politics, economics and similar branches, has been conceptually evaluated in very different definition approaches. When the general definition of tourism is made in line with the common aspects of these definitions, there is a definition of vacation, recreation, entertainment, culture, travel etc. outside the vital area where touristic consumers are constant. In the study, the tourism preference of Generations Y and Z, which will shape the tourism sector in the present and future, and its decisions were determined by a questionnaire applied to the selected sample. In the study, the tourism perceptions of the Y and Z generations, their expectations from the accommodation businesses, the reasons for choosing the accommodation business, the reasons for participating in tourism activities, the frequency of going on vacation; It was questioned whether there is a significant difference between the two generations based on their gender, education level and profession. In terms of touristic experience, searching for meaningful and spiritual experiences pushes people to go beyond the narrow form of spiritual tourism again and helps travellers to provide a spiritual integrity other than travel. (Bolton et al., 2013)

Looking at the studies of spiritual tourism, and the reasons that drive people to spiritual seeking, it can be stated that "seeking and finding new ways and methods to make themselves healthier by paying attention to the emotional and spiritual lives of people in addition to their physical care". According to Cohen (2002), this is. It has been expressed as spirituality and tourism in our new world where people remain strangers. In order to understand the travel motives of spiritual tourists, it is necessary to understand the motivations of spiritual tourists. It is seen for what purpose spiritual tourists travel, which emotions are at the forefront and in which areas they travel. According to the table in Cohen, it is seen that spiritual tourists seek spirituality in order to gain natural experiences by keeping their beliefs in the foreground and respecting the characteristics of the local people.

To summarize, spirituality is what most people feel internally, different from what they read and learn and what is imposed. It is thought that the universe gives every living thing a different power; spirituality is the awareness of this inner power. As long as spirituality remains in the language, it means nothing. Being connected with one's inner world means giving love to everything that exists in this world. Introversion, tolerance and acceptance, and mental, spiritual and emotional aspects must be peaceful and balanced.

## References


Ayhün, S. E. (2013), "Kuşaklar Arasındaki Farklılıklar ve Örgütsel Yansımaları", *Ekonomi ve Yönetim Araştırmaları Dergisi*, Cilt: 2, Sayı: 1, s. 93–112.

Bansal, H. and Eiselt, H. A. (2004). Exploratory Research of Tourist Motivations and Planning, *Tourism Management*, 25(3): 387–396.

Barton, C., Koslow, L., Fromm, J., & Egan, C. (2012). *Millennial passions: food, fashion, and friends*. Boston: Boston Consulting Group.

Beerli A. and Martin, D. J. (2004). Tourist Characteristics and the Perceived Image of Tourist Destinations: A Quantitative Analysis- A Case Study of Lanzarote, Spain, *Tourism Management*, 25(5): 623–636.

Behrstock-Sherratt, E. and Coggshall, J. G. (2010). Realizing the Promise of Generation Y, *Educational Leadership*, 67(8): 28–34.

Benckendorff, P. Moscardo, G. and Pendergast, D. (Eds.) (2010). *Tourism and Generation Y*. Wallingford, Oxfordshire: CABI International.

Bolton, R. N., Parasuraman, A., Hoefnagels, A., Migchels, N., Kabadayı, S., Gruber, T., Komarova, Y. and Solnet, D. (2013). Understanding Generation Y and Their Use of Social Media: A Review and Research Agenda, *Journal of Service Management*, 24(3): 245–267.

Cohen, A. B. (2002). The importance of spirituality in well-being for Jews and Christians. *Journal of happiness studies*, 3(3), 287–310.

Etlican, G. (2012), X ve Y Kuşaklarının Online Eğitim Teknolojilerine Karşı Tutumlarının Karşılaştırılması, Yüksek Lisans Tezi, Bahçeşehir Üniversitesi, Insan Kaynakları Yönetimi Bölümü, Istanbul.

Kyles, D., (2005), Managing Your Multigenerational Workforce, *Strategic Finance*, 87(6), 53–55.

Meltem, Altınay Özdemir, Acar, Ayşegül and Dinçer, Füsun İstanbullu, (2018), Dünden Bugüne Spiritüel Turizm, Uluslararası Sosyal Araştırmalar Dergisi, Cilt: 11 Sayı: 58 August 2018.

Phukan, H., Rahman, Z. and Devdutt, P. (2012). Emergence of Spiritual Tourism in India. *International Journal of Marketing and Technology*, 2(4), 300–317. Erişim Linki: http://www.ijmra.us.

Reeves, T. C. and Oh, E. (2008), Generational Differences, in *Handbook of Research On Educational Communications And Technology*, J. Michael Spector, M. David Merrill, Jeroen Van Merrienboer, Marcy P. Driscoll, (Eds.). 3rd ed., Taylor & Francis Group, New York.

Rinschede, G. (1992). Forms of Religious Tourism. *Annals of Tourism Research*, 19 (1), 51–67.

Vargheese, A. J. S. and Vargheese, P. S. (2012). Dimensions of Spiritual Tourism in Tuiticorin District of Tamil Nadu in India – A Critical Analysis. *Business Intelligence Journal*, 5(2), 245–251.

Williams, C. K. and Page, A. R. (2011), Marketing to The Generations. *Journal of Behavioral Studies in Business*, 3(3), 11.

# 4 Millennials and spiritual journeys

## The ways of Saint James in Northern interior Portugal as a contemporary pilgrimage

*Pedro Ricardo Coelho de Azevedo*

### 4.1 Introduction

Religion and spirituality are among the most common motivations for travel (Timothy and Olsen, 2006; Sharpley, 2009). Inserted in a phenomenon of contemporary pilgrimages to follow ancient religious paths, the Ways of Saint James are an excellent example of the symbiosis between religion, spirituality and tourism. In recent years, they have witnessed an exponential growth, being covered by an increasing number of tourists and pilgrims, especially by the Millennial generation, who look to challenge themselves and acquire a spiritual experience, because by travelling to places considered "sacred", they experience something different, achieving a deeper self-connection (Graburn, 1989). In addition, young people walk the Ways because they see it as a mystical route, allowing them a feeling of freedom, of authenticity, of sociability, of being in touch with nature and of escaping routine. In short, the Ways of Saint James is a way of knowing the world in a pure and authentic way.

The Millennials end up giving a new dynamic to the Ways of Saint James, as they travel the various itineraries with a variety of motivations (Lois-González and Santos, 2014), drawing on their very own identity characteristics. They make the trip either on foot or by bike, and mostly in groups. These indicators allow the Ways to be inserted in new forms of tourism that are different from conventional tourism. In this work we will adopt the designation of tourist-pilgrim, because this unites the concepts of tourist and pilgrim, since their motivations are very similar (Pereiro, 2017). In fact, there are many similarities between the pilgrim and the tourist, since both travellers are motivated by the experience of perceiving the trip and seek the same type of services and activities (Amaro et al., 2018). The Jacobean cult and the pilgrimages to Santiago de Compostela are a phenomenon that has several centuries of history, dating back to medieval times. In recent years, it has witnessed a profound revival.

In 2019, for example, another record for the pilgrimage was set, with the arrival of 347,578 pilgrims (https://oficinadelperegrino.com/), which indicates an exponential and constant growth in the number of pilgrims walking the Ways. Consequently, these data show that pilgrimage has intensified and diversified.

DOI: 10.4324/9781003170631-5

The United Nations World Tourism Organization (UNWTO) points out that youth tourism has become one of the fastest-growing segments in international tourism (UNWTO, 2016). The Millennials generation is highly regarded in today's world (Rainer and Rainer, 2011). Walking the Way has become popular among Millennials, both in Portugal and in other European countries. The Millennials interviewed have a specific profile and identity characteristics. Their motivations are also different from those of other so-called conventional tourists and pilgrims who walk the different Ways of Saint James. As will be shown, Millennials have visions and perceptions which are very adequate to the environment of the Ways of Saint James in the northern interior of Portugal: tranquility, history, nature, safety, and sociability, building a very specific image of these Ways. This study shows the growing importance of the Ways as they are increasingly used by Millennials and that the Ways go beyond their religious and cultural component, being walked by different types of tourists, as well as with different motivations.

Moving away from the religious motivations of the pilgrimage to Santiago de Compostela, the generation of Millennials who travel these Ways makes them fit into a phenomenon of new contemporary pilgrim motivations, a more spiritual one, linked to the search for new experiences, contact with nature and cultural heritage, to acquire moments of adventure, sociability and sport (cf. Granero Gallegos et al., 2007). Moreover, these new pilgrimages integrate various infrastructures and combine them with other tourist services. As will be shown, this pilgrimage experience has proven to be an enriching personal experience, especially for the youngest, who are part of the Millennial generation.

The structure of this work is divided as follows: there is a first section about the contextualization of the Millennial generation and a theoretical framework that explains its relation with the pilgrimages to Santiago de Compostela; the second section focuses on the geographical and historical contextualization of the existing Ways of St. James in the interior of northern Portugal; the third section explains the methodology adopted to carry out this study; the results obtained are then analysed; finally, the sixth section demonstrates the conclusions and their relevance to the topic.

## 4.2 Millennials: visions and speeches

### *4.2.1 Characterization of the Millennial generation*

The Millennial generation, also known as Generation Y or Gen Y, is generally characterized as regarding tourism and travel as fundamental activities for establishing their identities. This generation is defined in terms of the years of their birth, although authors vary in terms of where they set the divisions: some define Millennials as all those born between 1980 and 1994 (Yamawaki and Sarfati, 2019), whereas others extend this interval to between 1980 and 1999 (Garikapati et al., 2016), while other authors refer to millennials as the young people who belong to the new millennium, that is, those born between 1980 and 2000 (Lee and Kotler, 2016). It should be noted that the Millennial generation succeeds Generation X, who were born between the 1960s and 1980s; and also that it precedes Generation Z, which comprises those born between the late 1990s and the early 2010s.

Millennials travel more than any other generation (OECD, 2018). As Eran Ketter states, European millennials take between four and five trips per year (2020). Fundamentally, they end up playing an important role in boosting the economy and tourism in the regions. In recent years, there has been a growing connection between Millennials and spiritual travel, where they engage in meditation practices and seek to experience remarkable events, share moments from their lives and have convivial moments.

Millennials have a number of identity characteristics: they master new technologies, have environmental concerns and are sometimes members of environmental movements (Liu and Eastman, 2012), are used to technology (important for gathering and sharing information, including on sites and apps) (Kotz, 2016; Cunha, 2019) and are geared towards socialization. Because they are still in the process of self-discovery, Millennials choose and consume products or experiences that can inform this discovery (Ordun, 2016). They get their information about travel and accommodation mainly from the Internet. Most of those on the pilgrimage are single, being often either university students who about to finish their courses or young professionals.

### *4.2.2 Millennials in ways of Saint James*

The Millennial generation has the lowest expression of faith and religious behaviours (Nelson et al., 2010), going against the secularism present in society and in religious spaces (Turner, 2011). The Ways can be travelled in a mixed version, that is, one that combines the religious and spiritual versions, but also as a tourism experience, with a personal side that is linked to an initiation ritual (Prat, 2011). Table 4.1 presents the main differences between spiritual tourism and religious tourism.

In fact, the opening of sacred sites to tourism in recent years has led to the secularization of religion, and these sites are now visited not only for religious

*Table 4.1* Main differences between spiritual tourism and religious tourism

| *Spiritual tourism* | *Religious tourism* |
|---|---|
| Believe in something (plurality of manifestations) | Religions (Christianity, Buddhism, Islam, (...) |
| Paganism | dogmas/beliefs |
| Places connected to nature: parks, mountains (...) | Sacred places: chapels, churches, cathedrals (...) |

Source: Self-elaboration.

reasons, but also for other reasons. It is important to identify the main factors that lead the Millennial generation to walk the Ways of Saint James:

- Going through the Ways of Saint James marks a phase of growth and rediscovery of oneself. They end up being part of a trend in which increasing numbers of young people are travelling along the various Ways throughout Europe;
- Following secular paths combines with modernity in a reinterpretation, seeking to obtain unique and authentic experiences (McGettigan and Voronkova, 2016: 37);
- It allows moments of conviviality and sociability between tourists and pilgrims of various nationalities;
- It is understood as a form of liberation – the so-called "escapism", i.e. escape from routine (Cavagnaro et al., 2018). It also allows for moments of self-reflection, being with oneself, a preparation for life, autonomy (Seryczyńska, 2019), sometimes allowing to believe in or return to religion;
- It allows the establishment of contacts with local communities, their localities and traditions (Nisbett and Strzelecka, 2017);
- It provides an experience for all life. The pilgrim's own credential, the "Compostela", refers to the symbolic and material value of the Camino, since it serves as a material souvenir;
- It can be inserted into health tourism, since Millennials seek a healthy lifestyle and well-being;
- They use the route for sporting activities such as trekking and cycling;

The Pilgrim's Office provides some data to support that the Paths are travelled by Millennials taking into account the factors mentioned above, which should be analysed:

a) Young people are the second age group, up to the age of 30, who walk the Pathways the most, increasing in 2011 from less than 60,000 to more than 90,000 in 2019. Those aged between 30 and 60 make up the majority of these (Figure 4.1).
b) Without any doubt, the most frequent means of transport is walking. However, the use of bicycle to travel along the paths has remained stable, which can be explained by the fact that it is a mode of transport frequently used by young people and linked to new motivations such as leisure and sport (Figure 4.2).
c) In recent years, the Pathways have been widely travelled by students. In 2009 there were 26,844; by 2019, however, this figure had more than doubled to 62,681 students (Figure 4.3). This indicator is key because students correspond to the Millennial generation.

At a time when Covid-19 predominates, the Ways are seen as a safe place, integrated into nature, which gives only limited social contact. This idea of safety has triggered that the various paths have been considerably walked in recent times.

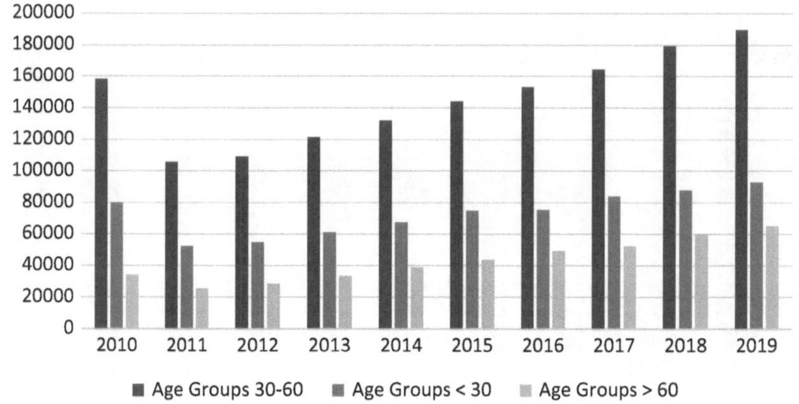


*Figure 4.1* Age groups of tourist-pilgrims between 2010 and 2019.

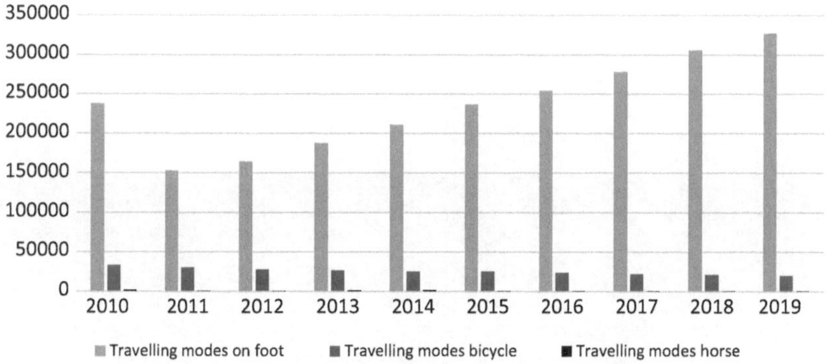


*Figure 4.2* Travelling modes between 2010 and 2019.

### ***4.2.3 Ways of St. James in the north interior of Portugal***

The Portuguese Way is the second most travelled way to Santiago de Compostela, after the French Way (https://oficinadelperegrino.com/estadisticas2/). The main existing Santiago Ways in the northern interior of Portugal are the Portuguese Interior Way of Saint James (PIWSC), the Portuguese Silver Way and the Torres Way. However, there are also other variants, such as the Terras de Basto Way and the Coração D'Ouro Way, which consist of transverse paths, as shown in Figure 4.4.

These are characterized by being routes that in recent years have seen a profound revitalization of their heritage and tourism.

The most walked path is the PIWSC, with thousands of tourist-pilgrims every year. The other paths have lower numbers because, on the one hand, they are still being revitalized and, on the other hand, they do not yet have a

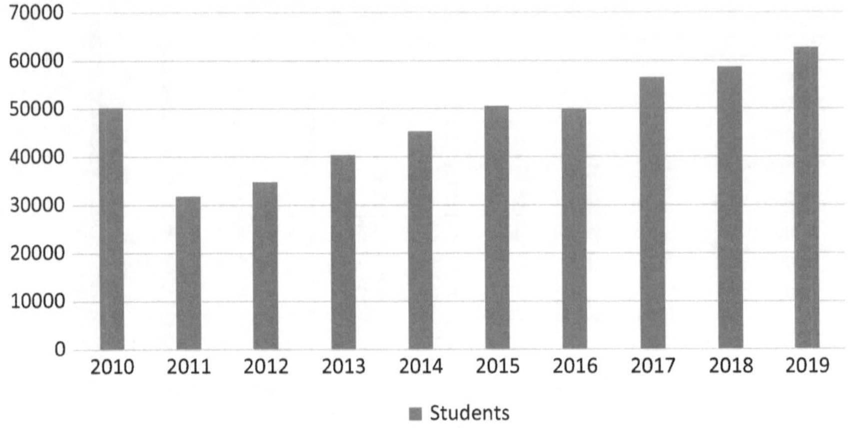


*Figure 4.3* Annual evolution of students between 2010 and 2019.

*Figure 4.4* Ways of Saint James in the northern interior of Portugal.

tourist dimension. The PIWSC has a length of about 205 kilometres between Viseu and the border between Portugal and Spain. It is also the most renowned route because it crosses areas of high historical and heritage interest, such as the UNESCO World Heritage site of the Alto Douro Vinhateiro. It also passes through countless cities and towns and features a number of other important tourist attractions.

Along with their recent revitalization, these paths share a number of features: their reduced tourist massification in comparison with other paths, ultimately allowing an intimate and pilgrimage experience; their location in predominantly rural territories and the environment, natural and landscape surrounding the various paths in the region, as shown in Figure 4.5, which shows the landscape framing of the Terras de Basto Way, situated in the middle of the Alvão Natural Park. In this way, it is sought after by specific segments of tourists, such as hikers and backpackers, among others.

The fact that these Ways are progressively being converted into cultural tourist routes makes them increasingly visible and sought after by national and foreign visitors. It should be noted that the cultural intimacy that the Ways provide make attract people from other countries to walk them, with particularly large numbers coming from Germany and Switzerland.

*Figure 4.5* Perspective of the Terras de Basto Way.

### 4.2.4 Methodology

The methodology applied in this study consisted in the use of quantitative and qualitative approaches to obtain broad and solid information, due to the complexity involved in studying the Millennials' traveling. In this way, interviews were conducted with tourist pilgrims, especially young people; in carrying out fieldwork; and also, fundamentally, in the application of participant observation. The methodological option of interviews and participant observation (Richards & Munsters, 2010), based on an empirical perspective, allowed us to analyze and characterize the importance of the Millennial generation for these Ways, framed in a contemporary pilgrimage.

In the first phase, we carried out a literature review, namely about the Millennial generation, in order to understand their relationship with pilgrimages from several perspectives. We also conducted a statistical survey related to the passage of tourist-pilgrims on the Ways of Saint James.

In a second phase, an open script with a semi-structured profile was built to support the interviews. The questions had two distinct sections: the first to record the interviewee's socio-personal information, including their name, age, studies or profession, and place of residence; the second section to record their experience on the chosen Way, their existing motivations, stages travelled, how they had travelled along the way, whether they were alone or accompanied, and also any other information considered relevant.

Two criteria were predefined for the interviews:

a) Interviewees had to fit the profile of Millennials, i.e. people between 19 and 35 years old. It is pertinent to define this sample in order to ensure uniformity in the identification of Millennials and that the sample shared these common characteristics (Kurrant, 2019).
b) To have travelled at least one or more existing Paths in the territory under study;

In view of this, the interviews were carried out during the months of January–November 2020, with tourist-pilgrims on the very Ways under study. We only considered 50 pilgrims, who constitute the study sample of this study. This was for two factors: first, only 50 of the interviewees fit the profile of Millennials, by age definition; secondly, because we consider that it represents a solid and representative sample of the number of tourist-pilgrims who travel these Ways.

At the same time, fieldwork and participant observation were adopted as a complement to the interviews, in order to dialogue and participate in the walks with the tourist-pilgrims, collecting information about the experience of Millennials considered relevant (Pereiro and Fernandes, 2018). At the same time, photographic records were taken. As we will see in the next section on analysis of the answers obtained during the interviews, we identified a set of motivations and perceptions that are largely identity of the millennial generation that walks these Ways of Saint James.

## 4.3 Analysis and results

### *4.3.1 Profile of interviewees*

Of the 50 interviews carried out with tourist-pilgrims who fall under the millennial designation, 8 are foreign tourists from a variety of countries; in addition, there are 6 Spanish, 2 Germans, and 2 Brazilians. The remaining 32 are Portuguese. There are 16 women and 34 men, all aged between 19 and 35.

Most of the interviewees live near the Ways: 29 live in localities crossed by the paths, 13 live in other parts of the country and 8 live in other countries. With regard to the occupations performed, 19 are students (especially university students), 30 have various professions such as administrative, teachers at different levels of education, businessmen, among others; only 1 interviewee did not have any occupation. 12 tourists reveal that they were walking the paths alone, while 38 did the paths accompanied, often in groups. Of the sample, 20 intended to walk the Ways as far as the end at the Santiago de Compostela Cathedral, whereas 30 revealed that they were walking the Ways in stages; that is, they returned to their starting point and then continued on the path from the next stage.

Another important indicator is the fact that 37 interviewees reveal that they are either religious or had some sort of belief or faith, while the remaining 13 declare they have no religion at all.

### *4.3.2 Interviews*

With the aim of maintaining a uniformity in the study of the paths, the interviews were applied to 34 tourist-pilgrims who have walked the PIWSC, as this is the most travelled route in the study area; in addition, 6 walked the Torres Way; 4 the Terras de Basto Way; 3 the Portuguese Silver Way; and 3 the Coração D'Ouro Way (Figure 4.6). However, it is worth mentioning that 12 tourist-pilgrims have already carried out two or more of the five itineraries under study and that 38 have already done other Ways of Saint James, either

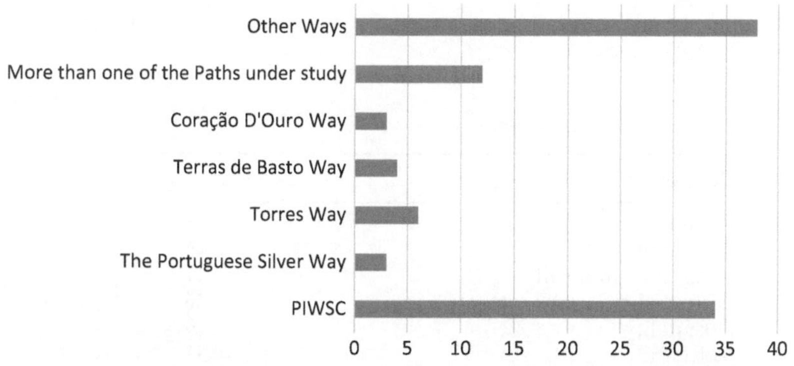


*Figure 4.6* Ways of Saint James in the northern interior of Portugal travelled by tourist-pilgrims.

in Portugal or abroad, which may indicate a certain "collectivism" in travelling several jacobean Ways.

In terms of motivations for walking the Ways of Saint James in the interior of northern Portugal, spiritual and multiple motivations both predominate, with 14 answers, followed by adventure (6) and by cultural and sporting motivations, both with 4 answers. Less expressive are the leisure (3), religious (3) and other motivations (2), as shown in Figure 4.7.

Regarding perceptions, the interviewees reported having several images in relation to these routes, having multiple perceptions, i.e. more than one, as shown in Table 4.2.

The Ways were done mostly on foot, by 42 pilgrims, while 8 did it by bicycle, as shown in Figure 4.8.

Regarding the activities performed during the trip, 42 had social moments, performed in proper places as we will see below; 14 respondents visited museums, 12 went to establishments to buy local and endogenous products; 22 visited gardens and parks; and 33 stopped in historical centres and visited the local heritage. 29 interviewees also stated that they did other activities, such as photographic records, thermal tourism, among others (Figure 4.9).

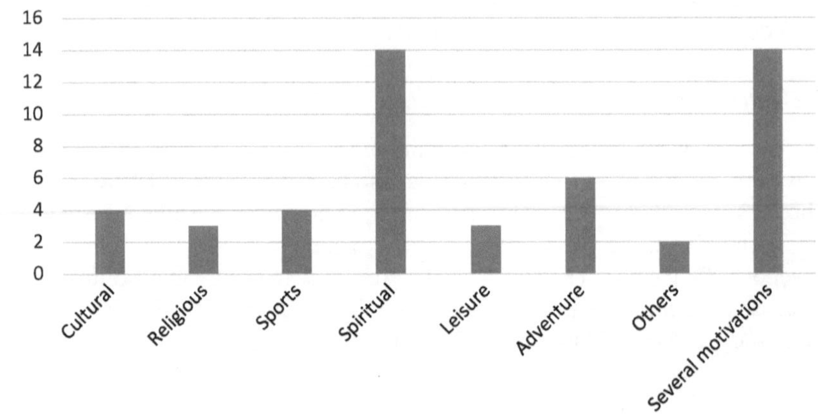


*Figure 4.7* Motivations.

*Table 4.2* Main perceptions

| *Perceptions* | *Number of responses* |
|---|---|
| rurality | 35 |
| tranquility | 46 |
| nature/environment | 47 |
| historical-cultural heritage | 26 |
| hospitality | 19 |
| Adventure | 32 |
| Mystical path | 40 |

Source: Own elaboration based on the interviews.

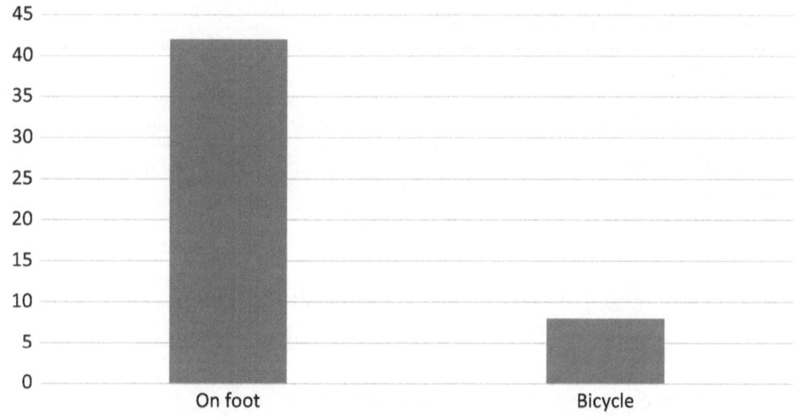


*Figure 4.8* Type of travel.

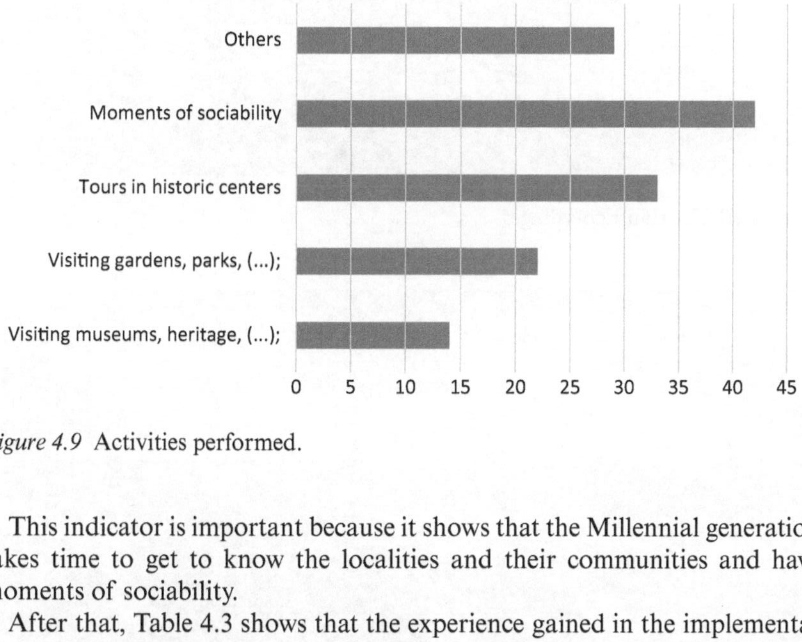


*Figure 4.9* Activities performed.

This indicator is important because it shows that the Millennial generation takes time to get to know the localities and their communities and have moments of sociability.

After that, Table 4.3 shows that the experience gained in the implementation of the Ways presents various responses, taking into account the internal and external perspectives:

The participant observation proved to be pertinent because it revealed that these tourist-pilgrims walk the Ways because they enjoy moments of conviviality and sociability. The eco-village of Cabrum, located in the PIWSC, is seen as a retreat that allows Millennials to find peace and their own path (https://www.ecolife-experience.com/), as shown in Figure 4.10.

They have specific meeting places, as shown in Figure 4.11, which represents a café on the Terras de Basto Way and is located in the Alvão Natural

*Table 4.3* Acquired experiences

| *Acquired experiences* | |
|---|---|
| *internal (personal)* | *external (tourist and cultural)* |
| personal growth | collection of paths taken |
| reflection | |
| better human being | |
| less materialistic | cultural enrichment |
| environmentally friendly | |
| proximity to religion and faith | academic learning |
| companionship/made friends, autonomy | |
| self-discovery | |

Source: Own elaboration based on the interviews.

*Figure 4.10* Cabrum Eco-village.

*Figure 4.11* Meeting point for tourists and pilgrims on the Terras de Basto Way.

Park. There are also other places, such as the hostels along the Terras de Basto Way. These spaces are meeting and socializing points between pilgrims of different origins.

Over the course of the study, a number of the participants provided insightful observations, as reflected below:

- "To do the Way is to be close to myself, it is to set out to discover, we detach ourselves from commitments and material goods" (male, 32 years old, Mondim de Basto, entrepreneur – Terras de Basto Way).
- "I acquired a new energy, I am a new me, I live the present" (male, 25 years old, student Chaves – PIWSC).
- "I do the Camino because I'm looking for confidence and personal growth, because the journey transmits tranquillity" (woman, 28, administrative, Vila Real – PIWSC).
- "The silence of the Way, the meeting with other pilgrims of various nationalities, the moments of sharing" (member of the association of walkers, 21 years old, student, Vila Real – several Ways). In short, these aspects clearly show the specific characteristics that each Camino de Santiago in the interior north of Portugal has, being an attraction for the Millennial generation.

## 4.4 Conclusion

It can be concluded that the young people who walk the Ways of Saint James located in the interior of northern Portugal are part of the generation of Millennials, because their motivations, perceptions and acquired experiences allow them to be integrated as such. It is notable that most of the tourist-pilgrims have a considerable connection with the Ways.

Being able to walk one of the Paths, whether in a line or in stages, is seen as a new experience and a sense of freedom that Millennials want to acquire. It ends up corresponding to a transitional phase in their lives, a break from their daily life. Travelling the Way in stages, which is a manifestation of slow tourism, they stop along the way to socialize, visit towns and monuments and get to know the local culture. In this way, they spend more time in the regions crossed by the paths and, in certain ways, they end up getting closer to conventional tourism and traditional pilgrims.

Thus, the three major premises detected in this study are: motivations, perceptions and acquired experiences. In view of this, the main arguments identified in this study regarding the young Millennials who travel these Ways are the following:

a) Moments of sociability;
b) Moments of authenticity;
c) Personal growth and autonomy;
d) Intimacy with nature;
e) The rural presence;

Despite the lower relevance given to religious motivations, some tourist-pilgrims reveal that they do seek to be close to religion. Yet even in this sense, we are not present in a pilgrimage in its broadest sense, but rather in the search for an approximation to something sacred.

The study discussed in this chapter can be replicated to other Ways and other regions, as it would be important to understand the growing presence of Millennials in the several itineraries of the Ways of Saint James in Portugal.

Although many Ways of Saint James are travelled annually by Millennials, the Ways analysed in this study gather a set of characteristics which give them their own identity features, allowing them to appeal to the Millennial generation, such as: the tranquillity, the natural landscape all along the Ways, and the issue of rurality, among other aspects mentioned above.

## Acknowledgements

This work is carried out within the scope of the Research Grant with the reference SFRH/BD/136459/2018 attributed by FCT – Portugal.

The Center for Transdisciplinary Studies for Development (CETRAD) is financed by national funds through FCT – Foundation for Science and Technology, I.P., under the project UIDB / 04011/2020.

## References

Amaro, S., Antunes, A. and Henriques, C. (2018) 'A closer look at Santiago de Compostela's pilgrims through the lens of motivations', *Tourism Management* 64: 271–280.

Cavagnaro, E., Staffieri, S. and Postma, A. (2018) 'Understanding millennials' tourism experience: values and meaning to travel as a key for identifying target clusters for youth (sustainable) tourism', *Journal of Tourism Futures* 4(1): 31–42.

Cunha, M. N. (2019) 'The tourism journey, from inspiration to post-travel phase, and the mobile technologies', *African Journal of Hospitality, Tourism and Leisure* 8(5), 1–17.

Garikapati, V. M., Pendyala, R. M., Morris, E. A., Mokhtarian, P. L. and McDonald, N. (2016) 'Activity patterns, time use, and travel of millennials: a generation in transition?', *Transport Reviews* 36(5): 558–84.

Graburn, N. Tourism: the sacred journey. In: Smith V. *Hosts and Guests: The Anthropology of Tourism*. Philadelphia: University of Pennsylvania Press, 1989.

Granero Gallegos, A. A., Ruiz Juan, F. and García Montes, M. H. (2007) 'Estudio sobre las motivaciones para recorrer el Camino de Santiago', *Apunts, Educación Física y Deportes* 3(89): 88–96.

Ketter, E. (2020) 'Millennial travel: tourism micro-trends of European Generation Y', *Journal of Tourism Futures* 7: 192–196.

Kotz, P. (2016) 'Reaching the millennial generation in the classroom', *Universal Journal of Educational Research* 4(5): 1163–1166.

Kurrant, C. (2019) 'Biographical motivations of pilgrims on the Camino de Santiago', *International Journal of Religious Tourism and Pilgrimage* 7(2): 11–18.

Lee, N. and Kotler, P. (2016) *Social Marketing: Changing Behaviors for Good*, California: Sage Publications.

Liu, J. and Eastman, J. (2012) 'The impact of generational cohorts on status consumption: an exploratory look at generational cohort and demographics on status consumption', *Journal of Consumer Marketing* 29(2): 93–102.

Lois-González, R. C. and Santos, X. M. (2014) 'Tourists and pilgrims on their way to Santiago. Motives, Caminos and final destinations', *Journal of Tourism and Cultural Change* 13(2): 149–164.

McGettigan, F. (2016) 'Walking Labyrinths: spirituality, religion and wellness tourism', *International Journal of Religious Tourism and Pilgrimage* 4(5): 37–50.

Nelson, M. F., James, M. S., Miles, A., Morrell, D. L., & Sledge, S. (2017). Academic integrity of millennials: The impact of religion and spirituality. *Ethics & Behavior*, 27(5), 385–400.

Nisbett, G.S. and Strzelecka, M. (2017) 'Appealing to goodwill or YOLO-promoting conservation volunteering to Millennials', *VOLUNTAS: International Journal of Voluntary and Nonprofit Organizations* 28(1): 288–306.

OECD (2018) *Tourism Trends and Policies 2018*, Paris: OECD Publishing.

Ordun, G. (2016). 'Millennial (Gen Y) consumer behavior their shopping preferences and perceptual maps associated with brand loyalty', *Canadian Social Science* 11(4): 40–55.

Pereiro, X. (2017) 'Portuguese touripilgrims in the Portuguese inside way to Santiago de Compostela', *Revista Turismo & Desenvolvimento* 1(27/28): 413–423.

Pereiro, X. and Fernandes, F. (2018) *Antropologia e turismo. Teorias, métodos e praxis*, La Laguna (Tenerife): PASOS.

Prat, J. (2011) '¿Por qué caminan? Una mirada antropológica sobre el Camino de Santiago', in Nogués-Pedregal, A. and Olmos, F. (eds) *La cultura sentida. Homenaje al profesor Salvador Rodríguez Becerra*, Seville: Signatura Ediciones, 495–529.

Rainer, T. S. and Rainer, J. (2011) *The Millennials*, Tennessee: B&H Publishing Group.

Richards, G. and Munsters, W. (coords.) (2010) *Cultural Tourism Research Methods*, London: CABI.

Seryczyńska, B. (2019). Creative escapism and the Camino de Santiago. *International Journal of Religious Tourism and Pilgrimage*, 7(5), 48–55.

Sharpley, R. (2009) 'Tourism, religion and spirituality', in T. Jamal and M. Robinson (eds) *The Sage Handbook of Tourism*, London: Sage, 237–253.

Timothy, D. J. and Olsen, D. H. (2006) *Tourism, Religion and Spiritual Journeys*, London: Routledge.

Turner, B. S. (2011) *Religion and Modern Society: Citizenship, Secularisation and the State*, New York: Cambridge University Press.

UNWTO (2016). *Spain*. UNWTO Highlights 2016, 16.

Yamawaki, M. and Sarfati, G. (2019) 'The millennials luxury brand engagement on social media: a comparative study of Brazilians and Italians', *Internext: Revista Electrônica de Negócios Internacionais Da ESPM* 14(1): 14–30.

# Part II

# Youngsters' inclination towards spirituality

The aim of Part II is to present research in the field of travel motivations and the inclinations of Millennials towards spirituality. This part amplifies the inclination of youngsters towards spirituality. Modern Millennials are inclined towards unexpected solo trips relating to interaction with nature, intercultural exchange, meditation, wellness, spiritual healing, active engagement and immersion in the local practices and lifestyle. The part also deliberates on linking spirituality to wellness, health and medical travel prospects among Millennials. It elaborates on the impact of spiritual tourism on work-life balance and the quality of life of Millennials and tourist behaviour and travel adoption among millennials. The major objective of this part is to provide insights into the growing inclination of Millennials towards spirituality.

Chapter 5, "Determining the Travel Motivations, Experiences and Expectations of Millennials: The Case of İstanbul, Turkey" by Sıla Karacaoğlu, aims to determine the key travel behaviours of Turkish Millennials aged 21-40 living in Istanbul. This chapter puts an emphasis on understanding the travel patterns of this generation and shows how catering for their changing needs is of vital importance for the current and future planning of the tourism industry over the years to come.

The next chapter, Chapter 6, "Millennials' Perceptions of Spirituality, Wellness and Travel" by Melanie Smith and Inge Yan Fong Chan, considers the role that spirituality plays in Millennials' everyday life in relation to other domains (e.g. social, economic, emotional, environmental) and in wellness (e.g. physical and mental health, lifestyle habits), as well as in travel motivations and benefits. The study confirms that Millennials mostly define spirituality in relation to the self and individuality, followed by connections to nature and humanism.

Chapter 7, "Millennial tourism: Myths, wellness, lifestyle, and practicality toward the economy" by Poon Wai Ching, uses the theory of cultural complexity and intends to identify how Millennial myths and lifestyles impact their mindset on the practicality of travelling. The chapter discusses that the

DOI: 10.4324/9781003170631-6

wellness tourism segment is another branch of Millennials' tourism. Millennials are more lifestyle-focused and target grander values on wellness experiences and health conditions than material possessions.

The final chapter in this part, Chapter 8, "Tourist behavior and travel adoption among Millennials" by Gül Erkol Bayram, evaluates tourist behaviour and travel adoption among Millennials and Millennials' travel characteristics, travel habits, and future travel trends are specified. This chapter puts an emphasis on the fact that it is important for businesses to increase their awareness about generations, understand their characteristics, develop personalized special products for them, adopt appropriate management understanding and leadership style in line with their wishes and expectations, and create the necessary working conditions.

# 5 A descriptive study on determining the travel behaviours of Millennials

## The case of Istanbul, Turkey

*Sıla Karacaoğlu*

## 5.1 Introduction

Gender, income, age, life standards and geography are among the determinants used by marketers to categorize consumers. Although the success of the methods mentioned above is certain, there exists another innovative categorization technique: to group consumers into generations (Pennington-Gray & Lane, 2002). Group segmentation based on generation has procured significant attention among researchers in marketing area due to the notion that the birthplace and life experience, particularly their coming-age years, have certain effects on behaviour, attitudes and values (Fountain & Lamb, 2011). People who show differences in terms of values, preferences, needs and expectations are often defined via generational segmentation theories (Davids, 2018). The reason behind applying the generational approach is based upon the argument that the values and consequent conduct shaped by these values are invariable, to some extent, and they separate the target generation from other generations. Hence, it is unbiased to state that market segmentation attained through generational cohorts is more fruitful compared to sole age segmentation and that generational approach renders the marketing of the target groups possible (Fountain & Lamb, 2011).

Currently, tourism marketers have realized the importance of targeting homogenous components of a heterogeneous market, that is segmentation, rather than considering the market as a whole. As a result, this efficient technique is used as a tool to define guest types aiming at envisaging who will react to promotions accordingly (Huang & Petrick, 2010). Understanding the effects of the social and demographic trends on traveller attitudes and behaviour is vital, therefore, applying multiple characterization and criteria; such as sex, attitudes, motives, age and activity preferences; while categorizing consumers holds great importance in comprehending the tourism market (Benckendorff & Moscardo, 2013). It should be borne in mind that using age as a demographic variable should imply chronological age, age groups and time periods rather than sole chronological age since marketing research, which has its focus point on alterations in consumption rates according to age cohorts, is inefficient in clarifying the effects of cohort succession (Huang & Petrick, 2010). In addition, factors such as beliefs, values and travel

DOI: 10.4324/9781003170631-7

experiences as well as age shape the tourism demands of generations owing to the difference in terms of socio-cultural and environmental background; each generation displays similar values, attitudes and conducts in themselves (Glover & Prideaux, 2008). The reason behind the importance of this generation is that Millennials were born in the age of information and constitute a major part of the workforce. Therefore, compared to preceding generations, Millennials have an increased motivation in terms of international tourism and they see travel as a way of life (Pendergast, 2010; Richards, 2007). With regard to generational cohort theory, generations are labelled with features which are the results of the events they witnessed in the years that shaped their characters; therefore, the information acquired via generational theory helps us in understanding visitor behaviour trends, attitudes and intentions in tourism (Benckendorff & Moscardo, 2013).

A comprehensive literature review has been conducted and it has been noted that there are very few studies which investigate the travel behaviours of Turkish Millennials (Albayrak & Özkul, 2013; Birdir, Toksöz, & Birdir, 2019; Caber, Albayrak, & Crawford, 2020; Çakar & Seyidoğlu, 2016; Kutluk & Avcıkurt, 2019). Therefore, it can be said that there is a gap in the area concerned. Istanbul was chosen as the research area in this study since it is one of the world's leading metropolises and it is the most developed city in Turkey. In this context, it offers important travel opportunities to those living there. Thus, in order to contribute to the development of the relevant literature, it is intended to analyze the travel and tourism preferences and behaviours of Millennials in Istanbul who are reckoned to be the core generation that shapes the contemporary travel movements in this study. In this regard, Millennial travelers are the subject of this research and the results of this study are expected to give ideas to the representatives of different sectors of the tourism industry, researchers and academics who want to have information about the tourism behaviour, demands and future expectations of Millennials.

## 5.2 Literature review

### *5.2.1 Generational theory*

From the studies, it is obvious that there is no agreed definition regarding the term "generation" and it is pointless to designate it with certain dates. In addition, to generalize a certain generation with the same views is impossible since it is obvious that people who constitute a generation are diverse in terms of beliefs, traditions, etc. (Greenberg & Weber, 2008). However, it has to be stated that a certain group of people holding common major characteristics are the people to whom the term generation refers. In this context, the term "generational group" (in other words a cohort) (Pritchard, 2014) is used to define people who were born at a certain time and experience the major changes in life, simultaneously being at a certain state of their lives at different times (United Nations [UN], 2001). From this point of view, it can be stated that there exists no sole and accurate type of generational theory.

Nonetheless, there are several types of theories which use the frame of it and which are reliable with regards to theorization (Huntley, 2006; Inglehart, 1997; Mannheim, 1928/1952; Strauss & Howe, 1991) and they are used to categorize generational cohorts of people according to expectations, needs, preferences and values (Davids, 2018).

Following the end of the Great War, (Mannheim, 1928/1952) and José Ortega y Gasset (1923 Ortega y Gasset, 1931, Ortega y Gasset, 1930 1932) focused on the exchanges among generations in order to find for explanations for social changes in the community (Bolin & Skogerbø, 2013). Mannheim first addressed the concept of generation in his controversial essay, "On the Problem of Generations", written in 1927/1928 and republished in 1952. Mannheim's study was the first study in the 20th century to develop the concept of generation with a systematic and detailed approach using social science methods (Jaeger, 1985). In his essay, Mannheim emphasizes that being a member of the same generation puts individuals on a common line in terms of their social and historical progression, and thus restricts them to probable experiences of a certain period, tailoring them to react in a certain characteristic physically and mentally. In addition, Mannheim argues that, apart from the effect of biological factors, social factors called generation location, generation as actuality and generation unit should all be taken into consideration when defining the generation (Mannheim, 1928/1952). To him, younger generations differ from older generations and this discrepancy lies behind the clash of values and belief systems which triggers the social change in the community (Pritchard, 2014).

The most well-known theory about generations was put forward by Strauss and Howe (1991) in the book *Generations: The History of America's Future, 1584–2069*. Although their theories were developed by taking into account only the history of the United States, the authors stated that generations and trends elsewhere in the world form similar cycles. For Strauss & Howe (1991: 60), a generation is "a cohort-group whose length approximates the span of a phase of life and whose boundaries are fixed by peer personalities". By determining the human life span as an average of 80 years, they stated that there exist 20 years between each generational intervals, meaning that the time length between birth and pregnancy is 20 years. They categorized these phases into four distinct segments: "youth", which lasting from ages 0 to 21; "rising adulthood", from ages 22 to 43; "midlife", from ages 44 to 65; and "elderhood", from ages 66 to 87. The focus point of the study reveals that there is a repetition after every four years in terms of generational social cycles (Pritchard, 2014). In addition, they stated that "epochal events" which a generation experiences between youth and adulthood are factors that shape a generation (Strauss & Howe, 1991). In other words, each generation is infused with a certain characteristic as a result of many conditions. These conditions can be counted as icons, changes in the physical and social environment besides financial and political incidents happen in the course of formative years such as the Civil War, the Watergate scandal, the bombing of Hiroshima, and the Great Depression. Alternatively, they may include rites of passage such as parenthood, marriage, retirement and so on (Dulin, 2008).

In this way, cohorts can be defined as people who are peers, who have alike experiences which may result in similar thoughts and emotions. This theory posits that common experience is a determinant of common fate (Rosow, 1978). These epochal events affect us at a deeper level and the things that we were doing at those times are crystal clear in our memories and, in time, we become aware of that they tailor us as a person.

### *5.2.2 The generations*

In the literature, it is seen that there are debates among the researchers about the age groups considered in the classification of generations, and there is no consensus due to the different turning points and stages experienced by societies in the historical process. Therefore, generations are classified closely to each other but with different dates and names (Howe & Strauss, 2000; Kyles, 2005; Lancaster & Stillman, 2002; Zemke, Raines, & Filipczak, 2000). In addition, geographical differences have to be taken into consideration since counterpart generations from Eastern and Western societies display clear differences (Knight, 2009).

As mentioned in the preceding section, it is the generation theory of Strauss and Howe (1991) which is the most known and accepted in the literature. As to this theory, they argued that there are seven generations in the USA, however, "The Missionary Generation" (1860–1882) and "The Lost Generation" (1883–1900) are aged above 100 and they no longer live in the society. The rest of the generations are aged between 1 year and 100 years. The first generation is the "Greatest (GI) Generation" (1901–1924). "Silent Generation" (1925–1942) comes second, followed by the "Baby Boomers" (1943–1960). Following the Boomers, we have "Generation X" (1961–1981) as the fourth generation. This is then followed by Generation X and "Generation Y" or the so-called "Millennial Generation" (1982–2002). Finally, those born in 2003 and later is called "Generation Z".

This generation theory does not comply map on to the situation in Turkey. When economic, social, political and technological developments are considered, generational distinction in approximately 100 years of existence of the Republic of Turkey can be stated as follows: the "Silent Generation" (1927–1945); the "Baby Boomers" (1945–1964); "Generation X" (1965–1979); "Millennials" (1980–1999); and "Generation Z" (2000 and later) (Kuran, 2018a). Since the study investigates the travel behaviours of Millennials, the next section will not mention the other generations. Only the general characteristics of Millennials will be summarized.

## 5.3 The characteristics of millennials

Millennials take place in the literature under many different names such as "Generation Y" (Gen Y), "Generation Why" (Pendergast, 2010), "Generation Zero" (Lepeyko & Blyznyuk, 2016), "Generation Me" (GenMe) (Twenge, 2006), "Echo Baby Boomers" (Eisner, 2005), "Nexters", "Generation Next"

(Gardner, 2006; Zemke, 2001), "iGen" (Twenge, 2006), "N Gen", "Net" or "Web Generation", "Digital Generation" (Benckendorff, Moscardo, & Pendergast, 2010). Millennials are the children of the Baby Boomer generation and the siblings of Generation X. In their formative years, they witnessed globalization and technological developments, especially in communication and wireless connectivity, and they were the first generation to have lived in a digital environment in their most formative years (Lepeyko & Blyznyuk, 2016). That is why both Millennials and the other generations that have come after them are also called "digital natives" (Prensky, 2001). Since the development of the Internet along with the computer took place during this generation, Millennials had more contact and relations with the outside world compared to other generations. This generation, which grew up during the years of incredible developments in world politics and technology, has also witnessed terrorism and some wars (such as the Iran–Iraq Gulf War, Afghanistan, Bosnia-Herzegovina wars, the September 11 attack) (Duygulu, 2018; Lepeyko & Blyznyuk, 2016). In addition, those in this generation, whose parents are very different from the parents of previous generations, have mothers and fathers (known as "helicopter parents") (Costanza & Finkelstein, 2015) who have more active roles in ensuring that their children can grow up in good conditions, receive a good education and stay safe (Özkan, 2017).

Howe and Strauss (2000) used seven terms to describe the characteristics of this generation: "special", "confident", "sheltered", "achieving", "team-oriented", "pressured" and "conventional". For Millennials, self-esteem, prompt results, spiritual serenity, flexibility, and fun factors in activities hold great importance. They tend to opt for goals which are reasonable in achievement and it can be stated that the main factors which shape this generation are developments in IT and digital technologies (Lepeyko & Blyznyuk, 2016). Millennials were born into the information age and this created a larger generation gap (in terms of the difference in values and character traits); it can be stated that it is the most important change in our society up to date (Pendergast, 2010). Having an "always connected" obsession (Ketter, 2020), Millennials utilize every means of technology for tourism activities such as planning trips, booking and checking reviews. They also use social media as a tool to document their experiences constantly as a result of being connected (Hamed, 2017). In addition, they can be defined as lavish with temporary desires. Spending money and following the latest updates of technological tools are the commonly observed behaviour among Millennials (Pendergast, 2010). It can also be stated that the brand obsession and lack of brand loyalty of this generation make them unreliable consumers (Goldgehn, 2004; Ting, Lim, de, Koh, & Sahdan, 2018). They have a nature that wants to be active, not passive, in decision-making and constantly asking the question of "why" (Pendergast, 2010). According to Martin (2005), they have an indecisive attitude towards authority; they both respect it and question it at the same time. Millennials, while they esteem the authority in setting task and making requests, believe that questioning the methods will end in enhanced results.

Kuran, who has carried out the largest number of studies on the generations in Turkey, defines people who were born in the period 1980–1999 in Turkey as Generation Y, with Millennials constituting about one-third of the country's population, the nation's highest demographic. With regard to Kuran, Millennials are the generation that rewrites the codes of both daily life and business life, believing that respect should be presented to those who deserve it, having a high tendency both to influence the community and to be influenced by them, and are expected to change jobs more than 10 times during their career (Milliyet Newspaper, 2018). Being born into a Turkey which is open to the world is the most prominent feature of this generation, in contrast to preceding generations (Arslan & Staub, 2015).

There are some studies which categorize Millennials into two separate cohorts: the old or early Millennials (birth year approximately 1980–1990); and the young or late Millennials (birth years approximately 1990–2000). Each of these has certain different characteristics. There are also, however, some studies which reject this type of characterization. The term "contradiction" is what we depict in the literature when we take the defining characteristics of Millennials into account (Pendergast, 2010). It can be stated that these opposite views about Millennials in the literature stem from the fact that their basic values are still forming (Lepeyko & Blyznyuk, 2016).

## 5.4 Travel behaviours of millennials

Recently, youths have become the latest focal point of tourism trends and Millennials who are defined as young tourists/travelers are the starring characters in travel movements (Expedia/Future Foundation, 2016). According to the United Nations World Tourism Organization [UNWTO] (2016), Millennials are estimated to make up about 23% of all international tourists. Recently, lots of countries have started to wake to the significance of Millennials in terms of their role in raising the value of the tourism destinations since they spend more time and money during their travels compared to the older generations. As to the ITB World Travel Trends, 2018/2019 Report, Millennials also form the 40% of the international tourism market and spending on travelling is a priority for them rather than purchasing goods. They prefer travelling to having possessions, such as car, home and so on (Expedia/Future Foundation, 2016; Glover, 2010; Sofronov, 2018). It is expected that the expenses of Millennials will constitute more than 50% of the travel cost in the whole world in the following years (Hamed, 2017). In this context, it can be stated that Millennials are to be both the fastest generation with regard to growing in population and the largest generation in terms of spending on expenses.

Since Millennials consider the major life milestones such as parenting and marriage as stages that can be postponed (Moscardo & Benckendorff, 2010), they have a higher potential travel budget than preceding generations, making travelling more important compared to setting a standard life. In the beginning phase of their professional lives, they want to find their true selves and they use travelling as a tool to discover what they really desire in life (Watson, 2020). The relation between Millennials and tourism can be summarized as follows: "Travel is a way of life. A certain level of risk is part of travel, even though it can be minimized through careful planning" (Richards, 2007:4). There is a clear distinction in the travel habits of Millennials and previous generations. As distinct from their parents, their vacation time has more importance (Hamed, 2017). It is certain that Millennials are experienced in travelling and are good at finding the best travel opportunities. Since they are accustomed to domestic tourism due to their travels with their parents, they are mostly interested in visiting domestic and foreign tourism destinations. In addition, the drop in the prices of transportation has increased the desire to travel abroad among Millennials (Glover, 2010). Nonetheless, they desire for more, owing to the perception of themselves as explorers not tourists (Starcevic & Konjikusic, 2018). Therefore, mass tourism attracts them less; they are more excited regarding the variety of opportunities in a tourism destination rather than stereotyped characteristics (Hamed, 2017). The logic behind the travel decisions of Millennials have been revealed through many studies (Chiu, Ramli, Yusof, & Ting, 2015; Giachino, Truant, & Bonadonna, 2020; Reisenwitz & Fowler, 2019). Experiencing new cultures, acquiring new languages, finding a job or studying abroad are the reasons of their travel preferences in addition to witnessing different lifestyles and meeting new people (şchiopu, Pădurean, Ţală, & Nica, 2016). In fact, for Millennials, socializing with new people during their travel is much more important than buying souvenirs (Watson, 2020).

During their travels, being active and adventurous, they want to live the authenticity of local life (Richards, 2007) with new and unusual experiences (Starcevic & Konjikusic, 2018) in unexplored destinations (Glover, 2010). Experience and authenticity are vital components of their travels and they prefer living like a local to being a mere tourist (Watson, 2020). This tourist type also prefers to take advantage of the broader experiences offered by the locals and interact with them on a more meaningful level, as they seek more local authenticity. At this point, Millennials commonly make use of the sharing economy, using apps such as Uber, Airbnb, Eatwith, LikeaLocal while interacting with the locals. Consequently, the sharing economy helps them to decrease their travel costs at the same time.

For Millennials, travelling is a top priority and they enjoy taking instantaneous decisions on weekend travels whenever they see a possibility to do so (Sofronov, 2018). Singles, single parents and singles and childless couples constitute a large proportion of Millennials who are travelling as tourists (Glover, 2010). "Carpe Diem" and "Change" are their mottos in life and they think of themselves as smart consumers; therefore, they want to get good service for the money they spend (Starcevic & Konjikusic, 2018). Expecting to see the value of the money is more common with young Millennials due to the fact that they have a limited travel budget (Faat, Mohd Nasir, & Nik Mohd Nor, 2017). Millennials have a tendency towards spontaneous travel. Thus, it is normal to label them as highly unpredictable in the tourism market

(Starcevic & Konjikusic, 2018). However, they are also the most well-informed generation that has ever lived (Sofronov, 2018). Hunger for experience and information is one of the most prominent features of Millennials; they truly make full use of their travel by any means (Richards, 2007). They are eager to acquire new travel information and buy travel goods and services (Ahn, Lee, & Lee, 2020). With the developments in technology, Millennials draw on information from the Internet, mobile devices and social platforms as key factors for their travels (Watson, 2020). With regards to travel planning, blogs and social media platforms that include travel reviews are more dependable as information sources compared to mass media marketing (Veiga, Santos, Águas, & Santos, 2017). Mobile travel applications and travel platforms in digital environment are commonly used by Millennials for planning travels and posting their experiences for everyone to see. During their travel, they also use interactive maps (Benckendorff et al., 2010; Hamed, 2017; Watson, 2020). They are in need of a confirmed experience from their peers to see a destination as desirable. On the other hand, the travel planning process can be extremely complicated for Millennials due to the variety of information sources (Starcevic & Konjikusic, 2018).

At last, sustainability, which is a mega trend, naturally affects Millennials. Sustainability is the factor that they seek for in choosing tourism products and travel destinations and the sustainability of tourism enterprises is significant for them (Bochert, Cismaru, & Foris, 2017). Sustainability and responsible tourism are priorities in their experiences – they try local dishes at local restaurants rather than eating in chain restaurants. Given this sensibility, they call themselves "World citizens" (Watson, 2020) who feel an obligation to make the world better and therefore volunteer tourism opportunities are particularly attractive to them (Veiga et al., 2017). This is the reason behind their activist impulse. Thus, travel and tourism suppliers should consider including environmental protection activities in their offers as this could help Millennials to achieve a better travel experience (Hamed, 2017; Starcevic & Konjikusic, 2018).

### *5.4.1 Methodology*

This study aimed to identify the travel behaviours and patterns of Millennials living in Istanbul and to develop suggestions for tourism academics and professionals in line with the results obtained. For this purpose, considering the exploratory feature of the research, the researcher preferred the semi-structured interview method for data collection. As to Rojek, Shaw, and Veal (2006), using qualitative research methods, enable us to reach deeper and richer data which results in a better understanding of behaviour, opinion and experience because specific responses, understandings and depth are lost when the sample grows. However, the main purpose of qualitative research is not to generalize the study universe, but to examine the views, experiences and interpretations of participants about subject, person or phenomenon in depth. The researcher benefited from the relevant literature while preparing interview questions (Chiu et al., 2015; Franzidis & Hritz, 2014; Guedes, 2017; Kattiyapornpong, 2009; Reisenwitz & Fowler, 2019; Richards & Morrill, 2020; şchiopu et al., 2016). The final form of the questionnaire was given after taking the opinions of expert academicians. It is known that qualitative analysis is used to review, interpret and synthesize the data collected (Fossey, Harvey, McDermott, & Davidson, 2002) and it is generally recommended that the sample size should be at least 15, although a clear number is not specified in qualitative data collection studies (Mason, 2010).

The reason behind choosing Istanbul as the target population in the study is due to the fact that Istanbul, being one of the few largest metropolises in the world as well as the largest in Turkey, provides many travel opportunities and possibilities for Millennials living in Istanbul. Since the study was designed to determine the travel behaviours of Millennials living in Istanbul, it was stated that both people living in Istanbul and between the ages of 21 and 40 could participate in the study within the scope of purposive sampling. In this direction, the researcher first sent the questionnaire to a few academicians between the ages of 21 and 40, who she believed would give information about the purpose and subject of the study, and then these people forwarded the questionnaire to other people and the participants increased like a chain and with the snowball sampling method, data were collected by email from 21 people between 1 October and 1 November 2020. After examining the interviews, it was seen that the responses of 21 participants were repeated and it was thought that the data saturation point was reached and the interviews were terminated. The data obtained were transferred to the computer environment by the researcher, arranged and subjected to descriptive analysis.

### 5.4.2 *Findings*

Findings obtained from the research consist of two parts. First and foremost, (1) the socio-demographic characteristics of the participants (Table 5.1) were examined. Then the data obtained for determining (2) the travel behaviours of Millennials living in Istanbul was evaluated. Accordingly, a comprehensive literature review was conducted and 19 questions were prepared to be addressed to the participants. While the participants could give more than one answer to some of these questions, they gave open-ended answers that they could comment on some of them, and closed-ended answers (yes/no) that were closed to interpretation for the rest.

### 5.4.3 *Socio-demographic characteristics of participants*

As it can be seen in Table 5.1 that 12 of the participants are women and 9 are men. The age distribution of Millennials ranges from 21 to 39, and the participants are predominantly over 30 (n = 14). According to the marital status column, 10 of them are single, 5 of them live together, 4 of them are married and 2 of them are divorced. In terms of education level, it is seen that 10 of the participants have a Bachelor's degree, 6 of them have a PhD degree, 3 of

*Table 5.1* Socio-Demographic Characteristics of Participants

| | *Gender* | *Age* | *Marital Status* | *Level of Education* | *Occupation* | *Level of Income* | *Frequency of Internet Usage* |
|---|---|---|---|---|---|---|---|
| **R1** | Female | 39 | Single | Ph. D | Academician | Medium | Daily |
| **R2** | Male | 36 | Married | Ph. D | Academician | Medium | Daily |
| **R3** | Female | 36 | Married | Ph. D | Engineer | Medium | Daily |
| **R4** | Female | 32 | Living with another | M.A | Ph. D student | Medium | Daily |
| **R5** | Female | 29 | Single | B.A | Dietitian | Medium | Daily |
| **R6** | Male | 39 | Married | Ph.D | Engineer | Medium | Daily |
| **R7** | Female | 32 | Single | Ph.D | Engineer | Medium | Daily |
| **R8** | Male | 28 | Single | B.A | Tour guide | High | Daily |
| **R9** | Male | 26 | Single | B.A | Tour guide | Medium | Daily |
| **R10** | Female | 24 | Living with another | B.A | Stewardess | High | Daily |
| **R11** | Male | 35 | Single | B.A | Psychologist | High | Daily |
| **R12** | Male | 33 | Single | M.A | Lawyer | High | Daily |
| **R13** | Female | 37 | Living with another | B.A | Manager | High | Daily |
| **R14** | Male | 38 | Living with another | B.A | Manager | High | Daily |
| **R15** | Male | 34 | Divorced | M.A | Sales Expert | Medium | Daily |
| **R16** | Female | 22 | Single | High school | Undergraduate | Low | Daily |
| **R17** | Female | 33 | Single | B.A | Engineer | Medium | Daily |
| **R18** | Female | 21 | Living with another | High school | Undergraduate | Low | Daily |
| **R19** | Male | 25 | Single | B.A | Not employed | Low | Daily |
| **R20** | Female | 34 | Divorced | Ph.D | Journalist | Medium | Daily |
| **R21** | Female | 39 | Married | B.A | Teacher | Medium | Daily |

them have an MA degree and 2 of them are high school graduates (who are currently continuing their university education). Nonetheless, in terms of income level, it is obvious that the majority (n = 12) stated that they are at the middle income level and the number of those with high income level is less than 1 in 3 of the participants (n = 6). Finally, it was determined that the frequency of internet usage of all participants is on a daily basis.

## 5.5 Travel behaviours of participants

After evaluating the socio-demographic data, the data on the travel behaviors of the participants were analyzed. In this context, it was determined that the participants take at least 2 (n = 6) and at most 10 (n = 1) holidays per year, and als that most of them (n = 12) spend 2 weeks per year on vacation. When the length of holiday preference is examined, it is noteworthy that all of them prefer short breaks, and that more than half of those who prefer both short and long holidays (n = 13) draw attention. Some of the participants who prefer short breaks (R5, R7, R13, R14) stated that the intensity of work life necessarily leads them to prefer short breaks. A participant who takes both short- and long-term vacations explained the reason for this preference as follows:

> ...To reduce my work stress, I take a long vacation for 2 weeks at least once a year. I usually prefer to travel abroad. While I am working, I plan my short-term holidays for nearby destinations that coincide with the weekend or public holidays...
>
> (R3)

It is understood that the majority (n = 10) stated that their vacation budget is €2000 and above. While the participants stated that accompanying holiday members are predominantly friends, partners and family, the number of those who prefer to go on vacation alone (n = 5) is lower.

It is noteworthy that in the booking channel of travel or vacation, all participants use digital platforms, and only two people prefer both digital and traditional methods, namely travel agency/tour operator. Participants who prefer digital platforms indicate the reasons behind their preferences as the ability to freely plan their own vacations (R4, R10, R18, R20), flexibility in realizing time and activities for their interests (R6, R8, R11, R16), fast, result-oriented detailed filtering, alternatives and add-ons (transportation, restaurants, etc.), integration of these platforms with tickets and other digital tools (calendar, e-mail, etc.) (R3, R6, R10, R12, R14, R20) and establishing an appropriate budget–benefit relationship (R2, R4, R5, R15, R16, R17, R18, R19). A participant who prefers digital platforms in travel and vacation planning explains the reason for this as follows: "I check every option from digital platforms comparatively at the same time; travel agencies/tour operators only offer the contracted locations to the consumers. This is not attractive at all..." (R14).

The most frequently preferred means of transportation by the participant Millennials are: airplane (n = 19), car (n = 13), train (n = 11) and bus (n = 10). A small number of participants stated that they go on vacation by bike, motorbike and cruise ship and ferry. The majority of Millennials participating in the study stated that they prefer to fly when travelling abroad and this provided them with advantages in terms of time, budget and comfort (R1, R3, R2, R8, R9, R10, R11, R12, R21). In addition, some of the participants stated that travelling domestic and closer destinations by car provides flexibility in terms of sudden plan changes (R13, R14, R15, R21). Some of those who prefer to go by train to the destinations which they have planned to visit have emphasized that they do it consciously due to concerns over their carbon footprint (R3, R4, R11, R20). The most preferred accommodation types are, respectively, hotel (n = 16), renting apartment (n = 14), friends/family house (n = 10). Homestay (n = 4), hostel (n = 3) and resort (n = 3) accommodation are the least preferred. Those respondents who choose "hotel" stated the reasons for their preferences as mostly hygiene (R1, R2, R3, R6, R13, R14, R20, R21), comfort (R2, R13, R14, R15, R21), security (R2, R6, R20, R21) and to earn bonus points (R3, R6, R13, R14). Participants stated that they prefer to stay in a renting apartment when they go on vacation with a large group of friends (R4, R10, R11, R16) and cannot find cheap hotels (R2, R4, R8). Those who prefer homestay stated that they make these stays when they want to get to know the culture of the destination more closely (R9, R18, R19), while those who prefer a hostel seek entertainment and adventure (R16, R18, R19).

According to vacation destination preference, almost all of the participants stated that they prefer to travel abroad (n = 20). On the other hand, those who take domestic and abroad vacation (n = 13) make up more than half of the participants. However, the number of participants who prefer only domestic holidays (n = 1) and only abroad (n = 6) is low. When the data are analyzed, it is seen that the participants travel to developed and undeveloped destinations at an equal rate (n = 17). Those who prefer to visit both developed and undeveloped countries (n = 9) constitute almost half of the participants. The number of Millennials who visit only developed (n = 4) and only undeveloped (n = 4) destinations is low. Those who prefer to go on a vacation in undeveloped destinations stated that they find the culture in these geographies more interesting and authentic (R1, R7, R8, R9, R17, R20) and that they have the opportunity to experience extraordinary and different experiences (R13, R14, R18, R19). More than half of the participants stated that they do not want to visit the places they have visited before (n = 14). A participant who said I would go to the same place again expressed the reasons as follows:

> ... If I establish a connection with that place, it gives me pleasure to go again. I would like to see the changes about the place and to see if the places in my memories remain the way I saw them last, and to see if I still get the same pleasure...
>
> (R6)

All of the participants (n = 21) stated that they regard vacation as a priority in their lives. They expressed these priorities as giving up other needs (R3, R6, R10, R11, R12, R18, R20) and saving money (R2, R4, R16, R18, R19, R21). R4 summarized her thoughts on this matter as follows:

> ...Going on vacation is really a big priority in my life. Seeing new places, trying different foods and drinks, learning about the history and culture of a destination I don't know... All of them make me very excited and happy. Even dreaming of vacation gives me energy of life...

R18 expressed the priority of the holiday in her life by stating

> ... I do not have a regular income because I am still a student. I save a large part of the money I receive from my family and part-time income to go on vacation. I save my money in social life in order to travel to the places I have planned to see, I don't do unnecessary shopping and waste my money...

Nonetheless, few participants said that they take risks in life for vacation (n = 4). Participants explained these risks as taking out loans for holidays (R8, R19), and travelling despite reasons such as war, terrorism (R18, R19), epidemic (R8, R9, R18, R19). R19 said "...There is a risk in all areas of life, so if I get a suitable vacation opportunity, I will definitely take it..."

To the question of criteria prioritization while planning a vacation, the answers given by the participants the most are budget (n = 12), places not visited before (n = 10) and new experiences (n = 9). R17 explained her thoughts on the budget as follows:

> ...The worsening economic conditions for the last one or two years, the decline in purchasing power, and the gap of the Turkish lira-exchange rate difference have caused me to review my travels and shape them according to these factors. My priority is to balance the budget and the benefits I will get in return...

Countries that do not require a visa, less explored places, authenticity, gastronomy and spirituality have followed these priorities. R5 reveals her thoughts on this issue;

> ...For me, the holiday means unforgettable memories that I want to explore, learn new things, and constantly remember places I have never been to before. I don't like to be stuck for a long time and I don't prefer to complete my trip without exploring different things. That's why I act accordingly while planning my vacation, I generally do road trip...

Another participant said:

> ...I plan my vacation to grow and develop spiritually at least once a year (such as yoga camps and volunteer works). These travels to places where I integrate with nature, with very few things, unknown places help me to discover myself, make me mindful, calm my soul. I feel that I am relieved, liberated and filled with peace...
>
> (R20)

Millennials answered Tripadvisor (n = 14), Booking.com (n = 13) and Skyscanner (n = 8) to their three most frequently used digital platform questions while vacationing. These answers were followed by Airbnb, Instagram, travelblogs and Kayak. One of the participants who preferred renting apartment states the reason for this as follows:

> Instead of an isolated accommodation experience in a luxury hotel, I rent houses from Airbnb that match the spirit of that city. Sometimes I do Couchsurfing. Thus, I can mingle with the local people and learn more about the culture of the destination...
>
> (R9)

Participants stated that in planning their vacation they mostly do research on accommodation (n = 18), cultural/historical/natural places (n = 13), food & beverage (n = 13). These top three were followed by transportation lines, local activities, and photographic locations. The most purchased product from digital platforms was determined as accommodation (n = 20), followed by transportation (n = 16), car rental (n = 7) and restaurant reservation (n = 6). Museum tickets, event tickets, local tours and outdoor activities are other touristic products purchased from the digital platforms by the participants. The main motivations of the participants in the research that led them to travel; are to learn about new things (cultures, places, people, etc.) (n = 20) to experience new things (different, authentic, adventurous, etc.) (n = 18) and to have unforgettable memories (n = 17). These are followed respectively by physical and mental refreshment, feeling free, being in touch with nature, and having a good time with their family and friends.

Finally, when Millennials' situation of reading others' comments about their travel experiences on digital platforms and sharing their experiences (photos, videos, recommendations, comments, etc.) during the travel planning process is analyzed, it was determined that all of the participants did at least one of these actions, and some more (R4, R8, R9, R10, R16, R18, R19). R3 conveys her thoughts on this subject as follows: "...Before I travel, I definitely research local food/drinks and try them when I arrive at the destination. It is very important for me to read other people's restaurant/cafe/pub recommendations and experiences..." Participants stated that it is caused by reasons such as the desire to read/make these posts on digital platforms; being inspired while planning travel routes (R1, R5, R11, R12,

R18, R20), exchanging ideas (R5, R8, R16, R17, R18), learning/transferring experience (R3, R4, R5, R8, R10, R13), R16, R19), helping others (R1, R3, R8, R10, R20), creating digital photo archive (R1, R2, R3, R6, R10, R13, R15, R17, R20, R21) and photo or feeling incomplete when not sharing videos (R16, R18, R19), etc. In this context, the participant who stated that Instagram is an important social media tool that affects travel preferences, expressed her thoughts as follows:"…Social media and especially Instagram photos have drastically changed our vacation choices. If your holiday in our age cannot be shared on social media, you are not considered to have made that holiday..." (R16). Another participant describes the effects of social media and digital platforms on vacation decisions with the following sentences:

> ... My friends and I plan our travels based on the lists of the must-go places created by photos or blogs on Instagram. Especially the experience-oriented sharing of others (experiencing rural life, participating in local cookery workshops, seeing the northern lights, hot air balloon tours, snowmobiling, etc.) can motivate me to live that experience...
>
> (R18)

However, another participant expressed her thoughts as follows: "...I think most people prefer some routes to share on Instagram. For myself, I sometimes see some places on Instagram and include them in my travel plan, but I cannot say that Instagram has affected my travel preferences completely..." (R5).

## 5.6 Conclusion

Millennials are among the largest generations in the history of the world (Expedia/Future Foundation, 2016). In terms of education and culture, it is fair to say that Millennials have a higher education level and are more diverse than previous generations (Bachman, Hull, & Haecker, 2020) and due to the technological advance in their era, they are more linked to information and the world (Ketter, 2020). In addition, Millennials will have constituted "75% of the global workforce by 2030" (Meister, 2012), yet they have high disposable income besides discretionary spending (Chen & Chou, 2019). These developments have led to the fact that this generation, which has grown in better living conditions compared to other generations, has high awareness, sees the meaning of life as travelling and experiencing, and constitutes the vast majority of the global workforce. They travel more each year and this helps the development of different sectors related to it by offering new opportunities to the travel and tourism industry. At the same time, as the next big key market (Internationale Tourismus-Börse Berlin[ITB], 2019), this generation will continue to rapidly transform the travel and tourism industry in line with their preferences, requests and expectations.

Due to the importance of Millennials in the travel and tourism industry, a lot of research has been done by tourism academics, tourism marketers and policymakers. In parallel with this increasing interest, the travel attitudes and behaviours of 21 Millennials living in Istanbul were examined in depth using the semi-structured interview method. In the light of findings obtained from socio-demographic data, it is seen that the aforementioned finding is greatly supported by the literature (Glover, 2010; Moscardo & Benckendorff, 2010); in other words, the Millennials participating in the study have a low attitude towards marriage. In addition, they have a stronger preference for living together or even for celibacy. It can be said that the education level of the participants is quite high and they have good professions parallel to this situation. This result coincides with the fact that Millennials are the most educated generation and, generally, young people who are receiving higher education grow in numbers as to the relevant literature (Dutzik, Inglis, & Baxandall, 2014; Kong, Wang, & Fu, 2015; Leask, Fyall, & Barron, 2014). Nevertheless, even if the earnings related to these professions are high, considering the expensive living conditions of Istanbul, it can be said that it is an expected result that people cannot evaluate their income level as high. It can be thought that those who see themselves in the low-income group are young Millennials; they are in this category due to the fact that they are either students or have only just graduated. Finally, it is unsurprising that all of the participants use the Internet on a daily basis (Reisenwitz & Fowler, 2019; şchiopu et al., 2016). This generation, which was introduced to the Internet and computers in adolescence, has a high rate and frequency of Internet usage.

When the travel patterns of the Millennials participating in the study are examined, it is understood that the participants have a tendency to take frequent holidays despite their intense working conditions or education. The fact that all of them stated that they prefer short breaks (Kattiyapornpong, 2009; Starcevic & Konjikusic, 2018) and the fact that more than half of those who prefer short breaks together with long holidays reflect the importance of vacation in their lives. As a matter of fact, all of the participants stated that they consider vacation as a priority in their lives (Internationale Tourismus-Börse Berlin[ITB], 2019). In this direction, giving up other needs and saving money in order to do annual travel and tourism activities (Ketter, 2020) are among the sacrifices made by the Turkish Millennials participating in the research. The findings also show that although the participants prioritize the travel and tourism phenomenon in their lives and make various sacrifices in order to achieve them, they also tend to avoid risks in the same issue. This result can be explained by the term "sheltered", which is one of the seven basic characteristics that Howe and Strauss (2000) stated about Millennials when introducing their theory. The risk aversion of Millennials, who were raised under protection by their "helicopter parents", can be interpreted as an expected result in this context. The research results of Expedia/Future Foundation (2016), carried out with 1000 Millennials from different countries, also reveal that personal safety is more important for this generation. The report in question supports the findings of this research. In addition, as another finding, it was determined in the study that the most preferred type of accommodation is "hotel" and this preference is based on reasons such as hygiene, comfort, security and score collection. The aforementioned finding can be explained by the association of value for money, cleanliness and safety as very important for Millennials (Jennings et al., 2010). On the other hand, a small number of participants stated that they take risks in life for travel and tourism. When the answers given by these participants to the questions are examined in detail, it can be interpreted that they are young Millennials and take risks in order to perform activities that they perceive as adrenaline-fuelled or physically/mentally challenging (Guedes, 2017; Xie, 2017).

This generation is also defined as "always connected" (Ketter, 2020). In line with the importance of technology and the Internet in their lives, Millennials are inclined to use technology (non-traditional media) to get information for travel and vacation planning (Reisenwitz & Fowler, 2019). In this context, it is seen that the data obtained from this study are compatible with the literature. Since all of the participants stated that they benefit from digital platforms while planning their vacation due to the various advantages (Xiang, Magnini, & Fesenmaier, 2015) and only two people use digital platforms and travel agency/tour operators. In addition, it was determined that the participants get well-informed before travelling and use a variety of sources of information for travel planning (Internationale Tourismus-Börse Berlin[ITB], 2019). They also benefit mostly from popular platforms such as Tripadvisor, Booking.com and Skyscanner (Liberato, Abreu, Alen-Gonzalez, & Rocha, 2018; Reisenwitz & Fowler, 2019). In addition, the finding which reveals that they utilize varied forms of digital technology (photos, videos, webcams, mobile phone apps, blogs, etc.), in which they research the experiences of peers in the digital platforms, read their comments, and perform these actions frequently, shows also a similarity to many studies in the literature (Guedes, 2017; Leask et al., 2014; Reisenwitz & Fowler, 2019). In addition, as to the research findings, learning about the experiences of others inspire (şchiopu et al., 2016) and motivate (Liberato et al., 2018) them while making their own travel and vacation plans. In this context, some of the participants' comments for Instagram were found dramatic. It can be predicted that Instagram will further influence the decisions of Millennials on travel in the coming years and will become a determinant social media platform for the travel and tourism industry.

In the study, it was determined that the participants mostly prefer to have a holiday with their friends, partners, and family members. This result is similar to some studies and research in the literature (Internationale Tourismus-Börse Berlin[ITB], 2019; Leask et al., 2014; Starcevic & Konjikusic, 2018). In the light of the answers given by the participants, it was determined that the reasons that lead Millennials to travel are mainly motivational factors such as learning new things, experiencing different things, having unforgettable memories, feeling free, being in touch with nature, resting physically and spiritually, and having a pleasant time with friends and families. In line with these data, which overlap with many studies in the literature (Guedes, 2017;

Jennings et al., 2010; Kattiyapornpong, 2009; Liberato et al., 2018; şchiopu et al., 2016; Starcevic & Konjikusic, 2018), it can be expected that the interest of Millennials towards sustainability, culture, local, authenticity, gastronomy, extraordinary experiences, nature and spiritualism-based tourism types will continue to increase in the coming years.

## 5.7 Limitations and future research

As in many studies, there are some limitations in this research. First of all, this study is limited to investigating the travel behaviours of Millennials in Istanbul, and the results cannot be generalized. Secondly, the researcher preferred to collect data online due to reasons such as time, cost and, of course, the COVID-19 pandemic. It is recommended to include Millennials who live in different cities in Turkey in the sampling and reach a higher sample size and design empirical research for those who will investigate this matter in the future. It is thought that a more in-depth examination of the travel behaviours of Turkish Millennials will enrich the current academic literature. In addition, it is considered that the researches in the related field are important in terms of giving ideas to travel and tourism industry professionals and public institutions to make tourism marketing and planning.

## References

Ahn, Y. J., Lee, B. C., & Lee, S. K. (2020). Analysis of Korean millennials' travel expenditure patterns: an almost ideal demand system approach. *Asia Pacific Journal of Tourism Research*, 25(1), 3–14.

Albayrak, A., & Özkul, E. (2013). Y kuşağı turistlerin destinasyon imaj algıları üzerine bir araştırma. *International Periodical For the Languages, Literature and History of Turkish or Turkic*, 8(6), 15–31.

Arslan, A., & Staub, S. (2015). Kuşak teorisi ve içgirişimcilik üzerine bir araştırma. *Kafkas Üniversitesi Iktisadi ve Idari Bilimler Fakültesi Dergisi*, 6(11), 1–24.

Bachman J. R., Hull J. S., & Haecker S. (2020) Millennials are not all the same: Examining millennial craft brewery and winery visitors' social involvement, self-image, and social return. *International Journal of Tourism Research*, 1–11. https://doi.org/10.1002/jtr.2400.

Benckendorff, P.,& Moscardo, G. (2010). Understanding generation-Y tourists: Managing the risk and change associated with a new emerging market, P. Benckendorff, G. Moscardo, & D. Pendergast In *Tourism and Generation Y* (pp. 38–46). Wallingford: CABI.

Benckendorff, P., & Moscardo, G. (2013). Generational cohorts and ecotourism, R. Ballantyne, & J. Packer In *International Handbook on Ecotourism*, (pp. 135–154). Cheltenham,: Edward Elgar.

Benckendorff, P., Moscardo, G., & Pendergast, D. (2010). Introduction, P. Benckendorff, G. Moscardo, & D. Pendergast In *Tourism and Generation Y*. (p. x), Wallingford: CABI.

Birdir, S. S., Toksöz, D., & Birdir, K. (2019). Yeşil yıldız uygulamaları: Y kuşağı örneği. *Journal of Tourism Theory and Research*, 5(2), 123–133.

Bochert, R., Cismaru, L., & Foris, D. (2017). Connecting the members of generation Y to destination brands: A case study of the CUBIS project. *Sustainability*, 9(7), 1197.

Bolin, G., & Skogerbø, E. (2013). Age, generation and the media. *Northern Lights: Film & Media Studies Yearbook*, 11(1), 3–14.

Caber, M., Albayrak, T., & Crawford, D. (2020). Perceived value and its impact on travel outcomes in youth tourism. *Journal of Outdoor Recreation and Tourism*, 31, 1–10.

Çakar, K. & Seyidoğlu, F. (27–28 May 2016). *Youth tourism: Understanding the youth travellers' motivations, Proceedings of the Multidisciplinary Academic Conference* (pp. 248–253), Prague, Czech Republic.

Chen, C. F., & Chou, S. H. (2019). Antecedents and consequences of perceived coolness for Generation Y in the context of creative tourism – A case study of the Pier 2 Art Center in Taiwan. *Tourism Management*, 72, 121–129.

Chiu, L. K., Ramli, K. I., Yusof, N. S., & Ting, C. S. (2015). Examining young Malaysians travel behavior and expenditure patterns in domestic tourism. *Asian Social Science*, 11(9), 77.

Costanza, D. P.,& Finkelstein, L. M. (2015). Generationally based differences in the workplace: Is there a there there? *Industrial and Organizational Psychology: Perspectives on Science and Practice*, 8(4), 308–323.

Davids, A. (2018). *Investigating the use of generational cohort theory to identify total reward preferences.* Doctoral dissertation, Faculty of Commerce, University of Cape Town, South Africa.

Dulin, L. (2008). Leadership preferences of a generation Y cohort: A mixed-methods investigation. *Journal of Leadership Studies*, 2(1), 43–59.

Dutzik, T., Inglis, J., & Baxandall, P. (2014). *Millennials in Motion: Changing Travel Habits of Young Americans and the Implications for Public Policy*. CoPIRG Foundation Frontier Group, USA.

Duygulu, S. (2018). Yeni medya teknolojilerinin K-kuşağının ebeveynleri ile olan iletişimine etkisi. *TRT Akademi*, 3(6), 632–652.

Eisner, S. P. (2005). Managing generation Y. *SAM Advanced Management Journal*, 70(4), 4–15.

Expedia/Future Foundation (2016). *Millennial traveller report: Why millennials will shape the next 20 years of travel*, available at: http://expediablog.co.uk/wp-content/uploads/2016/10/Expedia-Millennial-Traveller-Report.pdf (accessed: 12 October 2020).

Faat, F., Mohd Nasir, N. H., & Nik Mohd Nor, N. M. S. (2017). Domestic travel behaviour among generation Y. *Journal of Tourism, Hospitality & Culinary Arts (JTHCA)*, 9(2), 1–10.

Fossey, E., Harvey, C., McDermott, F., & Davidson, L. (2002). Understanding and evaluating qualitative research. *Australian & New Zealand Journal of Psychiatry*, 36(6), 717–732.

Fountain, J., & Lamb, C. (2011). Generation Y as young wine consumers in New Zealand: How do they differ from Generation X? *International Journal of Wine Business Research*, 23(2), 107–124.

Franzidis, A., & Hritz, N. (2014). Profiling the Gen Y tourist. *Turizam: međunarodni znanstveno-stručni časopis*, 62(3), 323–330.

Gardner, S. F. (2006). Preparing for the nexters. *American Journal of Pharmaceutical Education*, 70(4), 1.

Gasset, Joy (1932). *The revolt of the masses*. Routledge.

Giachino, C., Truant, E., & Bonadonna, A. (2020). Mountain tourism and motivation: Millennial students' seasonal preferences. *Current Issues in Tourism*, 23(19), 2461–2475.

Glover, P. (2010). Generation Y's future tourism demand: Some opportunities and challenges, P. Benckendorff, G. Moscardo, & D. Pendergast In *Tourism and Generation Y* (pp. 155–163). Wallingford: CABI.

Glover, P., & Prideaux, B. (2008). Using population projections to identify aspects of future tourism demand. *Advances in Hospitality and Leisure*, 4(4), 185–209.

Goldgehn, L. A. (2004). Generation who, what, Y? What you need to know about Generation Y. *International Journal of Educational Advancement*, 5(1), 24–34.

Greenberg, E. H., & Weber, K. (2008). *Generation We: How Millennial Youth are taking over America and Changing our World Forever*. Emeryville, CA: Pachatusan.

Guedes, L. C. (2017). *A descriptive study on the youth tourism market, based on generation Y travel experiences*. Doctoral dissertation, ISCTE Business School, Portugal.

Hamed, H. M. (2017). Marketing destinations to millennials: Examining the compatibility between the destination marketing organization website and the millennial tourist prospects. *Journal of Tourism and Recreation*, 3(1), 1–20.

Howe, N., & Strauss, W. (2000). *Millennials Rising: The Next Great Generation*. New York: Vintage.

Huang, Y. C., & Petrick, J. F. (2010). Generation Y's travel behaviours: A comparison with Baby Boomers and generation X, P. Benckendorff, G. Moscardo, & D. Pendergast In *Tourism and Generation Y* (pp. 27–37). Wallingford: CABI.

Huntley, R. (2006). *The World According to Y: Inside the New Adult Generation*, Sydney: Allen & Unwin.

Inglehart, R. (1997). *The Silent Revolution. Changing Values and Political Styles among Western Publics*. Princeton/Guildford: Princeton University Press.

Internationale Tourismus-Börse Berlin[ITB] (2019) World Travel Trends 2018–2019. *What are the trends to look out for? ITB and IPK travel trends report*, available at:https://www.itb-berlin.de/media/itbk/itbk_dl_all/ITB_2019_WTTR_Factsheets_ALLE_%20Web_4.pdf (accessed: 05 September 2020).

ITB World Travel Trends (2018/2019). Prepared By IPK International On Behalf Of ITB Berlin- The World's Leading Travel Trade Show.

Jaeger, H. (1985). Generations in history: Reflections on a controversial concept. *History and Theory*, 24(3), 273–292.

Jennings, G., Cater, C., Lee, Y.-S., Ollenburg, C., Ayling, A., & Lunny, B. (2010). Generation Y: Perspectives of quality in youth travel experiences in an Australian backpacker context, P. Benckendorff, G. Moscardo, & D. Pendergast In *Tourism and Generation Y* (pp. 58–72). Wallingford: CABI.

Kattiyapornpong, U. (30 November–2 December 2009). *Profiling generation Y – Where do they travel? Proceedings of the Australian and New Zealand Marketing Academy Conference, ANZMAC* (pp. 1–7), Melbourne, Victoria: Monash University.

Ketter, E. (2020). Millennial travel: Tourism micro-trends of European Generation Y, *Journal of Tourism Futures*, ahead-of-print(ahead-of-print). https://doi.org/10.1108/JTF-10-2019-0106.

Knight, Y. (2009). Talkin' bout my generation: A brief introduction to generational theory. *Planet*, 21(1), 13–15.

Kong, H., Wang, J., & Fu, X. (2015), Meeting career expectation: Can it enhance job satisfaction of Generation Y?, *International Journal of Contemporary Hospitality Management*, 27(1), 147–168.

Kuran, E. (2018a). *Telgraftan Tablete*. Istanbul: Destek Yayınları.

Kuran, T. (2018b). Islam and economic performance: Historical and contemporary links. *Journal of Economic Literature*, 56(4), 1292–1359.

Kutluk Bozkurt A., & Avcıkurt, C. (2019). Planlanmış davranış teorisi ile yerli Y kuşağı turistlerin destinasyontercih etme niyetlerinin belirlenmesi. *Afyon Kocatepe Üniversitesi Sosyal Bilimler Dergisi*, 21(4), 1294–1307.

Kyles, D.(2005). Managing your multigenerational workforce. *Strategic Finance*, 87 (6), 52–55.

Lancaster, L. C., & Stillman, D. (2002). *When Generations Collide: who they are. why they Clash. How to solve the Generational Puzzle at Work*. New York: Harper Business.

Leask, A., Fyall, A., & Barron, P. (2014). Generation Y: An agenda for future visitor attraction research. *International Journal of Tourism Research*, 16(5), 462–471.

Lepeyko, T. I., & Blyznyuk, T. P. (2016). Generational theory: Value-oriented approach. Бізнес Інформ, (11), 24–30.

Liberato, P., Abreu, A., Alen-Gonzalez, E. & Rocha, A. (2018), Generation Y: The competitiveness of the tourism sector based on digital technology, Antipova T., & Rocha Á. In *Information Technology Science* (pp. 227–240). New York: Springer International Publishing.

Mannheim, K. (1928/1952). The problem of generations, K. Mannheim In *Essays in the Sociology of Knowledge* (pp. 276–320), London: Routledge.

Martin, C.A. (2005). From high maintenance to high productivity: What managers need to know about generation Y. *Industrial and Commercial Training*, 37: 39–44.

Mason, M. (2010). Sample size and saturation in PhD studies using qualitative interviews. *Forum: Qualitative Social Research*, 11(3), 1–19.

Meister J. (2012). The future of work: Job hopping is the 'new normal' for millennials. *Forbes*. available at: https://www.forbes.com/sites/jeannemeister/2012/08/14/the-future-of-work-job-hopping-is-the-new-normal-for-millennials/?sh=1542dc1c13b8 (accessed: 18 September 2020).

Milliyet Newspaper, (2018). *Bir kuşağı anlamak, bir dönemi anlamaktır*. Available at: https://www.milliyet.com.tr/kariyer/evrim-kuran-bir-kusagi-anlamak-bir-donemi-anlamaktir-2687958 (accessed: 03 September 2020).

Moscardo, G., & Benckendorff, P. (2010). Mythbusting: Generation Y and travel. Y. P. Benckendorff, G. Moscardo, & D. Pendergast In *Tourism and Generation*, (pp. 16–26). Wallingford: CABI.

Ortega y Gasset, J. (1923). The idea of generations. *The Theme of Our Time, Complete Works*, 3, 145–156.

Ortega y Gasset, J. (1930). El error Berenguer. *El Sol*, 15.

Ortega y Gasset, J. (1931). La redención de las provincias. *Ortega y Gasset, José, Obras completas de José Ortega y Gasset*, 11.

Özkan, P. M. (2017). Generation Z – the global market's new consumers – and their consumption habits: Generation Z consumption scale. *European Journal of Multidisciplinary Studies*, 2(5), 150–157.

Pendergast, D. (2010). Getting to know the Y generation. P. Benckendorff, G. Moscardo, & D. Pendergast In *Tourism and Generation Y* (pp. 1–15). Wallingford: CABI.

Pennington-Gray, L., & Lane, C. W. (2002). Profiling the silent generation: Preferences for travel. *Journal of Hospitality & Leisure Marketing*, 9(1–2), 73–95.

Prensky, M. (2001). Digital natives, digital immigrants. *On the Horizon*, 9(5): 1–6.

Pritchard, J. C. (2014). *Inter-group communication between baby boomer leaders and generation Y followers: A cultural reasoning perspective.* Doctoral dissertation, Curtin University, St, Bentley, Australia.

Reisenwitz, T. H., & Fowler, J. G. (2019). Information sources and the tourism decision-making process: An examination of generation X and generation Y consumers. *Global Business Review*, 20(6), 1372–1392.

Richards, G. (2007). *New Horizons II: The Young Independent Traveller 2007.* Amsterdam, The Netherlands: WYSE Travel Confederation.

Richards, G., & Morrill, W. (2020). Motivations of global Millennial travelers. *Revista Brasileira de Pesquisa em Turismo*, 14(1), 126–139.

Rojek, C., Shaw, S., & Veal, A. (Eds.). (2006). *A handbook of leisure studies.* Springer.

Rosow, I. (1978). What is a cohort and why?. *Human Development*, 21(2), 65–75.

şchiopu, A. F., Pădurean, A. M., Ţală, M. L., & Nica, A.-M. (2016). The influence of new technologies on tourism consumption behavior of the millennials. *Amfiteatru Economic*, 18(10), 829–846.

Sofronov, B. (2018). Millennials: A new trend for the tourism industry. *Annals of Spiru Haret University Economic Series*, 18(3), 109–122.

Starcevic, S., & Konjikusic, S., (31 May–2 June 2018). Why millennials as digital travelers transformed marketing strategy in tourism industry, *3rd Tourism International Scientific Conference* (pp. 221–240), Vrnjačka banja, Serbia.

Strauss, W.,& Howe, N. (1991). *Generations: The History of America's Future, 1584 to 2069.* New York: William Morrow.

Ting, H., Lim, T. Y., de Run, E. C., Koh, H., & Sahdan, M. (2018). Are we baby boomers, gen X and gen Y? A qualitative inquiry into generation cohorts in Malaysia. *Kasetsart Journal of Social Sciences*, 39(1), 109–115.

Twenge, J. M. (2006). *Generation Me: Why Today's Young Americans are more Confident, Assertive, Entitled—and more Miserable than ever before.* New York: Free Press.

United Nations [UN], (2001). *Generations*, available at: https://www.un.org/esa/socdev/rwss/docs/2001/5%20Generations.pdf (accessed: 22 August 2020).

United Nations World Tourism Organization [UNWTO], (2016). *Annual Report 2016.* available at: https://www.e-unwto.org/doi/pdf/10.18111/9789284418725 (accessed: 01 September 2020).

Veiga, C., Santos, M. C., Águas, P., & Santos, J. A. C. (2017).Are millennials transforming global tourism? Challenges for destinations and companies. *Worldwide Hospitality and Tourism Themes*, 9 (6), 603–616.

Watson, C. M. (2020). *The authenticity of mass tourism attractions: Evidence from American millennials visiting the tower of London.* Master's Thesis, University of Lapland, Finland.

Xiang, Z., Magnini, V. P., & Fesenmaier, D. R. (2015). Information technology and consumer behavior in travel and tourism: Insights from travel planning using the internet. *Journal of Retailing and Consumer Services*, 22, 244–249.

Xie, J. (2017). The pursuit of freedom and its risks: The dreams and dilemmas of young Chinese backpackers. PhD Thesis, University of Glasgow.

Zemke, R. (2001). Here come the millennials. *Training*, 38(7), 44–49.

Zemke, R., Raines, C., & Filipczak, B. (2000), *Generations at Work: Managing the Clash of Veterans, Boomers, Xers, and Nexters in your Workplace*, New York: AMACOM/American Management Association.

# 6 Millennials' perceptions of spirituality, wellness and travel

*Melanie Kay Smith, Robert Kiss and Inge Yan Fong Chan*

## 6.1 Introduction

The main aim of this chapter is to explore Millennials' perceptions of the role that spirituality plays in their everyday lives, in their wellness and in travel. The work builds on the research of Mason, Singleton, and Webber (2007), who defined spirituality as ranging from traditional religion to secular beliefs, including new age therapies and humanism. It will also include relationships to nature and social connections after Brooks, Michaelson, King, Inchley, and Pickett (2018). Previous research has suggested a certain degree of hybridity amongst Millennials, where approaches to religion and spirituality are mixed (Halafoff et al., 2020) and not strictly binary, i.e. neither religious nor non-religious (Ammerman, 2013).The research will consider the role that spirituality plays in Millennials' everyday life in relation to other domains (e.g. social, economic, emotional, environmental) and in wellness (e.g. physical and mental health, lifestyle habits) as well as in travel motivations and benefits. These data could be useful when designing leisure and tourism products and services for Millennials. This includes knowing how important spirituality is in relation to other priorities in life, as well as which aspects of spirituality are deemed the most attractive and beneficial for this segment. There are also wellbeing implications of the research which could be used to encourage and provide health-enhancing activities, including travel.

The primary data collection takes the form of a questionnaire which was distributed among English-speaking Millennial students in Europe as well as Mandarin Chinese speakers in China, Taiwan and Hong Kong. The sample enabled the authors to reflect on cultural differences, which have already been observed in Australia and Canada (e.g. Halafoff et al., 2020; Watts, 2018), Italy (Palimisano & Pannofino, 2017) and Africa (Knoetze, 2018). Questions focused on respondents' definitions of and relationship to spirituality; how far religion or spirituality were present in their upbringing, culture and society; the role of religion and spirituality in their everyday life and wellness; degree of interest in religious or spiritual activities and experiences when travelling; and preferences for suggested spiritual trips (e.g. retreats, ashrams, pilgrimages, temple stays). Unlike previous studies, the research focuses more

DOI: 10.4324/9781003170631-8

closely on the relative role of spirituality in wellness compared to other domains and whether it is seen as an important dimension of life, health and travel for Millennials.

## 6.2 Millennials' attitudes to spirituality

Spirituality can be complex to define, especially in an age where many people would describe themselves as 'spiritual but not religious' (Ammerman, 2013). It can be argued that religion and spirituality are interconnected (Stausberg, 2014; Fedele & Knibbe, 2013), but it is also generally agreed that religion is now a *choice* (Percy, 2019) and that spirituality represents a more personal or individual approach than religion (Norman, 2012; Wilson, McIntosh & Wilson, 2013).

In their study of Australian Millennials, Mason et al. (2007) divided their definition of spirituality into three categories:

- Traditional: grounded in the tradition of a major world religion (e.g. Christianity, Judaism, Buddhism, Islam)
- Alternative: non-traditional religions or spiritual paths (e.g., New Age, partly derived from Eastern religions)
- Humanist: a worldview which affirms human experience and human reason (rather than religious traditions or spiritual paths)

The Pew Research Center (2018) defines Millennials as those born between 1981 and 1996. Their research shows that only four in ten Millennials say that religion is important in their lives (even though half believe in God) and older generations are much more likely to be religious. On the other hand, Millennials are just as likely to engage in spiritual practices. Waters and Bortree (2012) summarize some of the reasons for Millennials turning away from organized religion including scandals associated with religious leaders, the entanglement of politics in religion and disagreement with some socio-political issues, as well as the possibilities of engaging with spirituality online rather than in a church.

Waters and Bortree (2012) reiterate Howe and Strauss's (2000) opinion that Millennials prefer and need individual attention compared to previous generations. It could be argued that spirituality represents a more individual approach (Watts, 2018), as defined by Nelson, James, Miles, Morrell, and Sledge (2017:389), "Spirituality spreads beyond practices and rituals to encompass an individual's relationship to the transcendent and understanding of life." Collins (2000) argued that religion was not disappearing; it was instead mutating into personal spirituality of a more aesthetic, pleasurable and experiential nature. Horan (2017) also notes that Millennials are immersed in a world of postmodern thinking and multiple ways of knowing and understanding. Mason et al. (2007:161) describe this situation as more of a fate than a choice "Their eclectic mix of worldviews and the cautious relativism of their values result from facing an uncertain world with only fragile support for their identity, beliefs and values." Horan (2017) notes that, unlike previous generations, Millennials are more likely to question past beliefs and to reject top-down approaches from spiritual leaders. This has significant implications for organized religion as Millennials are exposed to so many different facets of spirituality. Bevins (2020) confirms this viewpoint, highlighting that we have moved to an age of uncertainty where people are no longer willing to place their trust in institutions and authorities. Watts (2018) noted that many Canadian Millennials perceived religion as restricting individual freedom and the potential to be one's authentic self. Percy (2019) also noted the craving for the 'authentic' among Millennials and their need for 'personal experience' of spiritual fulfilment.

In terms of the factors shaping their lifestyles, Nelson et al. (2017) suggest that Generation Y's lives have been shaped by economic instability, corporate scandals and worldwide tragedies. Percy (2019:165) aptly describes how Millennials "have many of the benefits of social media but are also keenly aware that the previous two generations dismantled the concept of free university education, relatively accessible prospects of home ownership, and jobs for life". Randle, Eckersley and Miller (2017) undertook a survey with more than 2000 people in four countries which examined levels of concern across 19 personal and 23 societal issues. They noted that country differences were small, whereas generational differences were substantial. Generation Y were the most concerned about personal issues, especially financial security and job-related issues. Generation X and Y also reported higher levels of stress than previous generations. Delbosc and Nakanishi (2017) summarized the delay in the life stage of Millennials because of their longer enrolment in higher education, spending longer in the parental household, later home ownership and postponement of marriage and child-rearing. Horan (2017) emphasizes the pressure that Millennials are under to perform well academically, to become more creative and self-actualized, as well as projecting a perfect profile on social media. All of these factors could have significant implications for wellness as well as spirituality.

There may be cultural differences too. In this chapter, we focus partly on Chinese-speaking Millennials. Sima and Pugsley (2010) describe Chinese Millennials as more affluent than their parents, they are often the only child in their family and consider individuality as a highly sought-after quality. This has led to a rise in a so-called 'me' culture and a shift away from the 'collective interest' mentality of the older Mao generation. This is supported by more freedom of self-expression through Internet-based technology. Tan and Tan (2016) imply that the Chinese understanding of spirituality can be quite deeply connected to the material world and lifestyle, especially the energy forces that help individuals to navigate their lives. Whereas the previous generation had limited opportunities to practice religion, there has sometimes been a revival of religion among Millennials (Mou, 2017). There is also a close connection between natural landscapes and spirituality in Chinese culture (Pong, Leung & Lung, 2020).

Some other studies have been undertaken among specific cultural or nationality groups. Mason et al. (2007) undertook research on spirituality with (then) young Millennials in Australia, among whom interest in traditional religion was declining, they had minimal interest in 'alternative spirituality' and there was an increasing indifference or growth in humanism. Watts (2018) carried out a study about the meaning of spirituality for Millennials in Canada. The consensus there was that spirituality was closely connected to the (true) self and the individual (rather than organized religion) as the guiding force. Palimisano and Pannofino (2017) included both Millennials and Generation Z in their study of young Italians and conclude that the boundaries between spirituality, church religion and secular viewpoints are rather porous. Even if they undertake their own spiritual journey, young people in Italy still define themselves in relation to their Catholic upbringing (although they often reject it). A small, but growing number of young people in Italy are partaking in alternative forms of spirituality, such as yoga or meditation or New Age activities. Knoetze's (2018) study of African Millennials compares the profiles of this generation to American Millennials, making the point that they are not all 'digital natives', especially in poor rural areas, but that many of them are still very active in church. Indeed, religion continues to play an important role in socio-economic and political life in Africa. There has also been some research in the Asian context, for example, Kim (2016) interviewed a number of Millennials in Seoul and found that Buddhist temples are now being used mainly as sanctuaries away from economic and social pressure and to discover one's true self.

Several authors suggest that young people may eventually be redirected back to religion through education in schools (Horan, 2017; Tauber, 2015). Tauber (2015: 497) recognizes the need to accommodate the individualistic and spiritual needs of young people within Judaism, for example. She argues that spirituality should be presented in partnership with religion rather than in opposition to it. Bevins (2020:17) argues that young people are interested in returning to "a truer Christianity, a more authentic Christianity" which gives them a sense of belonging and participation. Ross (2013) even suggests that young peoples' desire for intoxication, spectacle and transcendence at a music festival or concert may represent a misguided desire for communion with God.

However, it could be argued that conversion back to organized religion is not the only solution to providing a spiritual path for young people. Mason et al. (2007:256) refer to the 'spiritual supermarket' or 'spiritual marketplace', where consumers can purchase components of spirituality that do not form part of a complete package. This could be an Eastern form of meditation, doing yoga or Tai Chi or even reading Tarot cards or a horoscope. Some authors have cynically suggested that spirituality represents a 're-branding' of religion (Carrette & King, 2005), or a superficial balancing of body, mind and spirit (Modise & Johannes, 2016). However, there is often a greater commitment to the self, people and planet. The category of 'humanist' is also an important one, as self-proclaimed Generation Y humanists are unlikely to be religious and neither are they very attracted to alternative spiritual practices. They believe more in human and personal freedom, self-development and self-realization (Mason et al., 2007). Nature protection and new forms of environmentalism may also be viewed as non-religious spiritual practices (Taylor, 2010). Watts' (2018) research with Canadian Millennials also touched on spirituality as 'healing' from past hurt and trauma. This can be seen as one of the reasons that tourists would visit retreats (Kelly & Smith, 2017). Indeed, research has shown that spirituality is an important part of health and wellbeing (Thoresen, 1999; Bai & Lazenby, 2015) and can be as important as physical health (Ghiya, 2019). In Chinese culture, mental health is also thought to be affected by a lack of spiritual balance (Niu, McSherry & Partridge, 2020).

### *6.2.1 Millennials and the wellness industry*

An *Independent* newspaper article (2019) suggested that wellness is the 'Millennial' self-help, which tends to be spiritual in nature and soothes and calms rather than fixing what is wrong. However, popular media reports about health and wellness tend to be somewhat contradictory. Goldman Sachs (2020) suggested that Millennials are more likely to integrate wellness into their daily lives through healthy eating, exercising and smoking less than previous generations. Nermoe (2018) cites a report where Millennials rated different aspects of their life: 79% said family was important, followed by health and wellness at 53%, friends at 39%, spirituality at 31% and career at 27%. On the other hand, a *Time* article (Ducharme, 2019) indicated that although Millennials are often credited with driving the highly lucrative wellness industry, they are actually in worse health than Generation X-ers were. This includes most of the major lifestyle diseases, as well as depression. Many of these issues are connected to rising obesity, conditions like diabetes, the high cost of healthcare, increasing stress levels caused by financial crises, a divisive political and social climate and social media pressures. Cassata (2020) confirmed this finding, arguing that Millennials are not willing to invest in long-term health and wellness. They tend to prioritize and spend more money on experiences, travel, dining out and technology before personal wellness.

Popular media articles about Chinese Millennials suggest that this stressed and overworked generation is becoming increasingly attracted to wellness and health products and services (Meiling, 2020). Flora (2020) cites a 2019 survey where "living a healthy lifestyle" was listed as the top "key life priority" for both Millennial and Gen-Z Chinese consumers and 76% of affluent Millennial Chinese consumers said that they place more emphasis on wellness than was the case for their parents. However, Banta (2019) noted that there is not really a Chinese word to define wellness in the same way as in English, but that aspects of wellness can be divided into physical, mental and spiritual well-being, all of which are high priorities for Chinese consumers. Banta's study showed that spiritual health was ranked as high as other domains of wellness (e.g. mental, social, nutritional), especially for Millennials.

### *6.2.2 Millennials and travel*

Canavan (2018) emphasizes that leisure and travel are important to Generation Y, despite being affected by socioeconomic restrictions. The earlier-mentioned delays in Millennials' life stage, such as full-time jobs, marriage and children (Delbosc & Nakanishi, 2017) can actually support their propensity to travel. Liu, Wu, and Li (2019) emphasize that Millennials value experiences, authenticity, uniqueness and creativity when they travel. The findings of Canavan's (2018) study of Millennial backpackers shows them to be ambitious and adventurous travellers in pursuit of self-actualization while remaining closely connected to family and friends, as well as being embedded in online communities. This can include sharing travel experiences as well as being heavily influenced by word of mouth via social media (Liu et al., 2019). Floros, Cai, McKenna & Ajeeb (2019) note the difficulties that Millennials face when trying to disconnect from technology while travelling because of FoMO (fear of missing out) and 'nomophobia' (when they cannot use their device). However, they also suggest that Millennials are tired of constant connections and recognize the well-being benefits of digital-free tourism. Rita, Brochado and Dimova (2019) summarized UK and US Millennials' main motivational factors as the need 'to relax', the desire 'to escape from the ordinary' and the wish 'to experience different lifestyles'.

By contrast, Chinese travellers may place more emphasis on seeing famous sights and learning about different cultures and history (Xu, Morgan & Song, 2008). Luo et al.'s (2020) assessment of Chinese Millennial tourists is that they are experience-oriented, demand uniqueness and value social media presence. They are clearly more digital-friendly, but also more brand-conscious than previous generations. These authors' study shows that Chinese Millennials' destination loyalty is based firstly on emotional values, followed by novelty and then social values (i.e. making a good impression on others). Du, Liechty, Santos and Park (2020) suggest that social media engagement is an important leisure habit among Chinese Millennials, including during vacation, and it was found to enhance tourism experiences.

Spiritual travel has been defined as "subjective and individual travel for spiritual betterment and self-discovery" (Kato & Progano, 2017:245). It can include visits to retreats, yoga or meditation centres, ashrams, temples or monasteries to solve personal problems or for healing (Norman and Pokorny, 2017). It might also include elements of adventure or challenges (Cheer, Belhassen & Kujawa, 2017) or a pilgrimage (Kato & Progano, 2017). The natural landscape can also be important to spiritual tourists (Sharpley and Jepson, 2011; Bone, 2013; Gezon, 2018; Jiang, Ryan, & Zhang, 2018).

## 6.3 Research methods

The process of questionnaire design was based on a thorough search for previous studies of young peoples' perceptions of spirituality. Definitions of spirituality were constructed for the questionnaire based on articles by

Brooks et al. (2018), Collins (2000), Halafoff et al. (2020) and Mason et al. (2007). The search also included papers that had focused on specific generations, such as Green et al. (2013) and Percy (2019). Cultural differences were also taken into consideration based on studies of young people and spirituality in specific cultures (e.g. Knoetze, 2018; Palimisano & Pannofino, 2017). Studies about the relationship between health, wellness and spirituality were also consulted (e.g. Ghiya, 2019; Modise & Johannes, 2016; Thoresen (1999)). The tourism-related questions were inspired by the work of Cheer et al. (2017); Gezon (2018); Jiang et al. (2018); Kujawa (2017); Norman (2012); Norman & Pokorny (2017); and Wilson et al. (2013). The English version of the questionnaire was refined after discussions with Tourism Master's students from different religious backgrounds (e.g. Catholic, Muslim and Hindu) and the Chinese version was prepared after discussions with Chinese-speaking Tourism Master's students who were also studying subjects connected to wellness and culture. The questionnaire was prepared electronically to facilitate wider distribution and snowball sampling and closed-ended questions, mostly using the Likert scale, were included in order to make the comparison between the two language samples easier.

Both questionnaires were piloted among English and Chinese-speaking Master's students, who checked the clarity of the questions and the translation. The authors distributed the questionnaires in July and August 2020 to Masters and PhD students and young lecturers born between 1981 and 1996, using carefully controlled snowball sampling within their respective universities in Central Europe, mainland China, Taiwan and Hong Kong. 181 Millennials responded to the questionnaires: 86 in the English language and 95 in the Chinese language. The approach taken was based on non-probability, stratified sampling as it only targeted and allowed responses from Millennials who were attending one of the researchers' universities or were recommended by an already involved respondent. However, the snowball sampling approach allowed a certain degree of convenience and voluntary sampling to take place.

The English-language sample was more culturally diverse, with 30 nationalities represented from five continents; however, two-thirds were from Europe. Regarding the religion of the participants, in the English-language sample, 44 respondents identified as 'Not religious' (51.1%) and 34.9% were Christians, with a small number of individuals from other religions (e.g. only 7 Muslims, 1 Hindu but no Jewish respondents). The Chinese-language sample had 65 non-religious respondents (68.4%) and a small number of religious respondents who were mainly Christian, Buddhist or Taoist. Regarding the gender difference in the surveys, there was a significantly higher representation from women, with 126 female respondents and only 55 males. Smith (2008) highlights the gender bias in favour of women in online surveys, but it should also be added that there was also a much higher percentage of female students in the population sample because of the subject areas that they were studying (e.g. Tourism and Hospitality).

## 6.4 Findings

The Millennials' most popular understanding of spirituality was the strong connection to themselves (representing 16.4% of the total responses), which emphasizes this generation's 'self-importance', correlating with Watt's findings (2018). Connection to nature is also important (15.2%) (see Table 6.1). They also believed in body, mind and spirit activities such as yoga and meditation (12.5%), as well as showing strong cohesion towards family and friends (12.1%). Both belief in God and humanism gained the same score (9.9%). Only a very small number of respondents (2.4%) had no belief at all in spirituality.

Several significant cultural differences can be noted for English-language respondents compared to Chinese-speakers, for example: being humanist (64% compared to 36%), which includes helping others and self-development as defined by Mason et al. (2007). Differences can be seen in the connection to a force or energy outside of this world (62.8% compared to 47.2%) and connection to nature (61% compared to 39%). More Chinese-language respondents mentioned that they were not spiritual at all (83.3%), but more also believe in the supernatural (67.6% compared to 32.4%) or a religion (65.6% compared to 34.4%). Despite previous theories about Chinese society being more collective and less individualistic (Li, 2020), it can be seen that Chinese-speaking Millennials feel an even stronger connection to the self than English-speaking respondents. This re-emphasizes their sense of individualism and the increasing importance of "me" (Sima & Pugsley, 2010). Nevertheless, friends, family and society are still relatively more important for Chinese speakers. Body, mind and spirit activities such as yoga and

*Table 6.1* Which of the following comes closest to your own personal definition of 'spirituality'?

| *The respondents' personal definitions of 'spirituality'* | *Ch* | *Eng* | *Total* |
|---|---|---|---|
| A strong connection to myself | 47 | 36 | 83 |
| Being connected to nature and the natural world | 30 | 47 | 77 |
| Body-mind-spirit activities like yoga and meditation | 35 | 28 | 63 |
| Close connections with other people (Friends, family, society) | 39 | 22 | 61 |
| Being a humanist and helping others | 18 | 32 | 50 |
| A belief in God | 27 | 23 | 50 |
| A connection to a force or energy outside of this world | 16 | 27 | 43 |
| Belief in the supernatural | 23 | 11 | 34 |
| My religions (Christianity, Judaism, Islam, Buddhism, etc.) | 21 | 11 | 32 |
| I don't believe in spirituality | 10 | 2 | 12 |
| Total | 266 | 239 | 505 |

Source: Authors' own.

meditation are also more widespread in Asian cultures and there has been a revival of religion among this generation (Mou, 2017).

Most of the Millennials' self-evaluation of how they saw themselves spiritually showed that they believed themselves to be humanists who care about people and the planet (28.3% of total responses) (Table 6.2), echoing Mason et al.'s finding (2007). The second most common interest (although ranked much lower) was body-mind-spirit activities like yoga and meditation followed by some interest in holistic therapies (Chinese speakers were more likely to practise them with 62.9% compared to 37.1% of English speakers). The other responses reflect a certain degree of confusion about what spirituality is or hybridity, such as a tendency to be religious and spiritual (especially for Chinese-speakers) or 'spiritual but not religious' (more common for English-speakers with 72.7% compared to 27.3%). However, it is also important to note that a number of respondents claim neither to be religious nor spiritual. Overall, these findings suggest that religious interest is declining among Millennials, but is being substituted by 'alternative spirituality' (Mason et al., 2007).

When asked a question about which individuals or institutions have had the greatest influence on the respondents, parents, family, community and society were much more important than religious or spiritual leaders.

In terms of their lifestyle and wellness, Table 6.3 shows that their most frequently chosen responses were regular travelling followed by eating healthy food, connecting with nature and minimizing stress. Enjoying work and exercising were evaluated quite highly; however, the least popular statements were those relating to undertaking spiritual practice or religious activities which confirmed prior research again (Pew Research Center, 2018; Waters & Bortree, 2012). Doing exercise, eating healthily and undertaking spiritual practice were ranked higher for English-speakers, whereas Chinese-speaking Millennials seemed to be more concerned with socializing and minimizing their stress.

*Table 6.2* How would you best describe yourself?

| *Description of the 'spirituality' of the respondents* | *Ch* | *Eng* | *Total* |
|---|---|---|---|
| A humanist who cares about people and the planet | 36 | 52 | 88 |
| Interested in body-mind-spirit connection (e.g. through yoga, meditation) | 28 | 27 | 55 |
| Religious and spiritual at the same time | 30 | 8 | 38 |
| Curious about spirituality but unsure what it is | 17 | 19 | 36 |
| Interested in holistic therapies like Angels, Crystals, astrology, Tarot, etc. | 22 | 13 | 35 |
| Spiritual but not religious | 6 | 16 | 22 |
| Not interested in spirituality | 17 | 5 | 22 |
| A religious practitioner (Christian, Jew, Buddhist, etc.) | 10 | 5 | 15 |
| Total | 166 | 145 | 311 |

Source: Authors' own.

*Table 6.3* Which of the following statements are most true for you?

| *Lifestyle and Wellness* | *Ch* | *Eng* | *Total* |
|---|---|---|---|
| I have positive relationship with my family | 4.08 | 4.36 | 4.22 |
| I socialize with friends | 4.31 | 4.00 | 4.16 |
| I travel as often as I can | 4.04 | 4.01 | 4.03 |
| I try to eat healthily | 3.87 | 4.08 | 3.97 |
| I try to get out into nature and countryside | 3.85 | 3.91 | 3.88 |
| I try to minimize my stress levels | 3.95 | 3.79 | 3.87 |
| I enjoy my work or studies | 3.77 | 3.81 | 3.79 |
| I try to stay active and fit through exercise | 3.57 | 3.83 | 3.69 |
| I undertake some spiritual practice (e.g. yoga, meditation, body-mind-spirit activities) | 2.43 | 2.77 | 2.59 |
| I undertake religious activities (e.g. prayer, visiting church or temple, reading holy books) | 2.02 | 1.93 | 1.98 |
| Total | 3.59 | 3.65 | 3.62 |

Source: Authors' own.

*Table 6.4* Which of the following statements are true for you at the moment?

| *The level of religiousness of the respondents* | *Ch* | *Eng* | *Total* |
|---|---|---|---|
| I believe in a higher power, force or energy outside our world | 29 | 32 | 61 |
| Not many or none of my friends are religious or spiritual | 19 | 42 | 61 |
| I keep close connections to nature which feels spiritual to me | 19 | 41 | 60 |
| I do not practice religion or spirituality now | 25 | 26 | 51 |
| My friends and I do not discuss religion | 22 | 26 | 48 |
| I practice spirituality regularly through yoga, meditation or other body-mind-spirit practices | 18 | 24 | 42 |
| My friends and I do not discuss spirituality | 20 | 17 | 37 |
| Most of my friends are spiritual but not religious | 9 | 19 | 28 |
| I practice a religion regularly (e.g. through prayer, church or temple visits, reading holy books) | 13 | 15 | 28 |
| Most of my friends are religious | 19 | 7 | 26 |
| I grew up religious but I do not practice it now | 8 | 17 | 25 |
| I practice a different religion now from the one I grew up with | 11 | 0 | 11 |
| I feel comfortable practising my religion in a public place (e.g. praying) | 2 | 7 | 9 |
| If I do spiritual activities, it is for social or family reasons, not my personal belief | 0 | 8 | 8 |
| Total | 214 | 281 | 495 |

Source: Authors' own.

Regarding the respondents' current state of spirituality, they confirmed that they believed in a higher power's existence, but both language cohorts agreed that their peers were not practising religion or spirituality nor talked about religious or spiritual topics much (Table 6.4). Cultural differences showed that English-language respondents did not have many friends who were religious or spiritual, which correlated to studies about decreasing trust in religious organizations (Waters & Bortree, 2012). More important were close connections to nature that felt spiritual for them (68.3% versus 31.7%). Even if they grew up in a religious environment, they did not practice it now (68% compared to 32%), and had more spiritual friends who were not religious (67.9% versus 32.1%). On the other hand, a higher number of the Chinese-language sample stated that their friends were religious, but many of them were also practising a different religion from the one that they grew up with. Once again, this suggests a revival of religion in the post-Mau generation (Mou, 2017).

Figure 6.1 introduces the most popular spiritual travel destinations for Millennials which they would visit if they had enough time and money. It should be mentioned that the English-language Millennials were more enthusiastic about answering this question as they provided 58.7% of the responses; thus, all of the options were rated higher. The only question that was chosen by more Chinese people was 'None of the above' (61%), suggesting that they are much less interested in spiritual travel than their Western counterparts. However, both groups ranked the Thai massage course (15.4%) and Santiago pilgrimage (15%) highest, as well as a European monastery with meditation (10.7%). Ashton (2018) discusses the growing importance of spiritual retreat tourism in Thailand and Kim, Yilmaz and Ahn (2019) demonstrate that the Santiago de Compostela route has increasingly become part of secular or spiritual rather than religious tourism. The largest differences in favour of English-language Millennials were visiting a South Korean Buddhist temple (75.6%), an Ayurvedic centre (73.5%), and a meditation and yoga retreat in the Sahara Desert (68.8%).

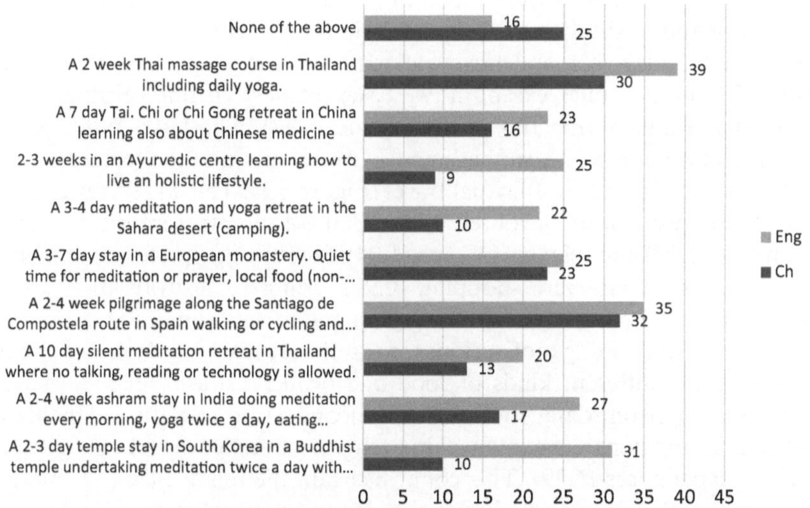


*Figure 6.1* Which of the following spiritual holidays would you find interesting in the next year or two if you had enough time and money?

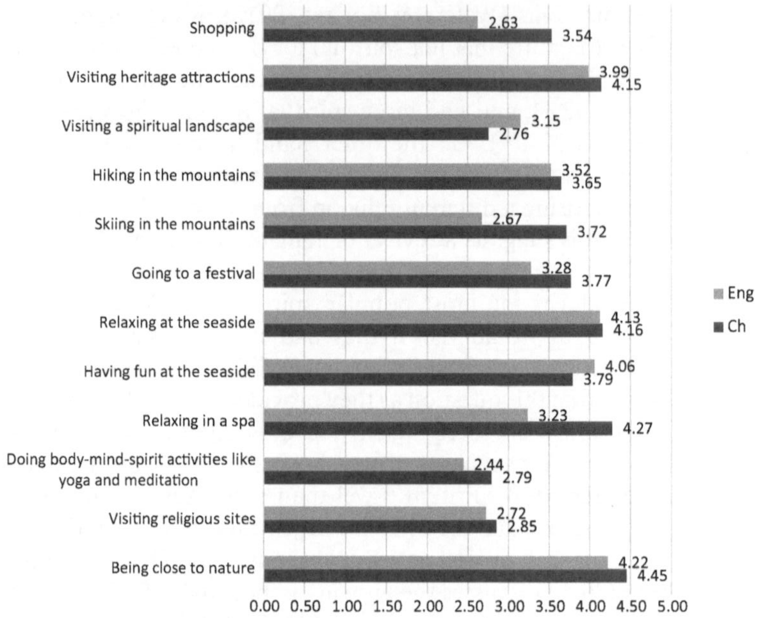


*Figure 6.2* Which of the following activities are the most interesting for you when you travel?

Two of the three most popular travel activities for the respondents related to switching off from daily routines: escaping to nature or relaxing by the seaside (Figure 6.2). The third most popular activity was visiting heritage attractions, followed by having fun at the seaside, relaxing in a spa, hiking in the mountains, skiing in the mountains. The lowest-ranked activities were doing yoga and meditation or visiting religious or spiritual landscapes.

Chinese-language respondents evaluated all of the given activities slightly higher (0.32) on average, except in two cases, namely visiting spiritual landscapes (0.39) and having fun at the seaside (0.27). The latter is based on behavioural difference, i.e. sunbathing for Westerners and the former on the relative lack of interest in spiritual travel (Figure 6.1). The largest differences between the two cohorts favoured by the Chinese sample were: skiing in the mountains (1.05) and relaxing in a spa (1.04). Significantly preferred activities for Chinese-speakers were shopping (0.91), going to festivals (0.49), and (although ranked relatively low) doing yoga and meditation (0.35).

The most outstanding motivations for travel were learning about other cultures, enjoying different kinds of food and drink, rest and relaxion, stress relief, escaping from daily routine and (re)connecting to nature. The least-favoured motivations were intensifying religious practice (1.86) and gaining spiritual experiences (2.39). This confirms again the relative lack of interest in spiritual and religious activities compared to others when travelling.

Regarding the cultural discrepancies, the Chinese-language respondents evaluated all activities slightly higher (0.22) on average, with the exception of

*Figure 6.3* Which of the following motivations and benefits are the most important for you when you travel?

escapism from everyday life. The largest anomalies can be seen in rest and relaxion (0.6), suggesting that Chinese travellers may be less active or need more recovery from stressful lives. The second-largest deviation between the two groups was (re)connecting to nature (0.45) and discovering myself (0.42). Pong et al. (2020) noted the importance of nature for wellbeing in Chinese culture and the latter point reconfirms the growing importance of the self in Chinese culture. Interestingly, there was a noticeable difference between the evaluation of health improvement, which seemed less important for English-language respondents. Indeed, it seems overall that travel affords more benefits for Chinese travellers than for English-speaking ones.

## 6.5 Conclusions

This chapter explored Millennials' definitions of spirituality, as well as perceptions of their own sense of spirituality. The data confirmed other studies which had shown the decline of religion among young people (Pew Research Center, 2018) in favour of more individual and experiential approaches to spirituality (Collins, 2000), including alternative therapies, connection to nature or humanism (Mason et al., 2007). Very few respondents did not believe in some form of spirituality (only 2.4%). The findings confirm that Millennials are mostly focused on the self and individuality (Watts, 2018), but one surprising finding is that Chinese speakers seem to be even more individualistic than English speakers, which confirms Sima and Pugsley's (2010) study but contradicts other studies that emphasize the collective nature of Chinese culture (Li, 2020). It seems that religion and spirituality are viewed as choices (Percy, 2019), however, although a higher percentage of Chinese

Millennials claimed to be 'non-religious' initially, deeper analysis suggested that a higher percentage are actually interested in religious practice than in the English-speaking sample, which supports the theory of Mou (2017). Chinese Millennials are also more likely to believe in the supernatural (interpreted as belief in spirits, ghosts or energy forces). In accordance with studies that show the unwillingness of Millennials to trust in top-down leaders and institutions (Bevins, 2020; Waters & Bortree, 2012), religious and spiritual leaders play a much less influential role in their lives than parents, family, friends or community. On the other hand, it is worth reiterating that Millennials may be more religious in certain parts of the world, e.g. Africa (Knoetze, 2018) which was under-represented in this study. Some authors have also emphatically suggested that young people would benefit from a return to collective religion (e.g. Bevins, 2020; Horan, 2017; Ross, 2013; Tauber, 2015).

Connection to nature is the second most significant response, as well as caring about the planet and people (humanism). This echoes Taylor's (2010) theories about nature and spirituality and confirms Pong, Leung and Lung's (2020) emphasis on the importance of natural landscape in Chinese spirituality. However, it should be noted that English-speakers are much more likely to define themselves as humanists in terms of the natural world and helping others. It can be seen that family, friends and social connections are still more important for Chinese respondents, despite the growth in individualism.

In terms of health, wellness and lifestyle, it is clear that religion and spirituality play a relatively minor role compared to other domains of life. Spirituality was not ranked as highly as other domains of wellness in this study, unlike that of Banta (2019). This finding is also slightly different from the study cited by Nermoe (2018), who suggested that wellness and spirituality are even more important than career for Millennials. The findings about travel activities and benefits confirmed Rita, Brochado and Dimova's (2019) study, which also emphasized the need 'to relax' and the desire 'to escape from the ordinary'. Interestingly, rest was more important for the Chinese and escapism for the English-speaking sample. Chinese Millennials tend to associate travel with more health benefits than English-speakers, which confirms the study cited by Flora (2020) who stated that a 'healthy lifestyle' was a major life priority for Chinese Millennials. This might be because this generation is especially stressed and overworked (Meiling, 2020). However, Chinese Millennials seem to be much less interested in spiritual travel as defined by Norman and Pokorny (2017), for example. Some interest was shown in the Santiago pilgrimage, but this is becoming increasingly secular (Kim et al., 2019).

It has to be reiterated that the least-favoured motivation for travel for both samples were intensifying religious practice and gaining spiritual experience. These findings suggest that these are very much niche activities or that travellers may not define their activities or benefits as 'spiritual'. If we return to the definition of spiritual travel by Kato and Progano (2017:245) as "subjective and individual travel for spiritual betterment and self-discovery", this fits better the Millennials' most-favoured definition of spirituality as being a connection to the self, followed by a connection to nature. The popular answer of 'being close to nature' when travelling is surely connected to a sense of spirituality even if it is not defined as such, echoing previous studies (Bone, 2013; Gezon, 2018; Sharpley & Jepson, 2011).

## 6.6 Future Research

Future research should clearly aim to explore the terminology that is used to define spirituality within the travel experience more clearly, especially with regards to self-development and the connection to nature. It is quite possible that the connotations of spirituality for many respondents in research is still inextricably connected to religion, which biases their responses and reactions. This sample is too small to make significant generalizations about cultural differences, but it already highlights some important differences between English and Chinese speakers which confirm other studies. It would be essential in future research to include a larger sample of respondents from different religious groups as well as nationalities to explore the meaning of spirituality also for Muslims, Hindus or Jews, for example (even those who are no longer practising). It would also be interesting to follow the changing travel habits and motivations of Millennials as they head towards older age to see if participants report higher levels of spirituality as they age as suggested by Green et al. (2013) and Moal-Ulvoas and Taylor (2014). It would also be useful to separate older and younger Millennials in a larger sample or to examine gender differences more closely.

## 6.7 Implications of the Research

This chapter offers some useful insights into the complexities of cross-cultural understandings of spirituality which have important social implications, especially for enhancing wellness, leisure and travel experience creation. The importance of self-development and nature connection should be emphasized. These findings could be used by tour companies for designing packages for young people; for agencies involved in planning educational visits or summer camps; as well as for spiritual destinations or individual sites when targeting segments of young travellers. The leisure and tourism industries and education sector can use such research to inform their product and service development to think carefully about how they design, label and promote their services to maximize wellness benefits and to improve the lives of this relatively disadvantaged generation of young people.

## References

Ammerman, N. T. (2013). Spiritual but not religious? Beyond binary choices in the study of religion. *Journal for the Scientific Study of Religion*, 52, 258–78. doi: 10.1111/jssr.12024.

Ashton, A. S. (2018). Spiritual retreat tourism development in the Asia Pacific region: investigating the impact of tourist satisfaction and intention to revisit: a Chiang Mai, Thailand case study. *Asia Pacific Journal of Tourism Research*, 23(11), 1098–1114. doi: 10.1080/10941665.2018.1526198.

Bai, M., & Lazenby, M. (2015). A systematic review of associations between spiritual wellbeing and quality of life at the scale and factor levels in studies among patients with cancer. *Journal of Palliative Medicine*, 18(3), 286–298. doi: 10.1089/jpm.2014.0189.

Banta, A. (2019). Caught between tradition and modernity: How Chinese Gen Z and Gen Y view wellness. *Global Wellness Summit*, Retrieved from https://www.globalwellnesssummit.com/wpcontent/uploads/2019/11/GWS2019_AmritaZanta_CaughtBetweenTraditionandModernity2.pdf.

Bevins, W. (2020). Young, restless, and liturgical: Young adults and the search for identity. *Liturgy*, 35(2), 18–24. doi: 10.1080/0458063X.2020.1739476.

Bone, K. (2013). Spiritual retreat tourism in New Zealand. *Tourism Recreation Research*, 38(3), 295–309. doi: 10.1080/02508281.2013.11081755.

Brooks, F., Michaelson, V., King, N., Inchley, J., & Pickett, W. (2018). Spirituality as a protective health asset for young people: an international comparative analysis from three countries. *International Journal of Public Health*, 63, 387–395. doi: 10.1007/s00038-017-1070-6.

Canavan, B. (2018). The complex cohort: a netnographic review of generation Y backpackers. *Leisure Studies*, 37(2), 184–196. doi: 10.1080/02614367.2017.1347699

Carrette, J., & King, R. (2005). *Selling Spirituality: The Silent Takeover of Religion*. London: Routledge.

Cassata, C. (2020). Why more millennials are willing to cut corners on their health, *Healthline*, Retrieved from https://www.healthline.com/health-news/why-arent-more-millennials-investing-in-their-health#What-influences-millennials?

Cheer, J. M., Belhassen, Y., & Kujawa, J. (2017). The search for spirituality in tourism: Toward a conceptual framework for spiritual tourism. *Tourism Management Perspectives*, 24, 252–256. doi: 10.1016/j.tmp.2017.07.018

Collins, S. (2000). Spirituality and youth. In M. Percy (Ed.), *Calling Time: Religion and Change at the Turn of the Millennium* (pp. 28–45). Sheffield: Sheffield Academic Press.

Delbosc, A., & Nakanishi, H. (2017). A life course perspective on the travel of Australian millennials. *Transportation Research, Part A*, 104, 319–336. doi: 10.1016/j.tra.2017.03.014

Du, X., Liechty, T., Santos, C. A., & Park, J. (2020). 'I want to record and share my wonderful journey': Chinese Millennials' production and sharing of short-form travel videos on TikTok or Douyin. *Current Issues in Tourism*, doi: 10.1080/13683500.2020.1810212

Ducharme, J. (2019). Millennials love wellness but they're not as healthy as people think, report says. *Time*. Retrieved from https://time.com/5577325/millennials-less-healthy

Fedele, A., & Knibbe, K. (2013). *Gender and Power in Contemporary Spirituality. Ethnographic Approach*. New York: Routledge.

Flora, L. (2020). China's wellness surge pushed into high gear by pandemic. 28 July. Retrieved from https://www.glossy.co/beauty/chinas-wellness-surge-pushed-into-high-gear-by-pandemic

Floros, C., Cai, W., McKenna, B., & Ajeeb, D. (2019). Imagine being off-the-grid: millennials' perceptions of digital-free travel. *Journal of Sustainable Tourism*, 9(5), doi: 10.1080/09669582.2019.1675676

Goldman, Sachs (2020). *Millennials: Coming of Age*. Retrieved from https://www.goldmansachs.com/insights/archive/millennials

Gezon, L. L. (2018). Global scouts: Youth engagement with spirituality and wellness through travel, Lake Atitlán, Guatemala. *Journal of Tourism and Cultural Change*, 16(4), 365–378. doi: 10.1080/14766825.2017.1310217

Ghiya, G. D. (2019). Promoting spiritual health and holistic wellness. *Journal of Health Management*, 21(2), 230–233. doi: 10.1177/0972063419835104

Green, M., Chavez, E. S., Carmody, M., & Duncan, P. (2013). Differences in religiosity and spirituality as a result of generation, gender and education. *The International Journal of Religion and Spirituality in Society*, 2(4), doi: 10.18848/2154-8633/CGP/v02i04

Halafoff, A., Shipley, H., Young, P.D., Singleton, A., Rasmussen, M. L., & Bouma, G. (2020). Complex, critical and caring: Young people's diverse religious, spiritual and non-religious worldviews in Australia and Canada. *Religions*, 11(4), 166. doi: 10.3390/rel11040166

Horan, A. P. (2017). Fostering spiritual formation of millennials in Christian schools. *Journal of Research on Christian Education*, 26(1), 56–77. doi: 10.1080/10656219.2017.1282901

Howe, N., & Strauss, W. (2000). *Millennials Rising: The Next Great Generation*. New York, NY: Random House.

Jiang, T., Ryan, C., & Zhang, C. (2018). The spiritual or secular tourist? The experience of Zen meditation in Chinese temples. *Tourism Management*, 65, 187–199. doi: 10.1016/j.tourman.2017.10.008

Kato, K., & Progano, R. N. (2017). Spiritual (walking) tourism as a foundation for sustainable destination development: Kumano-kodo pilgrimage Wakayama, Japan. *Tourism Management Perspectives*, 24, 243–251. doi: 10.1016/j.tmp.2017.07.017

Kelly, C., & Smith, M. K. (2017). Journeys of the self: the need to retreat. In M. K. Smith and L. Puczkó (Eds.), *The Routledge Handbook of Health Tourism* (pp. 138–151). London: Routledge.

Kim, H. M. (2016). Becoming a city Buddhist among the young generation in Seoul. *International Sociology*, 31(4), 450–466. doi: 10.1177/0268580916643089

Kim, H., Yilmaz, S., & Ahn, S. (2019). Motivational landscape and evolving identity of a route-based religious tourism space: a case of Camino de Santiago. *Sustainability*, 11, 3547. doi: 10.3390/su11133547

Knoetze, H. (2018). Analyzing transformational development as engagement with African millennials. *International Review of Mission*, 107(2), 483–499. doi: 10.1111/irom.12245

Kujawa, J. (2017). Spiritual tourism as quest. *Tourism Management. Perspectives*, 24, 193–200. doi: 10.1016/j.tmp.2017.07.011

Li, R. (2020). *Comparison of Chinese and Western Internal Spiritual Culture. Advances in Social Science, Education and Humanities Research, 415, 5th International Symposium on Social Science*, 177–180.

Liu, H., Wu, L., & Li, X. (2019). Social media envy: How experience sharing on social networking sites drives millennials' aspirational tourism consumption. *Journal of Travel Research*, 58(3), 355–369. doi: 10.1177/0047287518761615

Luo, J., Deyb, B. L., Cagri Yalkinc, C., Sivarajahd, U., Punjaisrie, K., Huang, Y., & Yen, D. A. (2020). Millennial Chinese consumers' perceived destination brand value. *Journal of Business Research*, 116, 655–665. doi: 10.1016/j.jbusres.2018.06.015

Mason, M., Singleton, A. & Webber, R. (2007). The spirituality of young Australians. *International Journal of Children's Spirituality*, 12(2), 149–163. doi: 10.1080/13644360701467451

Meiling, C. (2020, 7 October). For millennials, health is the new wealth. *China Daily*. Retrieved from http://www.chinadailyasia.com/article/145491

Moal-Ulvoas, G., & Taylor, V. A. (2014). The spiritual benefits of travel for senior tourists. *Journal of Consumer Behaviour*, 13, 453–462. doi: 10.1002/cb.1495

Modise, L., & Johannes, M. L. (2016). Well-being and wellness in the twenty-first century: A theanthropocosmic approach. *Journal of Religion and Health*, 55, 1876–1890. doi: 10.1007/s10943-015-0140-4

Mou, Y., (2017). Predicting the use of traditional Chinese medicine health maintenance approach from cultural and spiritual perspectives. *Journal of Religion and Health*, 56(3), 971–985. doi: 10.1007/s10943-016-0299-3

Nelson, M. F., James, M. S. L., Miles, A., Morrell, D. L., & Sledge, S. (2017). Academic integrity of millennials: The impact of religion and spirituality. *Ethics & Behaviour*, 27(5), 385–400. doi: 10.1080/10508422.2016.1158653

Nermoe, K. (2018, 12 September). Millennials: the 'wellness generation'. *Sandford Health*. Retrieved from https://news.sanfordhealth.org/sanford-health-plan/millennials-wellness-generation

Niu, Y., McSherry, W., & Partridge, M. (2020). An understanding of spirituality and spiritual care among people from Chinese backgrounds: A grounded theory study. *Journal of Advanced Nursing*, 76, 2648–2659. doi: 10.1111/jan.14474

Norman, A. (2012, June). The varieties of the spiritual tourist experience. *Literature & Aesthetics*, 22(1), 20–37. Retrieved from https://core.ac.uk/download/pdf/229423903.pdf

Norman, A., & Pokorny, J. J. (2017). Meditation retreats: Spiritual tourism and well-being interventions. *Tourism Management Perspectives*, 24, 201–207. doi: 10.1016/j.tmp.2017.07.012

Palimisano, S., & Pannofino, N. (2017). So far and yet so close: Emergent spirituality and the cultural influence of traditional religion among Italian youth. *Social Compass*, 64(1), 130–146. doi: 10.1177/0037768616683329

Percy, M. (2019). Sketching a shifting landscape: Reflections on emerging patterns of religion and spirituality among Millennials. *Journal for the Study of Spirituality*, 9(2), 163–172. doi: 10.1080/20440243.2019.1658268

Pew Research Center (2018, 13 June). Young adults around the world are less religious by several measures. Retrieved from https://www.pewforum.org/2018/06/13/young-adults-around-the-world-are-less-religious-by-several-measures

Pong, H.-K., Leung, C.-H., & Lung, C.-L. (2020). Validating the Chinese-translated version of the Spiritual Health and Life-orientation Measure (SHALOM) amongst the Chinese youth populations in 2010 and 2018. *Journal of Beliefs & Values*, 41(4), 489–508. doi: 10.1080/13617672.2019.1693823

Randle, M., Eckersley, R., & Miller, L. (2017). Societal and personal concerns, their associations with stress, and the implications for progress and the future, *Futures*, 93, 68–79. doi: 10.1016/j.futures.2017.07.004

Rita, P., Brochado, A., & Dimova, L. (2019). Millennials' travel motivations and desired activities within destinations: a comparative study of the US and the UK. *Current Issues in Tourism*, 22(16), 2034–2050. doi: 10.1080/13683500.2018.1439902

Ross, C. (2013). Spectacle, ecstasy, and music festivals: The spiritual longing of the lost generation. *Liturgy*, 28(3), 42–50. doi: 10.1080/0458063X.2013.774884

Sharpley, R., & Jepson, D. (2011). Rural tourism: A spiritual experience? *Annals of Tourism Research*, 38(1), 52–71. doi: 10.1016/j.annals.2010.05.002

Sima, Y., & Pugsley, P. C. (2010). The rise of a 'Me' culture in postsocialist China, youth, individualism and identity creation in the blogosphere. *The International Communication Gazette*, 72(3), 287–306. doi: 10.1177/1748048509356952

Smith, W. (2008). Does Gender Influence Online Survey Participation? A Record-Linkage Analysis of University Faculty Online Survey Response Behavior. *ERIC*. Retrieved from https://eric.ed.gov/?id=ED501717

Stausberg, M. (2014). Religion and spirituality in tourism. In A. A. Lew, C. M. Hall and A. M. Williams (Eds.), *The Wiley Blackwell Companion to Tourism* (pp. 349–360). New Jersey: Wiley.

Tan, C., & Tan, L. (2016). A shared vision of human excellence: Confucian spirituality and arts education. *Pastoral Care in Education*, 34(3), 156–166. doi: 10.1080/02643944.2016.1167109

Tauber, S. M. (2015). Spiritual not religious? The imperative of religious education as spiritual inspiration in Jewish education. *Religious Education*, 110(5), 495–498. doi: 10.1080/00344087.2015.1089721.495

Taylor, B. (2010). *Dark Green Religion: Nature, Spirituality and the Planetary Future*. University of California Press.

The Independent (2019, 19 August). *Why wellness is the millennial self-help*. Retrieved from https://www.independent.co.uk/news/long_reads/health-and-wellbeing/wellness-millennial-self-help-self-care-love-fear-a9056946.html

Thoresen, C. E. (1999). Spirituality and health: Is there a relationship? *Journal of Health Psychology*, 4, 291–300. doi: 10.1177/135910539900400314

Xu, F., Morgan, M., & Song, P. (2008). Students' travel behaviour: a cross-cultural comparison of UK and China. *International Journal of Travel Research*, 11(3), 255–268. doi: 10.1002/jtr.686

Waters, R. D., & Bortree, D. S. (2012). "Can We Talk About the Direction of This Church?": The impact of responsiveness and conflict on millennials' relationship with religious institutions. *Journal of Media and Religion*, 11, 200–215. doi: 10.1080/15348423.2012.730330

Watts, G. (2018). On the politics of self-spirituality: A Canadian case study. *Studies in Religion / Sciences Religieuses*, 47(3), 345–372. doi: 10.1177/0008429818764114

Wilson, G. B., McIntosh, A. J., & Wilson, A. L. Z. (2013). Tourism and spirituality: A phenomenological analysis. *Annals of Tourism Research*, 42, 150–168. 10.1016/j.annals.2013.01.016

# 7 Millennial tourism

## Myths, wellness, lifestyle and practicality toward millennial tourism

*Wai Ching Poon*

### 7.1 Introduction

According to the Pew Research Center, Millennials (a.k.a. Generation Y) are defined as those who were born between 1981 and 1996 (and therefore aged 24 to 39 in 2020) (Dimock, 2019). They are different from Generation X (born 1965–1980) and Baby Boomers (born 1946–1964) with respect to both their behaviour and lifestyle. Millennials are more technology-savvy, more innovative, more inclined to defer the completion of their education and getting married, while the Baby Boomers eagerly seek a steady job, get a house, a car, and form a family. The Millennials are experiencing increasing earnings, and with their greater spending power, they soon surpass Baby Boomers as the largest travel spenders.

According to Barroso, Parker, and Bennett (2020), Millennials have outshone Baby Boomers as the leading living adult cohort, rendering them crucial targets for largest travel spenders. Millennials have redefined various aspects of the marketplace, including social, demographic, technological, and economic ones, which clout in the many industries worldwide (Garikapati, Pendyala, Morris, Mokhtarian, & McDonald, 2016). This demographic group is the most profitable for businesses and stakeholders in the tourism industry.

Millennial tourism is one of the fastest-growing segments. This demographic prefer to spend money on consumer experiences (e.g., dream holidays) (Verissimo & Costa, 2018). They have increasing earnings with greater spending power, and their lifestyle has an association with the demand for tourism. The wellness tourism segment is another branch of Millennial tourism. Millennials are not wellness newbies, but it is not entirely clear what the nature of their wellness habits should be, i.e., what they expect and the mythical Millennials on tourism. Therefore, it is important to understand the behaviour of Millennial tourists as their character is reflective of the industry's need.

In this chapter, we employ the recent trends in global Millennials' travel statistics to understand their travel myths, lifestyle and practicality, wellness and sustainability. For instance, what do Millennial travellers want and need? What motivates them to travel? What is the role of social media in Millennial tourism? What is their priority in life? Are there any differences in the experience of Muslim Millennial travellers? We intend to understand

DOI: 10.4324/9781003170631-9

how Millennials' myths and lifestyles impact their mindset on the practicality of travelling. We then discuss the importance of sustainable tourism with its challenges and opportunities.

## 7.2 Millennials' travel trend statistics

Let us look at the trends in global Millennials' travel statistics to understand their travel myths, wellness and lifestyle. Not only can we understand their behaviour, but we can also comprehend the reasons they travel so much. eHotelier (2019 April 16) demonstrates that more than 80 per cent of Millennial travellers' favour paying extra for the convenience of flight schedule, and they prefer to reconnoitre international destinations rather than domestic trips. Moreover, they are more likely to lengthen their business trips and to extend their vacation expeditions in order to spend more time with friends and family. The key insights from these trends provide recommendations for the airlines and hotel industries, travel agencies and destination marketing organizations.

Recently, Condor Ferries (2020) carried out a survey of Millennials' travel trends. From the survey, the U.S. Millennials spent $1.036 billion on travel in 2017. The leisure travel and summer travel were reported $ 718.4 billion and $101.1 billion, respectively. They spent an average of $6,000 on an international trip. In terms of trip length, they took more short breaks. For example, 69 per cent took more trips over weekends. A study by New Horizons (2017) reported that the Millennials are happier when a scheduled trip is filled with more activities and experiences and more jovial if they travel for a longer time in different countries.

Some thrill-chasing Millennials travel for experience and seek out something dangerous for excitement, while others seek travel destinations with historical and/or cultural milieu contextual experiences. 62 per cent of them are prone to extend their business trips to the pursuit of cultural events. The majority (86 per cent) of Millennials elect new culture instead of party and shopping. 55 per cent took trips for relaxation, some visit people, some devote time with family, while others travel for a romantic getaway. One-third of Millennials travel to experience life, and a quarter of Millennials plan for family vacations overseas (Condor Ferries, 2020).

According to the AARP Travel Trends Report 2019, the average Millennial tourist took five trips per year in 2019; of these, three trips are international and the remaining two are local trips. Leonhardt (2019) assert that Millennials engage in more international trips than Generation X and more voyages than Baby Boomers.

### *7.2.1 Millennials' travel technology and booking statistics*

Millennials came of age during the period of the Internet explosion. Studies have found some correlation between Millennials and technology (e.g., Bolton et al., 2013; Nusair, Bilgihan, Okumus, & Cobanoglu, 2013; Nusair, Parsa, & Cobanoglu, 2011). Millennials travellers are extremely proficient users of digital technology, which mostly checks product offerings online. They can access information, check reviews and compare offers and prices within seconds from their keyboard. They are digitally obsessed and are prone to authentic and experiential travel experience, such as adventure, educational, cultural and/or culinary tourism, whereby they prefer to befriend and connect with other travellers. Having grown up in the digital age, Millennials rely on smartphones with apps. Among the most essential of the travel apps which Millennials use are TripAdvisor, Couchsurfing, Airbnb, Skyscanner, Hostelworld, Backpackr, Waze, Translate, etc.

In its survey, Condor Ferries (2020) reports that two-thirds of the Millennials book their trips using smartphones with travel apps installed. They rely extensively on apps to book flights, accommodation, events and activities. Millennial travellers are heavily dependent on smartphones, with more than 70 per cent using them to search for travel information, 66 per cent booking their travel and 74 per cent planning their trips online. Technology, undoubtedly, has changed the check-in and check-out channels for flights and hotels; it also assists the navigation process, makes restaurant bookings easier and helps with reshaping itineraries.

Condor Ferries (2020) report that 80 per cent of travellers prefer to use mobile booking because it is easier to get the information through a click of a button. More than 60 per cent wish to get a last-minute good hotel or flight deal. 70 per cent are inclined to book eco-friendly accommodation, while the majority (90 per cent) use apps when travelling to check for weather, maps, airlines info and post photos/videos in social media.

Millennials' booking preferences through online travel agents (57 per cent), hotel websites (36 per cent) and airline sites (32 per cent). However, Millennials still prefer to rely on the services of travel agents to get advice with regard to culture and experience-centred vacations. However, 13 per cent believe that using travel agents is not as safe as having a direct booking. More than 50 per cent use social media and research some review sites in the travel forums to decide on which option to pick and book (Condor Ferries, 2020).

## 7.3 Myth

Myth is part of traditions which record the people's lives as a collective memory in order to shape, socialize, and reconstruct people's culture. According to CBRE (2017) in a UK millennial survey, mythical Millennials have some common characteristics. Many Millennials are financially constrained. Millennials often still live in the family home; they both delay moving out, and will also return to live with their parents if their financial circumstances worsen. Many of them hope to relocate within two years, but they expect to rent to maintain a flexible lifestyle once leaving the family home, with no intention of owning a home or ever buying because of the unaffordability of homeownership. Many are not saving for the long term and depend on the parents when buying things. A more recent survey by Pew Research Center also concurs that Millennials are more likely to stay with their own family than preceding generations at the same phase of life. Over 50 per cent are not married (Barroso et al., 2020). Many Millennials decide to live with their parents at home so that they can save more disposable income to pay for experience services (Sachdev, 2020).

How do Millennials' myths and lifestyles impact their mindset on the practicality of travelling? The Millennial generation is the best educated in history (The Economist, 2016 21 January). They grew up in the gloominess of the financial crisis during a digital revolution and completely redefine what adventure means, choosing to spree on experiences. They view the travel paradigm differently compared to previous generations, such as Generation Z, Generation X and Baby Boomers, and perceive travel as an essential piece to their life experience and not a luxury good. According to Fromm (2016), Millennial travellers spend more than $180 billion per annum for the tourism industry, marking an upsurge of approximately 30 per cent since 2007. They are in their peak earning years and it is becoming increasingly affordable to pay for international travelling.

## 7.4 Lifestyle and practicality

### *7.4.1 What do Millennial travellers' want and need?*

According to Ooi (2019), "cultural complexity is the range of values, norms, beliefs, practices established in a society". The range of cultural appearances is a source for people to manage a multitude of social symptoms and to navigate the diverse situations they face in life. The Millennials' impact on the economy is huge across a range of tourist practices, behaviour, and habits when they travel. Hence, it is imperative that the tourism industry players, tourists from other generations and other stakeholders understand cultural complexity and respond to circumstances.

According to a recent survey by Bloomberg (2018), the Millennial generation made up 31.5 per cent of the world's population of 7.7 billion. Travel companies or online travel agencies should focus on this generation because of their potential spending power since their travel pattern focuses on experiences travel and cultural events or activities, and is heavily influenced by social media or user-generated content in drawing their itinerary decisions.

Millennials are curious and more agile than preceding generations. They place a strong emphasis on life experiences. Travel and adventure are ranked at the top of their priority list. If there is an option, they prefer to travel with friends or go on solo travels, rather than journeying with a family or a partner. Millennials believe that travelling may give them increased growth opportunities and experiences in employment- or educational-associated segments. They want a bit of everything and fancy to trail a deeper meaning of life and seek happiness in a career.

They are more interested in wanderlust and seeking out the hidden gems; they go undercover to where the locals hang out, and experience the local life. But they also seek out authenticated classic tourist hotspots: the Grand

Canyon, Niagara Falls, the Statue of Liberty, the White House, and Times Square in North America; Anne Frank's House (Amsterdam), the Parthenon (Athens), Sagrada Familia (Barcelona), Berlin Wall (Berlin), Atomium (Brussels), The Colosseum (Italy), Oia (Santorini), Buckingham Palace (London), the Louvre and the Eiffel Tower (Paris), and the Seven Sisters Waterfall (Norway) in Europe; The Great Wall of China (in China); and The Sydney Opera House (Australia). More importantly, they want the perfect Instagram photos opportunity. Hence, the impulse for a distinctive experience and the freedom of choice has changed the tourism industry.

There is a change in lifestyle preferences among Millennials. For example, grandparents and parents are now taught how to use smartphones and asked to send selfies to their grandchildren, parents watch movies online with young children, hotels and hostels are catching up with the demand created by the Millennials market, hotels are optimizing mobile apps to improve their services to customers. Forward-thinking resorts have introduced 'Selfie Stations' encouraging tourists to share experiences and tag location (Condor Ferries, 2020).

### *7.4.2 What motivates them to travel?*

#### *7.4.2.1 Millennials want a travel adventure*

Millennials are adventurous. They seek adventure pitching the hidden and uncommon things and are strong believers in hassle-free travelling. Millennials prefer consumer experiences and plan for a dream holiday. Millennials explore the spots they travel to. They generally endeavour independently to far destinations, extend their stay and engross in local cultures to shape their life experiences, including a taste test of the local food and unearth local hotspots and activities. For this, they prefer exploring remote destinations instead of major cities (UNWTO, 2016).

In terms of travel activities, the adventure-seeking Millennials heighten their experience, and some would like to enhance their experiential offering thrive with cultural activities. They are seeking more chances to diversify their lifestyle and experience the proficiency of new cultures. This cultural awareness has become the new indication of privileged circumstances. More importantly, whether it is digital or physical, they want a travel adventure that allows the creation of a shared experience with their friends.

#### *7.4.2.2 What is the role of social media in Millennial tourism?*

Social media is a key driver to record their daily lives. Social media platforms embrace social networks (e.g., Instagram, Facebook), microblogs/blogs/podcasts (e.g., Twitter, WordPress), sharing communities (e.g., YouTube), and reviews (e.g., TripAdvisor) (Cleffman, Feuerabend, Howald, & Kollmann, 2010). Enabling a commonplace platform for travellers' influencers maximize prominence and build more trust among their followers (Khamis, Ang, &

Welling, 2016). For travel influencers with a large number of social media followers signify a sign of popularity (Veirman, Cauberghe, & Hudders, 2017). In addition, content posted on social media by travel influencers is more effective than advertisements in influencing travel decisions (Sokolova & Kefi, 2020).

With the evolving technology, WiFi or a smartphone data plan plays a pivotal role in travel research and decision-making. As long as they stay connected, social media allow Millennials to receive infinite travel tips. This has become one of an integral part for Millennial travellers. Millennials are the generation who frequently surrounded by dynamic information from the influencers that have a strong impact on their travel decision-making process (Sokolova & Kefi, 2020). Drawing on social comparison theory, Liu, Wu, and Li (2019) find that benign social media envy has triggered Millennial travellers' aspirational tourism consumption after viewing travel experience sharing on social media, particularly those travellers with low self-esteem.

In a survey, Condor Ferries (2020) find that 87 per cent (50 per cent) used Facebook (Pinterest or Twitter) to make their booking. 97 per cent share travel experiences on social media, by posting photos or writing blogs to share their trip-spiration (Harrison, 2017) experiences, tips and tricks. Similarly, about 80 per cent grab friends and peers' advice on destinations and make the final booking decision based on their commendations (reviews, forums, comments) and 85 per cent still check for exclusive deals from different websites before confirming the booking. 97 per cent post their holiday daily, influencing friends' vacation choices of the trip.

Some travellers enjoy uploading their photos and videos in social media, which is motivated by their self-centred nature, while others just want to know who like their photos, and aware and recognise their experiences (Munar & Jacobsen, 2014). Through social media, Millennials build bonds with people who have similar interests by sharing their trips with the global audience. They interact and share the experience, which ultimately builds stronger trust among themselves (Nusair et al., 2013). They are enthused by posting and reviews shared by their influencers and friends on social media and treat that as a direction to strategize next trips. Therefore, stakeholders in the tourism industry should build a relationship with potential or existing customers, stay connected and interact actively on social media platforms (Parra-López, Bulchand-Gidumal, Gutiérrez-Taño, & Díaz-Armas, 2011).

### *7.4.3 Travel priority in life, travellers are on a budget and seizing short trips*

When the 9/11 terrorist attacks incident took place, the members of the Millennial generation were between the ages of 5 and 20. They entered the workforce broadly at the height of the economic recession, and thus their memory of adulthood have been moulded by the shadow of recession and wars in Iraq and Afghanistan (Dimock, 2019). They feel that life is short, and make travel a priority in life more than other generations.

Millennials are very budget-oriented and search for economical tourism services with good value for money (Verissimo & Costa, 2018). Millennials are swiftly becoming more affluent; they are cost-conscious consumers and very much on budget travel. Some Millennials would make travel a priority, reduce their daily spending, and increasing their savings just for the purposes of travel. Because of this, Millennials prefer to plan their itinerary conscientiously. Some of them work remotely during their holiday trips. Some apply for a credit card with travel rewards and disburses travel costs with rewards points. Notwithstanding, they have a different perspective on the purpose of money.

Millennials would spend more when they feel it is important to do so or if there is an opportunity to join a one-off luxury travel event or activity, such as a concert, a cultural show, or a festival. They first opt for deal-seeking to check if there is any opportunity to experience luxury travel. On account of Millennials being so technologically-savvy, it is easier for them to interlock in travel hacking and manipulate resource to find the best deal, search for cheap promotions, and hunt for a great bargain. Some Millennials who put travel as a priority adhere to a strict budget, and even boost their income by taking up multiple full-time jobs and side jobs.

Debt does not preclude Millennials from travelling. While they juggled debts that include car loans, mortgage, monthly expenses and student loans, Millennials opt to reduce their costs by budget travel. Even if they do take on more loan debt to travel, they focus on seeking out remarkable deals to visit more places given the budget constraints. They are perfectly happy to have long weekend trips rather than full-fledged vacations. Nevertheless, some Millennials are earning more, making travel affordable and manageable and opt for a wider range of leisure options (Bednall, Valos, Adam, & McLeod, 2012).

### *7.4.4 Convenience*

As technology-savvy digital nomads, Millennials are up-to-date travellers who desire the cultivation of messaging and they are happy to rely on the convenience of mobile phone apps. Not only are they looking for value-for-money services, but Millennials also consider a social environment that fits their needs. As mentioned earlier, Millennial travellers favour paying extra for the convenience of flight schedule. They also enjoy convenience when choosing accommodation. The accommodation they pick will allows them to share the freedom of being travellers, and they seek out value for money (Verissimo & Costa, 2018). They prefer to go for more economical Airbnb residences or the youth lodging hostels, instead of costlier hotels.

### *7.4.5 Muslim Millennial travellers*

Muslim tourists, both male and female, have risen rapidly and they now form a highly significant sector of the travelling population. As with other Millennials, young Muslims are more technology-savvy and prefer to book

Airbnb for free and easy travellers, and to strive for adventure travel. Muslim Millennials dare to challenge the status quo. They are likely to question and demand more in their travels. For example, they seek clarity when a location is not adequate to meet the needs of Muslim travellers, including food and praying facilities. They are more informed than previous generations and have a tendency to share information with their peers, almost expressing themselves in a collective voice.

## 7.5 Sustainable economy

Millennial tourism is closely linked to sustainability, i.e., how they contribute to the local community. Millennials are more open than any other demographic generation in giving back to the communities they visit (Travel Technology & Solutions, 2020). The idea of sustainable tourism is the aspiration to transmute authentic travel into an opportunity to make travellers and hosts develop an intercultural empathetic, create more quality jobs and play a part in building strong relationships within the national economy. When the income of local people increases, this would create more spending power and a healthier and livelier domestic economy. Their visit has a more significant impact on the locals and the environment (Haupt, 2019). As such, Millennials are gradually transforming the landscape of tourism and travel.

There are many ways to demonstrate the inherent connection the Millennials demand for sustainable tourism. They are always open to connect with local businesses, such as support re-cycling, re-using bath towels instead of replacing a new one from hotel staff, using waste treatment facilities to minimize carbon footprints, and preferring demand for socially responsible products. Tourists that enjoy nature and that prefer the natural environment may choose to travel to remote areas or suburbs; this is commonly known as ecotourism. Ecotourism is also perceived as securing a low-carbon footprint since Millennials prefer to walk, hike, or hire a bicycle, instead of taking transportation.

Millennials have been hit hard by the recession, which has had an impact on career advancement and may face job losses. These insecure circumstances may explain the trend towards shorter trips and the increasing popularity of higher demand for cheaper travel options (e.g., low-cost airlines, home-sharing, Airbnb, and ride-sharing platforms, hostels, and travelling in the off-season). Travel in low seasons could also curb environmental carbon footprints by opting for a sustainable holiday.

With the increasing purchasing power of Millennials, they enter the market, not only as travellers, buyers and users, but also as retailers, entrepreneurs, and creators of new online travel businesses. Travel businesses need to be aware of what Millennials like, their needs, their habits, and their tendencies in order to know how to cater trips to fulfil the desires of this generation. Therefore, it is extremely challenging to tackling sustainability in tourism. The future of sustainable tourism is the way forward, and it is important to understand the challenges and opportunities which this travel sector may face.

### *7.5.1 Millennials and wellness tourism*

The wellness tourism segment is another branch of Millennial tourism. Wellness is the new black and the wellness industry has been booming since 2017, from physical yoga to virtual mobile meditation apps. The Global Wellness Institute claims that the global wellness economy at more than $4.2 trillion (Cavill, 2019) and *wellness tourism* is estimated *to* hit $919 billion by 2022.

Millennials emphasize on lifestyle-focused and target grander value on wellness experiences over material, assets, or properties ownerships. Nutrition and diet are among the dominant segments in the wellness industry. Millennials incorporate more wellness values into their lifestyle, the wellness economy has been more integrative and becomes a dominant lifestyle, profoundly changing consumer behaviour and changing the markets.

The Millennials drive the growth of wellness tourism. Linking products to wellness and linking wellness to advertising strategies is a trend seen across many beauty brands and companies. Beauty currently accounts for around one-quarter of the wellness economy. Millennials are relatively more health-conscious and treat wellness daily. They exercise more often than in preceding generations. They use apps or smartwatches to track training data.

Among the many global wellness economies, the fastest-growing sectors are spas, tourism and real estate, but wellness ranges from beauty and personal care products, hemp and cannabidiol oil (CBD) to yoga by the sea and/or hiking in the mountains. Consumers use CBD products mainly to overcome pain and relieving anxiety. CBD has also been acclaimed to treat a range of medical ailments, from muscle spasms, depression and cancer to Alzheimer's and Parkinson's (Bryan, 2019).

Women use CBD for anxiety more than men. Female travellers may be more particular on yoga and spa or beauty-related wellness services, while male travellers are meticulous with regard to the quality of gym facilities. This growth can be attributed to cross-promotional branding and an uptick in sales innovations.

### *7.5.2 Wellness tourism and millennial women*

In more recent years, Millennial woman have increasingly traveled in search of wellness. According to the Travel Market Report, 66 per cent of the 4,600 women in their survey would opt for a wellness getaway, 40 per cent would prefer a fitness retreat, and 55 per cent would want to go to places such as a ranch or a farm (Dimon, 2018). When traveling, Millennial women seek out top fitness talent and healthy food. Most women prefer to experience something unusual or which they could not get at home, which prompts them to look into experience related to nature and to take the opportunities to pursue wellness in places which offer unique and comfortable experiences, such as hot springs, spas, facial treatments and sauna during their vacations.

### 7.5.3 *The dominant sectors: – nutrition and diet*

Millennials focus on a wellness approach to eating and altruism can prove an effective messaging techniques for organic or health food companies. Organic food can mean different diets of eating nutrition and healthy food. Globally, the retail sales of organic food have increased. The rise of eating beyond meat, in vegans and vegetarians, is another example of wellness making an impact. Evolving options in healthy eating, and the promotion of eating less meat because of health concerns, reflect travellers' perceptions about dieting. The time is right for marketers to shift their attention to this "sort-of-vegetarian" demographic. The trend changes in these sectors, in terms of both what millennial travellers want from the demand side and what is offered from the supply side, means that marketing strategies should stay agile for effective campaigns.

### 7.5.4 *Wellness tourism and over-tourism*

Wellness tourism is seen as a possible antidote to over-tourism, which is the extreme concentration of visitors to particular cities and landmarks. Wellness trips should be incorporated with wellness messages in order to draw tourists to cities which are less frequently visited, eco-friendly and health-focused.

### 7.5.5 *Direct-to-Consumer: the wellness market and brand marketers*

Direct-to-consumer (DTC) brands have been successful in the wellness category, particularly with regard to digital health or tapping into the telemedicine trend. The sky is the limit for wellness and health-focused DTC brands connecting with travellers. It is essential to find authentic ways to connect Millennial travellers who are engrossed in wellness (Cavill, 2019).

Travellers are generally making procuring decisions based on sentiment, passion or excitement in the wellness market. Marketers need to think of why the Millennials should choose to buy their brands of wellness service. Consumers pay a premium for wellness treatments so as to improve their lives both physically and mentally. Hence, companies should create an authentic brand that reverberate with customers' core values.

Brands focus on the result of the service they are serving or the product they are selling in order persuade travellers that they provide value for money. Millennial women feel revived and refreshed and build a sense of confidence with the instantaneous effect of any kind of wellness offer. Doing so could further amplify testimonials, and receive positive highlights or reviews in social media. When done authentically, this may yield trust, which can quickly transform customers into brand ambassadors in the wellness sector. For this reason, engage influencers could resonate with high-value millennial women segments. Therefore, creating this relationship with Millennial women travellers is crucial to position your brand to prevail in the wellness economy (Ackerman, 2019).

Brand marketers can develop campaigns fusing customer-centric focus by addressing consumer need-states. Marketers map the consumer experience to understand which touchpoints are most profound for Millennial travellers. And they can then brand them and invest by differentiating those touchpoints.

## 7.6 Millennials and the COVID-19 pandemic

The COVID-19 pandemic and the lockdown or closure of borders have had crippling effects on travellers' confidence worldwide. Once the restrictions are lifted, we believe that social-media influencers will continue to play a paramount role in reinstating confidence and contributing to the recovery footpath of the tourism industry. Social media remains long-standing as individuals depend on the updates to alleviate anxieties and influencers are the power of travel recovery strategies. GlobalData reported that YouTube, Twitter and TikTok are commonly used platforms for travel promotions amid the COVID-19 crisis, which hold marketing potential (GlobalData, 2020).

A GlobalData survey reports that 48 per cent of Millennials are still worried about the risk of contracting COVID-19 (Baratti, 2020a). However, in the American Express Travel Survey, which gauges consumer sentiment, reports that 32 per cent of respondents have travelled over the Memorial Day holiday weekend, 48 per cent of Millennials travelled over the holiday weekend, compared to only 27 per cent and 22 per cent of Baby Boomers and Generation X, respectively (Baratti, 2020b).

The Millennials, a generation often related to any sort of shifting paradigm, is expected to lead local travel recovery (Sachdev, 2020). But why Millennials? The reasons being they are paradoxically more anxious than previous generations, more price-sensitive, more craving for exploration experiences, and more adjusted with social media. A 'new normal' will emerge and Millennials are highly likely to lead the recovery. While they inevitably feel fear about the outcome of the COVID-19 pandemic, they are also the most enthusiastic to excuse from the family and start travelling once the restrictions are lifted.

## 7.7 What do Millennials want from the industry? Lessons for the tourism industry

The industry is more versatile than ever. The players in the tourism industry should always find an appropriate strategy to accommodate the Millennials' travel strategies, whether through travel agencies, websites or on-site or travel apps. The broader wellness tourism industry have seen increasing levels of spending. As Millennial travellers have increased their demand for wellness, businesses in the tourism and hospitality sectors to mobile app start-ups are shifting towards a health-conscious business model. Wellness should not be detached from gym or spa membership; they are not isolated, but connected. The massage and spa industry has seen dynamic growth in this area, with more travellers being drawn to a self-care oasis, across fitness, massage, and beauty brands, as a way to achieve relaxation and to reward themselves after hard work, as well as a way to escape their busy and overcommitted working life.

According to the survey by Condor Ferries (2020), holiday booking services that the Millennial generation prefers from the travel industry are through the website or travel apps. Moreover, more than half of them (68 per cent) prefer a service that could split the travel bill facilities with others upon booking. They also like instant customer service via instant messaging or live chat facilities that provide prompt replies. Furthermore, they long to get a service that customizes holidays based on their needs at a reasonable price. Only minority (16 per cent) willing to disclose personal data in exchange for discounts.

The stakeholders in tourism should be aware of some key Millennials' marketing takeaways. The Millennials have been close interactions in the social media; therefore, the retail industry can try to get your brand to mimic their desire and provide services or facilities that the Millennials are looking for. Loyalty plans or schemes offer VIP upgrades or discounts offer. These are precisely what Millennials want; they boast about it and keep revisiting. Moreover, since the majority use smartphones to research and do the booking, try to promote your products, packages, facilities make through mobile app or supplier websites. Update your travel agency or company website and apprise promotion information on social media channels in user-generated content. Communicate with the Millennials by feeding authenticity information with real photos and tag photos to groups can catch their eyes when searching travelling information (Condor Ferries, 2020).

According to a Consumer Content Report 2017 using targeting adult respondents in the US, UK and Australia, 90 per cent of Millennials value brand authenticity when deciding which brands to support (Cassidy, 2017). This means that Millennials place a high value on authenticity. Millennials influence the market and services as they are frequent travellers. That is why enticing Millennial travellers are indispensable to the hotel recovery plan (Dehler, 2020).

Therefore, Millennial travellers should be well-informed about remedies hoteliers have put in place to prepare for the COVID-19 recovery period, keep them connected and maintain excellent service. More importantly, explain the steps hoteliers are taking to alleviate risk and furnish consumers' insights into the steps the hotels are captivating to help the society navigate the current COVID-19 pandemic. Stay engaged and expand the hotel's social media presence and be responsive to any sort of communications related to enormous travel deals that tap into value-for-money weekend packages and/ or challenging trips with cultural experiences.

It is noteworthy that the phenomenon of over-tourism wears can impact the balance of local life. Airbnb has disrupted the traditional hotel industry and travel market and created more opportunities for Millennials to rent accommodation at an affordable rate. Travellers who stay in Airbnb can cause a rise in rent. Renting accommodations through Airbnb can also bring about a major loss for hotels.

## 7.8 Conclusion

The Millennial generation has some general characteristics, including mobility, connectivity, dependence on technology, vanity, a craving for experiences, and social awareness. There are five key characteristics of Millennials (Paris, 2017). First, Millennials perceive travel more as a commonality than a luxury. Secondly, Millennials grew up during remarkable innovations in information and communication technologies. They are both technology-savvy and technology-dependent. Technology allows them the chance to experience personalized and independent travel. Thirdly, Millennials focus and look for the authentic travel experience. They desire sociality and social settings. They often share and create experiences with their virtually proximate, syncing socializations onscreen through social media. They have a desire for unique virtual sharing experiences, the so-called "Instagram-able" moments. Fourthly, millennials value independence, and they strive for opportunities for educational learning travel. Fifthly, risk is perceived differently by millennials. They are more risk-averse in their spending. Millennial travel leans towards being oriented towards family and friends. When they see value travel with friends and family, they spend. They are flexible and resilient travellers. Finally, Millennials have greatly reflected how we live in and there is a change in travel behaviours, lifestyle preferences and priorities, either directly or indirectly.

There are a plethora of Millennials sharing daily wellness regimens. Millennials are a reflection of the globalized and connected. They are value-conscious, and well versed in connectivity as a normal necessity of daily life. They pursue expertise and request support to find authentic and personalized experiences, and prefer not to engage with travel agents.

The future of Millennial tourism is bright. Amid the COVID-19 pandemic, travellers' lifestyles and norms have changed and again proven that social media influencers are still effective and heightened. The Destination Management Group or tourism industry players should engage with influencers in order to plan a steadfast recovery. Travellers perceive the messages or posts via social media as more authentic, real, true or trustworthy than direct advertising campaigns, which inspire travellers to resume travelling and gain a grip within a new market.

## References

Ackerman, T. (2019). Position your brand to win in the wellness economy. Retrieved from https://www.forbes.com/sites/forbescommunicationscouncil/2019/12/04/position-your-brand-to-win-in-the-wellness-economy/#175cbf9f42e1.

Baratti, L. (2020a). *Gen Z and Millennials are key to travel industry's recovery*. Retrieved from https://www.travelpulse.com/news/features/gen-z-and-Millennials-are-key-to-travel-industrys-recovery.html.

Baratti, L. (2020b). *Survey shows 32 percent of Americans traveled over memorial day weekend*. Retrieved from https://www.travelpulse.com/news/features/survey-shows-32-percent-of-americans-traveled-over-memorial-day-weekend.html.

Barroso, A., Parker, K., & Bennett, J. (2020). As Millennials near 40, they're approaching family life, differently than previous generations. *Pew Research Center*. Retrievedfromhttps://www.pewsocialtrends.org/2020/05/27/as-Millennials-near-40-theyre-approaching-family-life-differently-than-previous-generations/.

Bednall, D.H., Valos, M., Adam, S., & McLeod, C. (2012). Getting Generation Y to attend: Friends, interactivity and half-time entertainment. *Sport Management Review*, 15(1), 80–90.

Bolton, R., Parasuraman, A., Hoefnagels, A., Migchels, N., Kabadayi, S., Gruber, T., Loureiro, Y., & Solnet, D. (2013). Understanding Generation Y and their use of social media: a review and research agenda. *Journal of Service Management*, 24(3), 245–267.

Bryan, K. (2019). *The CBD market: Who, what, where, why & how*. Retrieved from https://insights.digitalmediasolutions.com/articles/cbd-market.

Cassidy, P (2017). *Survey finds consumer crave authenticity and user generated content delivers*. Retrieved from https://www.socialmediatoday.com/news/survey-finds-consumers-crave-authenticity-and-user-generated-content-deli/511360/.

Cavill, S. (2019). *Trend update: The wellness industry continues to dominate across sectors*. Retrievedfromhttps://insights.digitalmediasolutions.com/articles/growth-in-wellness-industry.

CBRE (2017). UK Millennials: Myths and realities. UK Overview. *CBRE*. Retrieved from https://www.cbreresidential.com/uk/sites/uk-residential/files/Millennialsper cent20inper cent20theper cent20UK.pdf.

Cleffman, L., Feuerabend, A., Howald, F., & Kollmann, C. (2010). *ECCO Social Media Report: Die Integration von Social Media in die Unternehmenskommunikation*. London: ECCO International Public Relations Ltd.

Condor Ferries (2020). Millennials Travel: Statistics & Trends 2020. Retrieved from https://www.condorferries.co.uk/Millennials-travel-statistics-trends.

Dehler, N. (2020). COVID-19 and hospitality: Why attracting Millennial travellers is essential to your hotel's recovery plan. Retrieved from https://www.hotel-online.com/press_releases/release/covid-19-and-hospitality-why-attracting-millennial-travelers-is-essential-to-your-hotels-recovery-plan/.

Dimock, M. (2019). Defining generations: Where Millennials end and Generation Z begins. *Pew Research Center*. Retrieved from https://www.pewresearch.org/fact-tank/2019/01/17/where-Millennials-end-and-generation-z-begins/.

Dimon A. (2018). How Millennials are disrupting the wellness travel market. Retrieved from https://www.travelmarketreport.com/articles/How-Millennials-Are-Disrupting-the-Wellness-Travel-Market.

eHotelier (2019 April 16). *2019 trends in global Millennial travel*. Retrieved from https://insights.ehotelier.com/global-news/2019/04/16/2019-trends-in-global-millennial-travel/.

Fromm, J. (2016). What you need to know about Millennial travelers. *Millennial marketing*. Retrievedfromhttp://www.millennialmarketing.com/2016/05/what-you-need-to-know-about-millennial-travelers/.

Garikapati, V. M., Pendyala, R., Morris, E., Mokhtarian, P., & McDonald, N. (2016). Activity patterns, time use, and travel of Millennials: A generation in transition? *Transport Reviews*, 36(5), 558–584.

GlobalData (2020). Millennial influencers will be indispensable in travel sector recovery. *GlobalData*. Retrievedfromhttps://www.globaldata.com/millennial-influencers-will-be-indispensable-in-travel-sector-recovery/.

Harrison, V. (2017). Why Millennials are the next tourism frontier. *BBC News*. Retrieved from https://www.bbc.com/news/business-40701934.

Haupt, T. (2019). Millennial tourism trends – Internet and sustainability. *Tourism ReviewNews*. Retrievedfromhttps://www.tourism-review.com/millennial-tourism-needs-internet-connection-and-reviews-news10904.

Khamis, S., Ang, L., & Welling, R. (2016). Self-branding, 'micro-celebrity' and the rise of social media influencers. *Celebrity Studies*, 8(2), 191–208.

Leonhardt, M. (2019). Millennials are making travel a priority more than previous generations – that's not a bad thing. Retrieved from https://www.cnbc.com/2019/08/30/Millennials-making-travel-a-priority-more-than-previous-generations.html.

Liu, H., Wu, L., & Li, X (2019). Social media envy: How experience sharing on social networking sites drives millennials' aspirational tourism consumption. *Journal of Travel Research*, 58(3), 355–369.

Munar, A. M., & Jacobsen, J. K. S. (2014). Motivations for sharing tourism experiences through social media. *Tourism Management*, 43, 46–54.

Nusair, K. K., Bilgihan, A., Okumus, F., & Cobanoglu, C. (2013). Generation Y travelers' commitment to online social network websites. *Tourism Management*, 35(1), 13–22.

Nusair, K. K., Parsa, H. G., & Cobanoglu, C. (2011). Building a model of commitment for Generation Y: An empirical study on e-travel retailers. *Tourism Management*, 32(4), 833–843.

Ooi, C-S (2019). Asian tourists and cultural complexity: Implications for practice and the Asianisation of tourism scholarship. *Tourism Management Perspectives*, 31, 14–23.

Paris, C. M. (2017). Understanding Millennial Travelers. *Travel Bulletin*. Retrieved from http://travelbulletinme.com/2017/11/understanding-millennial-travelers/.

Parra-López, E., Bulchand-Gidumal, J., Gutiérrez-Taño, D., & Díaz-Armas, R. (2011). Intentions to use social media in organizing and taking vacation trips. *Computers in Human Behaviour*, 27(2), 640–654.

Sachdev, E. (2020). Hospitality matters: Why Millennials will lead the local travel recovery. Retrieved from https://www.bangkokpost.com/life/social-and-lifestyle/1987871/hospitality-matters-why-Millennials-will-lead-the-local-travel-recovery.

Sokolova, K., & Kefi, H. (2020). Instagram and YouTube bloggers promote it, why should I buy? How credibility and parasocial interaction influence purchase intentions. *Journal of Retailing and Customer Services*, 53, 101742.

The Economist (2016 January 21). Generation uphill. https://www.economist.com/special-report/2016/01/21/generation-uphill.

Travel Technology & Solutions (2020). How sustainable tourism is impacting the travel industry. Retrieved from https://www.tts.com/blog/how-sustainable-tourism-is-impacting-the-travel-industry/.

Veirman, M. D., Cauberghe, V., & Hudders, L. (2017). Marketing through Instagram influencers: the impact of number of followers and product divergence on brand attitude. *International Journal of Advertising*, 36(5), 798–828.

Verissimo, M., & Costa, C. (2018). Do hostels play a role in pleasing millennial travellers? The Portuguese case. *Journal of Tourism Futures*, 4(1), 57–68.

World Tourism Organization (UNWTO) (2016). Affiliate members global reports, volume thirteen – The power of youth travel. Retrieved from https://www.e-unwto.org/doi/book/10.18111/9789284417162.

# 8 Tourist behaviour and travel adoption among Millennials

*Gül Erkol Bayram*

## 8.1 Introduction

Technological developments, transportation and communication systems and the change of touristic awareness, along with generational differences, have highlighted the need for businesses to adapt to this change rapidly and respond to the changing consumer needs instantly. In today's world, where global competition, change and generational differences are intense, the concept of tourism has increasingly associated with having experiences and achieving self-development, through the interconnected processes of travelling, having fun, and resting. Today's tourists want to discover new places and to experience the culture, traditions and cuisine of that place. They believe in the importance of continuous education and development and show a high degree of awareness of social problems (Benckendorff, Moscardo, & Pendergast, 2010). Businesses which want to stay one step ahead of the competition should adopt the open system approach and continuously receive energy, information, and materials from the environment or other systems in which they operate, and present them to their environment or other systems as business, goods or services. Those enterprises which are unable to update themselves or which cannot keep up with the changes, are running out of energy over time, their activities are deteriorating, their balance is lost, and system activities come to a halt with the emergence of confusion and troubles (Mahmutoğulları & şahin Perçin, 2018).

Tourism has been an important aspect of human lives and it is seen that demographic factors have a significant effect on tourism, in terms of both domestic and foreign customers. In particular, the "age" factor can affect the generation of tourists, the wishes and expectations of consumers, their perspectives on life, their travel habits, and their way of working. For this reason, it is important for businesses to increase their awareness about generations, to appreciate their distinctive characteristics, to develop personalized products for them, to adopt appropriate a management and leadership style that is in line with their wishes and expectations, and to create necessary working conditions. In the present day, the Millennial generation is one of the most well-travelled tourist groups. Millennials are one of the busiest traveling tourist

DOI: 10.4324/9781003170631-10

groups today. They do not just visit a place, but also contribute significantly to the economy of the region and the country through the money that they spend there. The importance of this group is confirmed by much of the academic research carried out around this subject (Cavagnaro, Staffieri, & Postma, 2018; Haddouche & Salomone, 2018; Li, Li, & Hudson, 2013; Puybaraud, 2010; Valentine & Powers, 2013; Wiastuti, Lestari, Bejo Mulyadi, & Masatip, 2020). This study aims to provide insights into tourist behaviour and travel adoption among Millennials. Also, this study will try to seek address questions concerning different aspects of generations according to travel behaviour.

## 8.2 The concept of tourist behaviour, theories, studies

Tourist consumer behaviour can be defined as the actions or behaviours people take to research, buy and evaluate products, services, and ideas that meet their needs while travelling. Although touristic consumer behaviour is an autonomous and private consumption type in its own right, it can also be seen as an important part of general consumer behaviour. The theoretical and practical knowledge presented for general consumption behaviour can generally be regarded as equally valid when considering tourist consumption behaviour. There are different definitions regarding the concept of the touristic consumer. For example, according to one research study, the tourist consumer behaviour is the interaction process created by the tourism enterprises to meet the needs of the tourist products. According to Torlak and his friends (Torlak & Altunısık, 2002), the purchase decision defines the touristic consumer as a group in which the products and services they purchase differ from other sectors. The use of travel and tourism products in the tourism sector is the focus of tourism research. It is also very important for marketing research. To conduct marketing research correctly, it is necessary to have a proper understanding of consumer behaviour. Consumer behaviour examines the decision-making process of a touristic product by people with many choices. Tourists make many decisions before, during and after their vacation. These decisions are about transportation, route, accommodation, attractions and activities. A consumer's intention is reflected in their future purchasing behaviour and is an indicator of his or her behaviour. He decides whether or not to buy after evaluating his perception with internal and external resources. Marketers facilitate the purchasing decision, making it more attractive for potential consumers (Özdipçiner, 2009). Considering the tourism structure and characteristics, it is seen as a labour-intensive service sector. For this reason, seeing tourism marketing as a service marketing will make it easier to determine the factors that affect the purchasing behaviour of consumers, especially due to changes in product characteristics. It should not be forgotten that the products offered in the context of tourism mostly do not consist of a single product. They are often thought of as compound and interdependent products. How the production process takes place in the industrial sector is not important for the consumer, but this process has great importance in services based on intense customer relations. This is because the simultaneous production and consumption of the product in the service sector (Yıldırım & Tayfun, 2010), the labour-intensive feature (Hacıoğlu, 2005) and the complementary products (Rızaoğlu, 2004) require the perfect time, relations with consumers and physical environment. In this context; In the tourism sector, which is service-oriented, the importance is given to time, human and physical environment factors that can be the determinant of the satisfaction of consumers' expectations and their subsequent preferences. In this respect, although the purchasing decision process of the consumers is the same, the tourism sector differs from the industrial sectors as a service sector due both to the nature of the product and to the nature of the factors affecting the decision process. Therefore, many factors play a separate or coordinated role in shaping the interest, expectations and preferences of tourists towards touristic products (cited in Ilban, Akkılıç, & Yılmaz, 2011). According to Mathieson and Wall (1982), four related factors affect the consumer's decision-making process. It is possible to express them as follows (Kangasmäki & Koskelainen, 2005):

- Tourist profile; age and education
- Travel information; using different sources of information
- Destination sources and characteristics; destination attractions
- Travel features; distance and duration of the vacation

The consumer decision-making process shows similar features to the classical consumer decision-making process. These include: determining the need; making a decision about the need to complete the requirement; and evaluating the satisfaction after the decision. Moutınho (1993) stated, in his research, that the tourist behaviour process consisted of four elements: the decision-making process of tourists; the pre-purchase phase; the post-purchase evaluation; and the decision-making phase for the next purchase. Correia (2002) researches the purchasing process of the travellers with the pre-decision phase, the decision-making phase, and the post-decision phase. The consumer behaviour process in tourism begins with the purchasing decision process. What begins this process is the occurrence of the need. Although there are many factors involved in the emergence of a holiday need, it is difficult to determine individuals' needs and desires. Subjects such as recreation, entertainment, religion, education and work encourage individuals to travel (Içöz, 2001). With the emergence of these needs, individuals have begun to be influenced by stimulants such as advertisements, posters, friendly advice and social media. According to McDowall (2010), obtaining pre-vacation information causes individuals to be more satisfied with their travels and leaves them happier from their travel. After collecting and evaluating the information on touristic services and products, the decision-making phase comes. However, tourists can change their decisions at the last minute and choose to travel to a different region. In this context, the elements that shape a tourist's travel can be considered. It is possible to express the quality of the product, ease of payment, availability of the product, the reliability of the product, the consumer having free time for vacation, and political conveniences, and trust in the travel business (Hayta, 2008). Many studies in the literature indicate that tourists who are satisfied with their vacations prefer these same services and products in their future (Opperman, 1998). We can express that the experiences of tourists are possible with a comprehensive process (Yazgan, 2012).

Research on consumer behaviour in tourism has not only been an academic guide, but has also contributed greatly to development of the tourism industry. Middleton (1988) is an important researcher investigating tourist consumer behaviour. Middleton (1988) developed the tourist consumer behaviour model in his research and pioneered many researchers. According to this model, the tourist consumer behaviour model consists of four basic elements: stimulating inputs as communication channels; consumer characteristics; decision process; and purchasing output (Swarbrooke & Horner, 1999). Another important research was carried out by Moutınho (1993). According to this research, tourist behaviour consists of several fundamental factors that affect each other. To understand this system, there are some determining behaviours and these need to be addressed systematically. The research conducted by Mayo and Jarvis (1981) is on an individual travelling tourist. In this study, it is emphasized that the behaviour and decision-making factors involved in tourists travelling individually are more comprehensive and different than those travelling collectively. Ndubisi (2007) stated that concepts such as geographical location, cultural structure, religion, and ethnic identity are effective in the decision-making of tourists travelling as a family (Ilban et al., 2011).

Parmar and Gupta (2007) stated that family income, work and age are influential on the selection of products. Kumar (2012) stated that family culture, social status, work, education, experience, economic status, lifestyle, personality and attitude all provide motivational aspects to the purchasing behaviour of consumers. According to Decrop and Kozak (2009), people decide what kind of holiday they will buy; it affects economic, cognitive and simple decision-making situations. Since the holiday budget depends on the relationship between income and vacation expenditure, a person who decides to go on vacation acts by taking into account his economic situation in choosing a holiday destination. This situation causes the price of the holiday to affect the purchasing decision while choosing among the vacation alternatives (Demir & Kozak, 2011), a simple decision-making strategy to save money, energy, and time (Decrop & Kozak, 2009), destination attractiveness is described as a cognitive decision-making strategy that people have learned about their feelings and destination (cited in Yazgan, 2012).

The characteristics of touristic consumers and their participation in tourism are complex. The value given to the consumer by hotel businesses, which can analyze this well and fulfill a special service understanding for every tourist consumer, is reflected as a brand value. When the characteristics and motivation reasons of the tourist consumer are added to the characteristics of tourism and the complex structure of consumer behaviour in general, purchasing behaviour becomes more complex. Characteristics of tourism require an explanation of tourist consumer behaviour about the complex features of the industry. The products and services produced in the tourism industry differ from the output produced in other industries. Outputs produced in the tourism industry are intangible and they involve service delivery. Therefore, the qualities of these services, which serve to meet the needs of consumers, are expected to have a quality level that can meet tourists' consumer expectations. Consumers can only get an idea about the level of quality as a result of an experience. As the consumer is also an element of the production stage in evaluating the quality of the services due to the inseparability feature, it is very difficult to measure the quality. Touristic consumers evaluate the products and services of the hotel business according to their personal experiences and judgements, and individual evaluations are effective on their purchasing behaviour.

## 8.3 Influential factors of tourist behavior and motivations

When the literature regarding consumer behaviour is examined, the reasons affecting consumer behaviour are broken down in terms of internal-external and personal-non-personal or social, cultural, demographic, and psychological reasons (Cömert & Durmaz, 2006: 353). According to Özsungur and Güven (2017), consumer demands and needs change over time and cause changes in consumer behaviour for several reasons. As one of the most important consumer groups in the world, tourists have a slightly different feature than other product users. It is observed that, in particular, the products offered in tourism are not intangible, are consumed in the place where they are produced, cannot be stocked, and are subject to a rapid change. It is observed that consumers have difficulties in choosing touristic products (Demir & Kozak, 2011).

Motivation has a very important place in the decision-making process of touristic consumers. Individuals have different characteristics in terms of their qualifications, needs and expectations. This situation causes them to be affected differently in the process of being motivated in their working life. Motivation takes place in the process where all motivating elements are integrated with the individual. It has been expressed the meanings of motivation such as encouragement, orientation, motivation, but he cannot fully explain the English word motivation from which it originates. Motivation is a general word that corresponds to the occupation of a person, which is explained with words such as wishing, needing, and impulse (Luthans, 1992). Motivation is regarded as the main reason underlying human behaviour (Guay et al., 2010); Gredler et al., (2004) defined motivation as a factor that activates a behaviour to do or not to do. On the one hand, Badu (2005) defined it as the characteristic element of human psychology that strengthens the degree of commitment of individuals (Erkol Bayram, 2017). When the travel motivations of touristic consumers are mentioned in tourism literature, this reminds us of the motivation concept of what constitutes the factors that affect the travel of touristic consumers and why they travel (Torun & Çetin, 2015). They are influenced by many factors during the decision-making process. The most significant travel motivations are as follows (Kocatürk, 2018);

- Personality (individual/group, cautious/adventurous, confident/shy)
- Lifestyle (varies according to interests, those who prefer to travel alone, those who want to make new friends)
- Experiences (positive and negative experiences)
- Past (nostalgia for certain places)
- Perceptions (strengths and weaknesses)
- Status (how others are perceived).

According to Crompton and McKay (1997), these motivations come under the label of socio-psychological motivations: as escape, self-discovery, relaxation, dignity, longing for the past, increasing relations with relatives, and social interaction, cultural motivations are ranked as innovation and education. Swarbrooke and Horner (1999) classified the travel motivations of tourists under six points:

- Personal development: Desire to improve oneself, increase knowledge and acquire new skills.
- Gaining status: Willingness to stay fashionable and gain dignity.
- Cultural factors: Willingness to travel and experience different cultures.
- Physical factors: Sun, health, relaxation, exercise, and desire for sex.
- Emotional factors: Adventure, nostalgia, romance, escape, fantasy, and the desire for spiritual renewal.
- Personal Factors: It is the desire to visit relatives and friends, make new friends, and make others happy.

Travel motivation refers to some requirements that cause individuals to participate in touristic activities. Travel motivation has been studied by a large number of researchers in different fields, including psychology, sociology, and anthropology. Researchers who want to clarify the concept of travel motivation have researched Maslow's hierarchy of needs theory and used this theory as a guide to explain the concept of travel motivation in tourism literature. Researchers working on tourism began to shape their work in the light of this model, with Pearce, Morrison, and Rutledge (1998) mentioned about a tourism motivation model composed them, but this model was based on Maslow's model and not dependent on assumptions. When studies on travel motivation are examined, it is seen that Dunne's (2009) research is remarkable. Dunne (2009) researched the reasons for tourists' travels and stated that great success in tourism can be achieved by determining these reasons. Crompton and McKay (1997) mentioned some of the benefits of understanding travel motivations in their research on travel motivations. According to this research, understanding travel motivation contributes to the following issues (Crompton & McKay, 1997 cited by Arslan, 2019);

- Understanding the tourist motivation will pave the way to create higher-quality products and services.
- Better understanding of tourists' motivations is closely related to the satisfaction they gain from touristic experiences.
- It will bring huge gains to tourist businesses, as the motivation of the tourist comes before their satisfaction.
- Estimating the motivations of tourists will make a great contribution to destination marketers.

Understanding the preferences and motivations of tourists plays an important role in tourism movements. Issues such as product development, the preparation of marketing strategies, and the determination of targets, which are important parts of the tourism system, are necessary for sustainable tourism activities. However, it is not easy to predict tourists' travel motivation (Harman, 2014). Different marketing activities are carried out to identify the changing demands and needs of tourists. Different tourism stakeholders, such as hotel management, restaurants and travel management, examine and use consumer behaviour research in their marketing activities for tourists. These studies are cultural (culture, subculture, social class), social (reference groups, family, roles and status), personal (age and life period, profession, economic characteristics, lifestyle, personality) and psychological (motivation, perception, learning, beliefs, and attitudes). Touristic consumers realize the purchasing behaviour under the influence of each of these factors (Yükselen, 2008).

Another factor that affects the behaviour of touristic consumers is the culture. The culture of a society or a country has an impact on the determination of the products/services to be produced in that society or culture, as well as on the consumer behaviour displayed. For this reason, to adopt tourist consumer behaviours, it is necessary to analyze the cultural effects of the society very well and it should be known that they can display buying behaviours accordingly (Odabaşı & Ve Barış, 2007). The lifestyles of touristic consumers are affected by the groups in which they belong to the society. Each tourist consumer can be included in more than one sub-group. These memberships can be delineated in a number of different ways: according to a certain age group, according to ethnic and racial background, or according to a specific place of residence (Solomon, 1996). In addition to having common features with other cultures, each subculture also has certain differences that distinguish it from the cultures they belong to and other cultures (Onkvisit & John, 2004). The needs, consumption patterns, and purchasing behaviours of touristic consumers are affected by the characteristics of the social class to which they belong. The social classes are highly influential in many choice behaviours such as product/service selection and business preferences, causing them to be closely related to marketing strategies (Onkvisit & John, 2004). Social factors that affect the purchasing behaviour of touristic consumers consist of reference groups, family, roles and status. Another factor that affects the purchasing behaviour of an individual is personal factors. Personal factors consists of age and life period, profession and economic status, lifestyle and personality characteristics. The age of a tourist consumer causes differences in the individual's holiday decision, purpose, transportation and accommodation preferences (Hayta, 2008). Another factor that affects purchasing behaviour is personal factors. Personal factors consists of age and life period, profession and economic status, lifestyle, and personality characteristics. The purchasing behaviour of touristic consumers is affected by psychological factors, which are formed by motivation, perception, learning, and beliefs and attitudes. It varies depending on the size of the effort made to meet the existing needs of the tourist consumer and the size of the motive that causes him to realize that need (Onkvisit & John, 2004). While the needs of individuals change over time, certain changes occur in their behaviour. Therefore, these ongoing changes in needs and behaviours form the basis of motivation.

## 8.4 The classification of generations, destination choice and touristic consumption

Although there are similarities between the generations, they have different characteristics from each other due to technological developments and political and economic events. While the generation is philosophically expressed as the community of people who were born in the same years and shared the conditions of the same age and had similar obligations, it is defined as "the group of individuals which are umbilical generation, abdomen". The literature is examined; it has been determined that the search for generations is not very clear and there is some confusion. The complex structure of generations is based on historical events. For example, an important political event in that period separates generations from each other. Auguste Comte stated, as a result of the scientific researches on generation studies between the years 1830–1840, that generational changes are forces acting in the historical process (Ekiz, 2017). The importance of generations has made it an area worthy of exploration. Generation differences are of great importance in business life, which is one of the areas where the concept of generation is used most frequently. Because the correct implementation of a strategy, the superior performance of the employees, and the correct determination of the mission and vision of the enterprises are possible with the correct perception of generation differences (Akgemci & Kalfaoğlu, 2018).

With a general definition, generational individuals come together with individuals born in the same period and define themselves as a class. Especially when the classification of people is examined, it is seen that they are grouped by date of birth. This classification contains many features such as experience, preferences and habits (Kian, Wan, & Sivan, 2013). In their study, Reeves & Oh, 2007stated that the cohorts cover a certain age range and this age range covers a period of 15–20 years. Individuals in this group can understand and perceive each other more easily because they have similar qualities. Solomon (1996) emphasized that generations consist of individuals belonging to a certain age group and that these people have similar characteristics. For example, people born in the same period love the same superheroes or have witnessed the same wars and peace. The sentence people define as "my generation" comes from this point of view. When the literature is examined, it is seen that the generations have similar birth dates and similar generation intervals. In this context, it is seen that the generations are divided into three generations. The researchers evaluated the generations according to certain date intervals and stated that the generations could have different names in each period. So far, generations have been examined in five groups. These generations can be listed as follows (Kuyucu, 2014):

- Traditionalists/The Silent Zone (born between 1925 and 1945)
- Baby Boomers (born between 1946 and 1964)
- Generation X (born between 1965 and 1979)
- Generation Y (born between 1980 and 1999)
- Generation Z (born after 2000).

However, there are also classifications of different researchers regarding basic generations. Although the names are similar, the dates are not the same. Another the classification concerning the business groups is as follows:

- Traditionals (born between 1925 and 1946)
- Baby Boomers (born between 1946 and 1963)
- Generation X (born between 1963 and 1981)
- Generation Y (born after 1980)

According to Bergh and Behrer (2016), Generation Y (the Millennial Generation) consists of people born in 1996. Generation Y is a general name given to those born between 1978 and 2000. This generation is also defined as the internet belt or nexters, nationally. In another study, this age group, also known as "Millennials" or "Eco Boomers", shows that they are hopeful for the future and people in this age group agree that it is important to look positively on life (Solomon, 2011). Generation Y has a unique feature, different from all generations. The people of this generation are the first to appear in the digital age, and this is one of their main characteristics. With its dependence on technology, Generation Y has a very different character from the previous generations. It is possible to express the characteristics of Generation Y as follows (Mohsen, 2016):

- They are well educated
- They like to travel
- It had a technologically advanced feature
- They can use technology very well
- They prefer to enjoy the process rather than being able to see the result.

As Sweeney (2006) stated in his research, there are serious differences between Generation Y and Generation X. Among these differences, Generation Y is more sincere and social, more abstract thinking, more emotionally harmonious and mature, more honest and courageous, more sensitive and emotional, giving importance to sceptical and objective realities, more open to change, more perfectionist and disciplined. There are quite different features according to its generation. It can take more risks and be more productive than previous generations in doing business or participating in an activity. Generation Y, which has a different character from its parents, is transformational (Puybaraud, 2010). Generation Y is a generation that does not grow under difficult financial conditions. It is a quality that deals with technology and social media intensively and has instant access to all information on the agenda. The travel trends of Generation Y, which follow bloggers and social media phenomena, are also quite extensive (Valentine & Powers, 2013).

Although the term Millennials is used for Generation Y, the scope of the book also includes Generation Z, because Generation Z is now in a position to make travel choices and purchase tourist services. Looking at the general definitions, it is seen that there are different groupings regarding Generation Z. Schawbel (2014b) defined Generation Z as those individuals born between 1994 and 2000. Singh (2014) states that individuals born between the years of 1995 and 2012 belong to Generation Z. Generation Z, which is dependent on technology like Generation Y, carries out many jobs with technology, and carries out its lessons, work, food and beverage activities with technology. Generation Z, whose attention span is short, is a process-oriented structure that loves speed, adventure and also seeks out individual activities. Unlike the Y generation, the Z generation, which does not like socializing, is defined as the "New Silent Generation" . Generation Z, who uses technological equipment such as phones, computers and tablets in almost every moment of their lives, has difficulty communicating on a face-to-face basis because they constantly communicate over the Internet. Generation Z living in big cities usually has elderly families in developed societies, so their parents are much older than themselves. In this case, it is trained by very old teachers. For this reason, its communication with the internet can be quite high. Considering the relationship between Generations Y and Z and tourism, it is possible to say that there are serious differences, although there is a general similarity. For example; Generation Y is free-spirited, adventurous, takes fewer risks, and looks more environmentally friendly, but does not compromise on comfort. Generation Z is more interested in 'green' matters and nature, has doubts about itself and its identity, loves spending less than Generation Y, that is, more economical, likes to focus on individual activities instead of spending time with family and friends. While Generation Z emphasizes cause and effect, it is seen that it has a balanced structure. Generation Y is more emotional. While the Y generation prefers messaging and phone calls as a communication tool, Generation Z communicates only with symbols and symbols (Akten, 2016).

Generations defined as the consumption segment have a very important place in tourism marketing. Determining the desires, needs and expectations of generations for tourism activities is a research area that is needed in predicting the purchasing behaviour of tourists and repurchasing. When the travel trends of Generation X are examined, it is seen that they are quite curious and enthusiastic about travel. Generation X, who wants to explore different destinations, travels through technology as much as possible. Generation X prefers nature-based tourism types, but also seeks entertainment during their holidays. Generation X, who does not like risk and adventure, buys its holidays as a package as possible and wants to know every detail before the tour (Li et al., 2013). Individuals of Generation X want to do something with their families and children as much as possible and are happy about it. Generation X tourists with high brand addiction prefer certain brands as much as possible. When the travel trends of the Y generation are examined, it is seen that they are in a close relationship with technology. Generation Y, born in technology, inevitably trusts technology. For example, negative news about a hotel business on social media also has an impact on the tourism preferences of Generation Y. Or a new tourism trend seen on social media attracts the attention of Generation Y. Generation X prefers package tours with the recommendations of travel businesses, while Generation Y prefers them considering their social media and internet comments. Generation Y book tours online, take into account the comments on TripAdvisor sites and considers the suggestions of their friends and friends about a destination and travel business (Özel, 2017). Generation Y travels frequently. This generation, who is not thrifty about spending money, has been observed to frequently buy products from souvenir shops in the region they visit. When traveling preferences are examined, it is seen that they prefer low-budget businesses and buy affordable products despite spending a lot (Avcıkurt, 2018; Ertürk, 2019).

According to Cavagnaro et al. (2018), Generation Y wants to get away from the mundanity of their daily life; therefore, they want to see new places, try new foods, and meet new people. When the tourism activities of the Generation Y tourists who prefer to travel for rest are examined. In addition to nature-based tourism activities, it is observed that they prefer small-group touristic activities such as sport fishing, ecotourism and bicycle tourism. Generation Y seeks calmness and prefers to stay for a long time in the destinations they find. Generation Y, who attaches importance to local culture, wants to experience this as much as possible. In this respect, they have a close relationship with responsible tourism. This generation, which protects the environment and culture, strives to pass on natural and cultural heritage to future generations. Generation Y has witnessed many important environmental events, such as climate change. For this reason, the education they have received regarding the protection of rapidly depleting resources declared Generation Y being called environmentally friendly tourists in the tourism industry. When the travel preferences of Generation Z are examined, it is seen that they prefer digitalization-oriented travel. For example, they make their travel preferences through the use of digital media. When the adventurous structures of Generation Z and their social media habits come together, the trend tourism trends of recent years emerge (Haddouche & Salomone, 2018). Generation Z spend their free time on short vacations. Their preferences for hotel managements are in favour of boutique hotels. In restaurants, they usually prefer local managements that offer local products. Cleanliness and trust are in first place in terms of tourist services (Wiastuti et al., 2020). The travel trends of Generation Z are generally oriented towards destination-oriented regions where there is no mass tourism industry. Smart tourists in Generation Z are a comfortable and stress-free tourist group, and they make their holidays by making use of all the possibilities of technology. They tend towards individual tourism activities as much as possible. Due to the strict and standard rules on travel agency tours, Generation Z generally does not prefer travel agents. Instead, he organizes a tour programme and finds a tour guide with his means (cited in Güçlütürk Baran, Özoğul, & Noyan, 2020).

## 8.5 Travel adoption of millennials

When the research is examined, it is seen that the Y generation, which has a great command of technology and knowledge, places importance on reliability and transparency. Generation Z, born into technology, emerges as a generation that constantly questions and researches. Generation Y decides to go on vacation at least twice a year. This is also true for the Z generation. Social media, websites, search engines, friends' suggestions, and forum websites are among the factors that will affect the decisions of both generations. For Generations Y and Z, the concept of tourism activities and vacation is generally accepted as a social activity rather than resting. For this reason, it can be suggested that tourism professionals and agencies offer opportunities with entertainment content to Generations Y and Z. Generation Y both pays attention to the suggestions of their surroundings and also affects their purchasing behaviour. The same will be true for Generation Z. He points out that while the Internet usage rate of Generation Y is over 55 per cent, this rate will reach 90 per cent in Generation Z. This causes hotels and agencies to pay more attention to their place in the online media and to the comments made about them.

Considering the economic and social effects of tourism, ensuring its sustainability is important for businesses, states and societies. At the point of ensuring the sustainability of tourism, it is seen that there has been a serious change in tourism trends, from buying tickets and tours to accommodation and holiday habits. With the technology change, there is a change in the habits of people who take a vacation to discover new places, relax and renew. This situation is reflected in tourism trends. In the past, vacationers who went to an agency and bought their holiday or plane tickets can now choose the hotel they will vacation with one click on the Internet and then buy their flight ticket. Those who prefer package tours can choose from comparison sites and plan their vacation in just a few minutes. The reflection of technology on tourism trends is not limited to this; the prominence of social media is reflected in tourism. When choosing a holiday destination, tourists now prefer the places that influencers visit, they want to see the places they share and make memories. Generation Y, which will have a large part of spending power by 2020, and Generation Z, which is expected to change the dynamics of the tourism sector in 2030, also affect the trends in tourism. It will be the key to success for tourism professionals to touch these generations (Indigo Dergisi: Y-Z Kuşağı turizm sektörü, 2020).

It is known that generations are defined as Generations X, Y, and Z in the field of tourism marketing. In this context, the tourism sector needs to know the touristic preferences and needs of the Y and Z generations and also their characteristics as tourists. It is important to pay attention to some issues in the process of offering touristic products to Generations Y and Z. In this context, it is possible to express the tourism preferences of Generation Y and Z as follows (The Brandage.com, 2020):

- Generation Y, who has a great command of technology and information, attaches importance to reliability and transparency.
- Generation Z, born into technology, will appear as a generation that constantly questions and researches. In this context, it is necessary to add relevant information from social media pages and to make all promotions related to tourism.
- Studies indicate that Generation Y has decided to go on vacation at least twice a year. This will also apply to Generation Z. Social media, websites, search engines, friends' suggestions and forum websites will be the leading factors that will affect the decisions of both generations.
- For the Y and Z generations, vacations appear to be more of a social activity rather than an opportunity for resting. Therefore, it will be important for the tourism sector and agencies to provide opportunities with entertainment content to Y and Z generations.
- A majority of the Generation Y members state that they will not buy a service of which their friends do not approve. Members of Generation Y will both influence, and be influenced by, the suggestions of the people around them. The same will be true for the Z generation.
- The rate of Generation Y in the Internet population is over 55%. This ratio will be around 90% for Generation Z. Based on these rates, hotels and agencies will have to pay more attention to their place in the online media and the comments made about them.
- Generations Y and Z prefer communication through social media such as Facebook, Snapchat and Instagram. Therefore, personalized messages and campaigns will enable them to communicate more easily with these generations.
- Versatile, simple, clear, easy to understand, honest, transparent, sincere, and utilitarian approaches that make you feel free and included, will stand out among the points that touch the Y and Z generations.

## 8.6 Conclusions

Although the date range varies according to different lifestyles and the depth of social events in different geographies, it is accepted that Generation Y, in other words, the Millennium generation, generally includes individuals born between the years 1980 and 2000. Those born after 2000 are called Generation Z and attract attention as the important explorers of the future in tourism. Generation Y, who is more self-centered than other generations, is a self-confident, conservative and educated generation who is fond of brands, friends, entertainment and digital culture, adopts informal communication styles, adopts a stance, relaxed but boring attitude. It should be noted that the tourism industry also attracts the attention of a new generation that seems very determined to change the organization, business life rules and social relations. Generation Y, who prefer to live "experience" to have it, also call themselves "Y" explorers. This generation wants to explore the world and innovations around them. They seek pleasure rather than achieve results. During their vacation, they take pictures of everything, the lobby, the view of the room, and the room itself, and share them with their friends and family. This search conducted by Generation Y, who was born with the rising standards of living thanks to the advancing technology, started to work in business life and tries to search for the meaning of life by visiting and seeing new places. This search creates a series of opportunities for the travel and hospitality sector. Vacation is an indispensable expense item in the personal budgets of the Y generation. Generation Y, who prefer travel programmes and even holiday packages for different experiences in their travel and travel preferences, uses technology very carefully in this area (Avcıkurt, 2018).

In recent years, the issue that has been on the agenda is the travel preferences of Generation Z. Generation Z is the label given to those born in 2000 and after. They like to use the Internet and mobile technologies and prefer to access their products through mobile technologies. They are active in every field with smartphones, I-Pads or tablet computers, which are common today. They especially prefer socializing and marketing products via the Internet. Young people of this generation prefer adventure tourism and experience local products in rafting, canoeing, paragliding and safari. However, it also shares all its activities via social media. Social media is their identity and Generation Z develops alternative tourism in this way. Popular destinations of Generation Z are the most frequently visited destinations by influencers, so tourism businesses should first meet with tourism phenomena to direct this audience. Shares and labels are very popular, Generation Z certainly adds the brand of that service to the tourism products. In this respect, knowing the demands and needs of both Generation Y and Generation Z is considered an important factor for the success of tourism stakeholders.

This chapter attempts to determine the travel preferences and tourist behaviours of Millennials, based on the existing research. This chapter summarizes that the travel preferences of Millennials change according to current conditions, political and political instabilities, fluctuations such as pandemics and crisis. Factors such as tourist profile, travel information, the use of different information sources, destination selection and structure, destination selection elements, travel characteristics and travel and duration change and shape, according to generations. Based on the literature, it is possible to say that the differences between generations stem from the travel behaviour and touristic choices of Millennials. This research also sheds light on future research. The section examined travel behaviours and choices, elaborating on the factors affecting the motivation of tourists. It should not be forgotten that personal characteristics, such as intrinsic and extrinsic motivation factors, personal, lifestyle, experiences, past, perception and the current situation, may be effective in changing the travel behaviours and preferences of generations. Finally, it should be kept in mind that the scope of Millennials can be constantly changed and updated, as well as the factors that are effective in the travel preferences and behaviours of Millennials

## References


Akgemci, T., & Kalfaoğlu, S. (2018), X Ve Y Kuşaklarının Kariyer Uyum Yetenek Düzeylerinin Belirlemeye Yönelik Bir Araştırma: Lise Öğretmenleri Örneği, *Nevşehir Hacı Bektaş Veli Üniversitesi Sbe Dergisi*, 8(2), 231–247.

Akten, M. (2016). *Y Ve Z Kuşaklarının Turizm Algıları ve Konaklama* I*şletmelerinden Beklentileri*, Master's Thesis, Balıkesir University, Balıkesir, Turkey.

Albayrak, A., & Özkul, E. (2013). Y Kuşağı Turistlerin Destinasyon Imaj Algıları Üzerine Bir Araştırma, *International Periodical For The Languages, Literature And History Of Turkish Or Turkic*, 8 (6), 16–31.

Avcıkurt, C. (2018). Y Kuşağı Turistler Veya Y "Kâşifleri". [Online] Https://Www.Turizmajansi.Com/Haber/Y-Kusagi-Turistler-Veya-Y-Kasiflerih28067/ [Accessed in: 10. 12. 2020].

Aydın, G. Ç., & Başol, O. (2014), X Ve Y Kuşağı: Çalışmanın Anlamında Bir Değişme Var Mı?, *Kırklareli Üniversitesi Meslek Yüksekokullarının Elektronik Dergisi*, 4(4), 1–15.

Badu, E. E. (2005). Employee motivation in university libraries in Ghana: A comparative analysis. *Information Development*, 21(1), 3846.

Benckendorff, P., Moscardo, G., & Pendergast, D. (2010). *Tourism and Generation Y*. Cambridge: CABI International.

Bergh, V. D. B., & Behrer, M. (2016). *Markalar Nasıl Genç Kalır? Y Ve Z Kuşakları Için Markalama* (Çev: Canan Vaner). Istanbul: Mediacat.

Broadbridge, A., Maxwell, G.A., & Ogden, S.M. (2007), Experiences, perceptions and expectations of retail employment for Generation Y, *Career Development International*, 12, 6.

Cavagnaro, E., Staffieri, S., & Postma, A. (2018). Understanding millennials' tourism experience: values and meaning to travel as a key for identifying target clusters for youth (sustainable) tourism, *Journal Of Tourism Futures*. [Online] Https://Www.Emerald.Com/Insight/Content/Doi/10.1108/Jtf-12-2017-0058/Full/Html [Accessed in: 10. 12. 2020].

Cömert, Y., & Durmaz, Y., (2006), Tüketicinin Tatmini Ile Satın Alma Davranışlarını Etkileyen Faktörlere Bütünleşik Yaklaşım Ve Adıyaman Ilinde Bir Alan Çalışması, *Journal of Yasar University*, 1(4), 351–375.

Correia, A., (2002), How do tourist choose – A conceptual framework, *Tourism An International Interdisciplinary Journal*, 50(1), 21–29.

Crompton, J. L., & McKay, S. L. (1997). Motives Of Visitors Attending Festival Events. *Annals Of Tourism Research*, 24(2), 425–439.

Schawbel, D. (2014a). Why 'Gen Z' May Be More Entrepreneurial Than 'Gen Y. Http://Millennialbranding.Com/2014/High-School-Careers-Study/ [Accessed in: 20. 11.2020].

Decrop, A., & Kozak, M., (2009), *Decision Strategies in Tourism Evaluation, Handbook of Tourist Behavior: Theory & Practice*, United Kingdom: Routledge.

Demir, ş. ş. & Kozak, M. (2011). Turizmde tüketici davranışları modelini oluşturan aşamalar arasındaki ilişki. *Anatolia: Turizm Araştırmaları Dergisi*, 22(1), 19–34.

Dunne, G. (2009). *Motivation And Decision Making in City Break Travel: The Case Of Dublin*. VDM Publishing.

Ekiz S., (2017), *Kamu Kurumlarında Y Kuşağının Liderlik Beklentilerine Etki Eden Demografik Faktörler Üzerine Bir Çalışma: Adana Büyükşehir Belediyesi Örneği*, Master's thesis, University of Çağ , Mersin, Turkey.

Erkol Bayram, G. (2017). Motivasyonun Demografik ve Mesleki Özelliklere Göre Farklılaşması: Turist Rehberleri Üzerine Bir Araştırma. *Journal of Yasar University*, 12(48), 257–271.

Ertürk, M. (2019). Measurement of Service Quality of Hotels in TRNC: Servqual Model. *Journal of Management and Economic Studies*, 1(3), 35–42.

Gredler, M. E., Broussard, Sheri, Coates, ve Garrison, & Betsy, M. E. (2004), The relationship between classroom motivation and academic achievement in elementary schoolaged children. *Family And Consumer Sciences Research Journal*, 33(2), 106–120.

Guay, F., Chanal, J. Ratelle, Catherine, F., Mars, H. W., Larosa, S., & Boivin, M. (2010), Intrinsic, identified, and controlled types of motivation for school subjects in young elementary school children, *British Journal Of Educational Psychology*, 80, 711–735.

Güçlütürk Baran, G., Özoğul, G., & Noyan, E. (2020). Yeni Tüketiciler Arasındaki Kuşak Z'nin Tatil Tercihleri: Üniversite Öğrencileri Örneği, *Türk Turizm Araştırmaları Dergisi*, 4(2), 927–945.

Hacıoğlu, N. (2005). *Turizm Pazarlaması. 5.* Baskı, Ankara: Nobel Yayın Dağıtım.

Haddouche, H., & Salomone, C. (2018). Generation Z and the tourist experience: tourist stories and use of social networks, *Journal Of Tourism Futures*, 4(1), 69–79.

Harman, S. (2014). Bağımsız Seyahat Eden Yerli Gezginlerin Seyahat Motivasyonları Üzerine Bir Araştırma. *Uluslararası Yönetim Iktisat ve Işletme Dergisi*, 10(21), 107–128.

Hayta, A. B., (2008), Turizm Pazarlamasında Tüketici Satın Alma Süreci Ve Karşılaşılan Sorunlar, *Kastamonu Eğitim Dergisi*, Cilt: 16, 1, 31–48.

Içöz, O., (2001), "*Turizm Işletmelerinde Pazarlama Ilkeler ve Uygulamalar*", Ankara: Turhan Kitabevi.

Ilban, O. M., Akkılıç, E., & Yılmaz, Ö. (2011). Termal Turizmde Tüketici Satın Alma Davranışını Etkileyen Faktörlerin Belirlenmesi: Gönen Örneği, *Öneri Dergisi*, 9(36), 39–51.

Indigo Dergisi: Y-Z Kuşağı turizm sektörü, [Accessed in: 08.12.2020], Https://Indigodergisi.Com/2017/03/Y-Z-Kusagi-Turizm-Sektoru/.

Kangasmäki, H., & Koskelainen, M., (2005), "How Important Is An Image? Dutch Travel Agencies' Perception Of Sweden As A Travel Destination", Master's Thesis, Karlstads Universitet, Division For Business Administration, Karlstad.

Kian, T. S., Wan, W. Y., & Sivan R. (2013). Relationship between motivations and citizenship performance among generation X and generation Y, *International Journal Of Academic Research in Business And Social Sciences*, 3(11), 53.

Kocatürk, Emine (2018). *Yabancı Turistlerin Çekici Seyahat Motivasyonlarının Tekrar Ziyaret Etme Niyetlerine Etkisi*, Master's Thesis, University of Muğla, Muğla, Turkey.

Kumar, A., (2012), The Changing Buying Behavior of Customers in Organized Retail Sector Of Pune City, *International Journal Of Research In Social Sciences*, 2(1), 247–263.

Kuyucu, M. (2014), Y Kuşağı Ve Facebook: Y Kuşağının Facebook Kullanım Alışkanlıkları Üzerine Bir Inceleme, *Elektronik Sosyal Bilimler Dergisi*, 13(4950), 55–83.

Li, X., Li, X. R., & Hudson, S. (2013). The Application Of Generational Theory To Tourism Consumer Behavior, *An American Perspective, Tourism Management*, 37(3), 147–164.

Luthans, F. (1992), *Organizational Behavior*, McGraw Hill Book Co: New York

Mahmutoğulları, D., & şahin Perçin, N. (2018). *Y Kuşağının Turizm Endüstrisindeki Etkisi Turizmde Güncel Konu Ve Eğilimleri*, Detay Yayıncılık

Mathieson, A., & Wall, G. (1982). *Tourism: Economic, physical and social impact*. Harlow, London: Longman.

Mayo, E, & Jarvis, L. (1981). *The Psychology Of Leisure Travel*. Boston: CABI Publishing.

McDowall, S., (2010), International Tourist Satisfaction And Destination Loyalty: Bangkok, Thailand, *Asia Pacific Journal Of Tourism Research*, 15(1), 21–42.

Middleton, V.T.C. (1988). *Marketing* In *Travel and Tourism*. London, UK: Heinemann.

Mohsen, M. A. M. S. (2016). Committed generations: A case study on Generations X and Y employees in Saudi hotels. *Anatolia*, 27(4), 456–467.

Moutınho, L. (1993). Consumer Behavior in Tourism. *European Journal of Marketing*, 21(10), 5–44.

Ndubisi, N. O. (2007). Impact of joint product usage and family structure on joint decision to purchase a vacation by Malaysian spouses. *Journal of Vacation Marketing*, 13(2), 135–147.

Odabaşı, Y. Ve Barış, G. (2007). *Tüketici Davranışı*. (7. Basım). Istanbul: Mediacat Kitapları.

Onkvisit, S., John S. (2004). *International Marketing: Analysis and Strategy* (4th Edition). New York: Routledge.

Opperman, M. (1998) Destination threshold potential and the law of repeat visitation, *Journal of Travel Research*, 37, 131–137.

Özdipçiner, N.S. (2009). Turistlerin Satın Alma Kararlarındaki Kültürel Farklılıklar: Türk – Alman Karşılaştırması, *Ege Akademik Bakış*, 9(4), 1295–1311.

Özel, Ç. H. (2017). Kuşak Kavramı ve Turizme Yansımaları, Özkoç, H. H. & Bayrakdaroğlu, F. (Ed), *Kuşak Kavramın Disiplinler Arası Bakış*, Ankara: Nobel Akademik Yayıncılık.

Özsungur, F., & Güven, S. (2017). Tüketici Davranışlarını Etkileyen Sosyal Faktörler Ve Aile. *International Journal Of Eurasian Education and Culture*, 2(3), 130.

Parmar, J. S., & Gupta, Y., (2007), "Consumer Behavior Towards Cosmetics: An Empirical Analysis", *Journal Of IPM*, 4(4), 4 Meerut, India.

Pearce, P., Morrison, A. M., & Rutledge, J. L. (1998). *Tourism: Bridges Across Continents, Motivational Influences in Tourism Demand*, Sydney: McGraw-Hill.

Puybaraud, M., (2010). *Generation Y and The Workplace: Annual Report 2010*. Global Workplace Innovation.

Reeves, T. C. and Oh, E. J. (2007). Generation Differences And Educational Technology Research. In J. M. Spector, M. D. Merrill, J. J. G. Van Merriënboer, & M. Driscoll (Eds.), *On Educational Communications And Technology* (pp. 295–303). Mahwah, NJ: Lawrence Erlbaum Associates. 4(2), 945–944.

Rızaoğlu, B. (2004). *Turizmde Tanıtma*. Detay Yayıncılık.

Schawbel, D. (2014b), Gen Y and Gen Z global workplace expectations study, http://millennialbranding.com/2014/geny-genz-global-workplace-expectations-study [Accessed in: 12.07.2019].

Seçilmiş, C., & Köz, E. N. (2015). Sürdürülebilir Kültür Turizmi Algısının Nesillere Göre Tekrar Gelme Niyetine Etkisi, *Journal of Tourism and Gastronomy Studies*, 3 (3), 69–78.

Singh, A. (2014). Challenges And Issues Of Generation Z. *Iosr Journal of Business and Management*, 16(7): 59–63.

Solomon, M. R. (1996). *Consumer Behaviour*. Englewood Cliffs, NJ: Prentice Hall.

Solomon, M. R. (2011). *Consumer Behavior – Buying, Having And Being*, (9th Edition), Upper Saddle River, NJ: Pearson Education, Inc.

Swarbrooke, J., & Horner, S. (1999). *Consumer Behaviour in Tourism*, Butterworth-Heinemann.

Sweeney, R. (2006). *Millennial Behaviors And Demographics*. Newark, NJ: New Jersey Institute Of Technology, 12(3), 10.

The Brandage, [Accessed in: 10.12.2020] Https://Www.Thebrandage.Com/Y-Ve-Z-Kusagi-Turizm-Sektoerunu-Nasil-Etkileyecek

Torlak, Ö., & Altunısık, R. Özdemir, (2002). *Modern Pazarlama*, Istanbul: Degisim Publishing.

Torun, Y., & Çetin, C. (2015). Örgütsel Sinizmin Kuşaklar Bazında Değerlendirilmesi: Kuşaklara Göre Örgütsel Sinizmin Hefdefinde Ne Var?, Iş *ve Insan Dergisi*, 2(2), 137–146.

Valentine, B. P., & Powers, L. T. (2013), Generation Y values and lifestyle segments, *Journal of Consumer Marketing*, 30(7), 597–606.

Wiastuti, R. D., Lestari, N. S., Bejo Mulyadi, N., & Masatip A. (2020). The Generation Z Characteristics and Hotel Choices. *African Journal of Hospitality, Tourism and Leisure*, 9(1), 1–14.

Yazgan, ş. (2012). *Bilgi edinme aracı olarak blogların turistik satın alma davranışına etkisi*, Master's Thesis, Adnan Menderes University, Aydın, Turkey.

Yıldırım, M., & Tayfun, A. (2010) Turistlerin Tüketim Davranışları Kültüre/Milliyete Göre Farklılık Gösterir Mi? Alman Ve Rus Turistler Üzerine Bir Araştırma, Iş*letme Araştırmaları Dergisi*, 2(2), 43–64.

Yükselen, C. (2008). *Pazarlama Ilkeler Ve Örnek Olaylar*. (7th Edition). Ankara: Detay Publishing.

# Part III

# Holistic tourism

## A Journey to the self

Part III of this book focuses on holistic tourism and explores spiritual tourism as a pathway for learning sustainability among Millennials. Further, this part aims to find out determinants of spiritual tourism rejuvenation among Millennials, examines the role of digitization in spiritual tourism adoption and explores the marketing and management of spiritual tourism destinations with a special emphasis on the Millennials.

Chapter 9, "Holistic Tourism: a Pathway towards Eudaimonic Well-being" by Zahra Rahmani, Maree Thyne and Anna Carr, presents the case that COVID-19 is a wake-up call to focus on wellness and preventive interventions. This conceptual study takes a positive psychology approach and attempts to understand how retreat centres can meet millennials' quests for meaningful experiences and an authentic sense of self – eudaimonic well-being.

Chapter 10, "Spiritual tourism and heritage site marketing: The case of Indian UNESCO world cultural heritage sites" by Raymond Rastegar & Zohreh Zarezadeh, discusses approaches for the effective use of social media marketing to promote cultural attractions and successfully attract younger travelers. The findings of this study provide the opportunity to gauge the ability of UN World Heritage Sites to engage successfully with tourists and particularly secularly inclined Millennials over social media.

Next Chapter 11, "Millennials seeking spiritual enlightenment through psychedelic plant medicine in Mexican indigenous communities" written by Diego Hannon Ovies, examines the motivations of Millennial travelers to visit the indigenous Mexican town of Huautla de Jimenez (HDJ) to participate in a shamanic mushroom ritual. Initially, the descriptive framework displays the correlation between Millennial travel trends, New Age spirituality, indigenous shamanism, and psilocybin mushroom rituals. The chapter aims to encourage academic discussion on the touristic appropriation of indigenous spiritual practices and the lack of inclusion of indigenous communities in the process of the scientific validation of psychedelic plant medicines.

DOI: 10.4324/9781003170631-11

The final chapter in this part, Chapter 12 "From Gen Xers to Millennials: Transformation of Self through Tourism" by Reni Polus, Neil Carr and Trudie Walters, provides a conceptual clarification of the difference in how Millennials, compared with their Generation X siblings, identify and define the transformation of self through tourism. The findings of this chapter, in general, indicate clearly that the different lifestyles and values that exist in the different generations lead to differences in consumption patterns in tourism.

# 9 Holistic tourism

## A pathway towards Eudaimonic well-being

*Zahra Rahmani and Anna Carr*

### 9.1 Introduction

Millennials, those born during the period 1982–1999 (Howe & Strauss, 2000), are now adults who seek to establish their social identity (Suar, Jha, Das, Alat, & Patnaik, 2020). This group of change agents (O'Malley & Williams, 2012) is reported to be unable to be their authentic selves (Deloitte Global, 2020). Despite being well educated (Pew Research Center 2019) and cooperative (Howe & Strauss, 2000), millennials have been observed to be highly 'fragile' (Eddy, Charles, & Richard, 2016) as many of them are not used to failure compared to their predecessors (Zainal Badri, Mun, & Ramos, 2020). As Winn and Dykes (2019) have noted, Millennials are not resilient in face of life's turning points and work difficulties. The way they have grown up has made them so vulnerable to emotional adversities (Buzza, 2017). Accordingly, Millennials are the most pressured generation suffering from anxiety and depression (American Psychological Association, 2014).

Given the reported increase in their stress levels and the fact that, together with Generation Z, Millennials supply most of the global workforce, their psychological and mental health issues present an enormous challenge to be solved (Deloitte Global, 2019, 2020). The current COVID-19 pandemic and the rapid pace of life transformation have thrown their psychological distress into sharp relief. Bearing in mind that Millennials are highly socialized with a greater desire for a sense of bonding and belonging (Hart, 2006), the throes of dealing with social distancing and self-isolation to cope with the COVID-19 pandemic have heightened their feelings of boredom, depression, stress and anxiety.

We remain hopeful that, sooner or later, the pandemic will come to an end. Nevertheless, crises and turning points seem to be the constant reality of life (Glouberman & Cloutier, 2017). In the face of such life challenges, people normally seek to leave their problems behind and relax on holiday, which tends to afford them hedonic pleasures and episodic happiness. However, there are times when more long-lasting, transformative, or self-developing outcomes are desired (Smith & Diekmann, 2017). In that case, people look for alternate sacred and spiritual spaces of reflection and contemplation

DOI: 10.4324/9781003170631-12

away from the distractions of their hectic world, where they can be themselves to be able to turn inward for an explanation (Fu, Tanyatanaboon, & Lehto, 2015). In response, holistic tourism has been developed specifically to provide time, space, and activities for self-reflection, personal growth, identity-building, and the development of psychological well-being (Gill, Packer, & Ballantyne, 2019; Thal & Hudson, 2019).

However, research has yet to deeply examine how holistic tourism may foster eudemonic well-being. Considering the unprecedented COVID-19 pandemic as a wake-up call to focus on wellness and preventive interventions, this study attempts to understand how wellness retreats can enable Millennials to achieve higher levels of well-being, helping them recover from the current crises and cope with the future transitions.

To answer this question, after an elaboration on the concept of 'wellness', the overlaps and differences between 'holistic tourism' and 'spiritual tourism' will be clarified, and then, through the lens of positive psychology and based on Ryff's (1989) conceptualization of eudaimonic well-being, this chapter goes on to explore how a holistic experience at retreat centres can lead to optimum levels of well-being.

## 9.2 From health to wellness and towards holistic tourism

> The part can never be well unless the whole is well.
>
> -Plato

Similar to the concept of health tourism, there are diverse definitions or understandings in the literature upon the terminology and typology of 'wellness tourism'. As referred to by Voigt (2010), this term is employed to cover a wide range of leisure activities like spa tourism, New Age tourism, volunteer tourism, adventure tourism, yoga tourism, spiritual and religious tourism, and holistic and retreat tourism. This lack of consistency becomes more obvious when it comes to characterizing 'wellness tourism', 'holistic tourism' and 'spiritual tourism'; insofar as even throughout a single piece of work, these concepts are used synonymously and/or distinctively, without discussing the variety of perspectives in this arena (Rahmani & Carr, forthcoming).

Although there is no consensus over a single precise definition for these terms s (Mueller & Kaufmann, 2001; Voigt, 2010; Voigt, Hall, Brown, & Howat, 2011), it is more generally agreed that as the non-water-based subsectors of 'wellness tourism', both holistic tourism and spiritual tourism tend to take place in purpose-built retreat centres. These two markets can be differentiated by their motivational approaches to well-being. While spiritual tourists focus purely on the spirit and its transcendence, holistic tourists aim for a balance of body-mind-spirit (Rahmani & Carr, forthcoming). 'Wellness Tourism' targets specific aspects of an individual's body, mind, and spirit episodically, whereas 'holistic tourism', as a new and alternative form of tourism (Rocha, Seabra, Silva, & Abrantes, 2016), aims at balancing, maintaining,

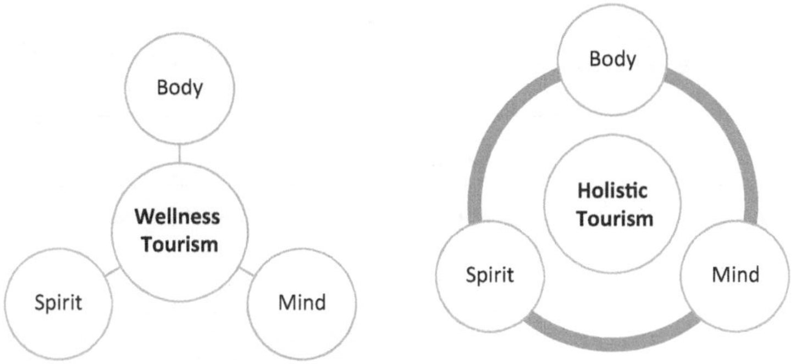


Figure 9.1 'Episodic' versus 'Holistic' approach towards well-Being.
Source: Adapted from Smith and Kelly (2012; 2006).

improving and developing an individual's 'body-mind-spirit' as a whole (Kelly, 2012; Kelly & Smith, 2009) (Figure 9.1).

Accordingly, it can be presumed that the core product of holistic retreats is 'body-mind-spirit integration'. If we consider 'wellness' as a multi-dimensional concept (Voigt, 2010; Voigt, Howat, & Brown, 2010), holistic tourism, which conveys notions of 'integration' and 'wholeness', attempts to incorporate all dimensions of wellness into the experience it offers (Kelly & Smith, 2009). This balance and integration can contribute to optimum levels of well-being, which is deeply meaningful and self-actualizing (Smith & Diekmann, 2017; Voigt, 2017; Voigt, Howat, and Brown (2010).

Collating various viewpoints in the aforementioned studies, Figure 9.2 is an attempt to illustrate exactly where the phenomenon of 'Holistic Tourism' lies in the Wellness Tourism Spectrum. As depicted here, meditation and yoga retreats form the core pathways of holistic tourism as they are meant to integrate an individual's body, mind, and spirit (Smith & Puczkó, 2009). These retreats are also categorized as sub-sectors of spiritual tourism (Sibi & Abraham, 2017). In addition to meditation and yoga retreats, the secular spiritual quest may also be pursued in pilgrimage centres like ashrams (Kelly, 2010, 2012; Lea, 2008; Narottama, 2016; Sharpley & Sundaram, 2005; Smith & Puczkó, 2009; Suresh, Ravichandran, & Ganesan, 2011; Voigt, 2010), temples (Sibi & Abraham, 2017; Voigt, 2010), monasteries (Bone, 2013; Ouellette, Kaplan, & Kaplan, 2005; Voigt et al., 2010), and sanctuaries (Bone, 2011, 2013).

Retreat centres are specifically established to serve holistic and secular spiritual tourists, whereas the primary mission of places such as temples, ashrams, and monasteries is to cater to traditional spiritual tourists with a religious affiliation. These pilgrimage centres have recently started to host holistic tourists as well (Kelly & Smith, 2009; Voigt, 2010). Given the complexity and multilayered notion of holistic tourism, the distinction between the context of holistic and spiritual tourism is not always as clear-cut. Recognizing this ambiguity, 'purpose-built holistic retreat' and 'ashram, temple, monastery, and sanctuary' are categorized separately, but with a dashed line to convey the overlaps (Figure 9.2).

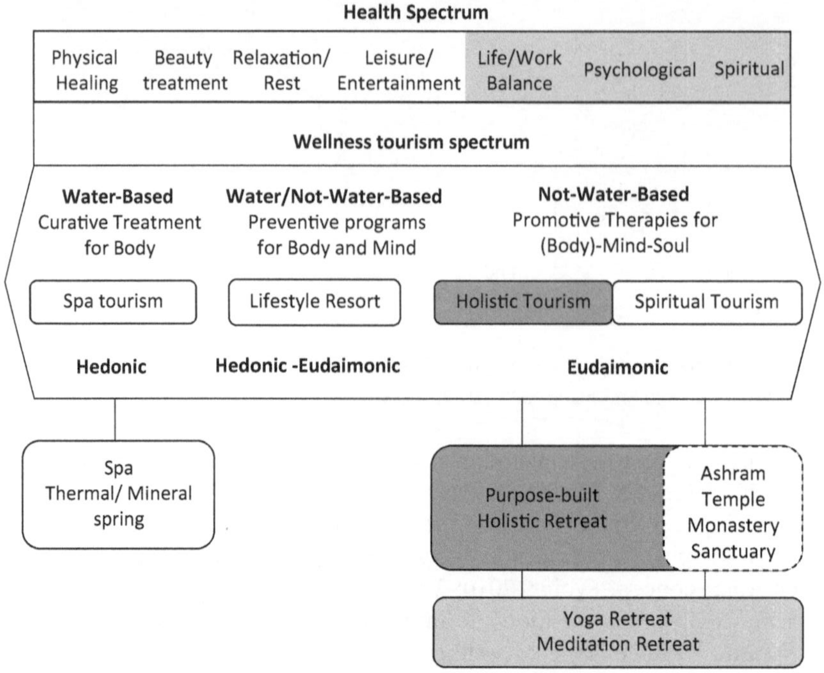


Figure 9.2 Holistic tourism in the bigger frame of health and wellness tourism.

Source: Adapted from Rahmani and Carr (forthcoming); Smith and Puczkó (2009); Voigt (2017); Voigt, Howat, and Brown (2010).

'Holistic tourists' are those who stay at least one night at a purpose-established and owner-run retreat, where they can take advantage of collective and/or individual classes, therapies, treatments and programmes to improve their physical, psychological, spiritual, and/or social well-being through achieving a state of mind-body-spirit balance or spiritual enlightenment (Gill et al., 2019; Heintzman, 2013; Kelly, 2010, 2012; Kelly & Smith, 2017). The majority of retreat visitors (holistic tourists) are 35–55-year-old women with an average age of 45 (Kelly, 2012) and higher levels of income and education (Heintzman, 2013; Smith & Puczkó, 2009). Considering Millennials' social challenges and behavioural concerns and the above-mentioned demography, sooner or later, Millennials are going to constitute a major share of the holistic tourism market.

## 9.3 Towards a conceptualization of eudaimonic well-being

Through integrating body, mind, and spirit, the ultimate mission of a holistic holiday is to facilitate tourists' spiritual enlightenment and psychological development (Kelly, 2010). However, only a few studies have covered the psychological side of the retreat experience; even in those studies, improved psychological well-being is assumed rather than being assessed (Thal &

Hudson, 2016). A brief contextualization of 'eudaimonic well-being' would help to understand the reason behind this research gap.

From the earlier Aristotelian perspective, eudaimonia is interpreted as happiness; today's philosophers, however, interpret it as flourishing and optimal functioning (Huta & Waterman, 2014). Whether as happiness or flourishing, the core concept of eudaimonia is 'knowing the true self' (Huta & Waterman, 2014; Waterman, 1993). Accordingly, self-discovery, self-realization and self-development can lead one to a higher state of flourishing and happiness (Ryff & Singer, 2008). Holistic tourists' experience of self-fulfillment and an authentic sense of self can be primarily considered as eudaimonic well-being (Smith & Diekmann, 2017; Voigt, Howat, & Brown, 2010). However, this concept, in tourism studies, is predominantly characterized by physical and social challenges and adversity (Nawijn & Filep, 2016). Accordingly, self-actualizing experiences have been mainly studied in contexts where mastery of challenge and skills is required and tourists attempt to accomplish their lifelong aspirations through pushing their body and mind to the limit (e.g. Holm, Lugosi, Croes, & Torres, 2017; Huta & Ryan, 2010; Matteucci & Filep, 2017). Examples include highly social activities like volunteer tourism (e.g. Alexander & Bakir, 2013; Coghlan & Weiler, 2018; Kontogeorgopoulos, 2017), extreme sports, physically challenging and risk-related activities such as white water rafting, and rock/mountain climbing (Boudreau, Mackenzie, & Hodge, 2020), and adventure tour-guiding (e.g. Houge Mackenzie & Brymer, 2018; Houge Mackenzie & Raymond, 2020).

To fully understand eudaimonic experiences, researchers (Diener, 2009; Medeiros, Gonçalves, Veiga, & Caraciolo, 2020; Nawijn & Filep, 2016) have called for the conceptualization of eudaimonia in diverse tourism activities. Recognizing the significance and dearth of research about eudaimonic experiences in still and slow contexts such as spiritual and holistic tourism, the next section elaborates on how a holistic experience on retreat can provide individuals with a deeply meaningful experience of well-being.

### *9.3.1 The intersection of holistic tourism and eudaimonic well-being*

Through the lens of positive psychology –the study of what makes life worth living (Seligman & Csikszentmihalyi, 2000), and, according to Steiner and Reisinger (2006) and also Dunn's definition of (high-level) wellness (1959), central to holistic tourism as a deeply meaningful and transformative experience (Voigt, 2017) is the concept of 'eudaimonic well-being' – the higher state of 'flourishing' and optimal functioning that can be achieved through self-development and realization of one's unique potential (Annas, 1993; Gable & Haidt, 2005; Ryff & Keyes, 1995; Ryff & Singer, 1996, 2008; Tiberius & Hall, 2010).

One of the best-known theoretical foundations to examine eudaimonic well-being is Ryff's (1989) conceptualization of 'psychological well-being', which includes six substantive determinants: personal growth, purpose in life, autonomy, positive relations with others, environmental mastery, and

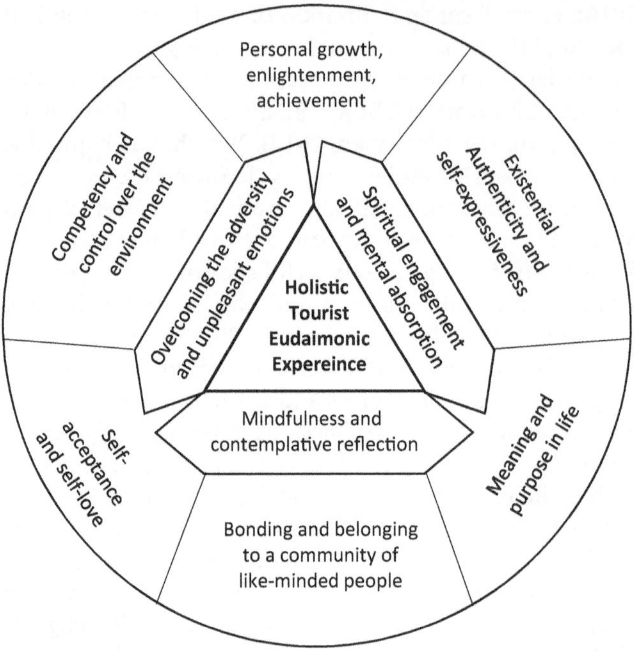


*Figure 9.3* Key dimensions of holistic tourist eudaimonic experience.

self-acceptance. Through an extensive review of various eudaimonic approaches, Huta and Waterman (2014) realized that these dimensions are further acknowledged by researchers who have tried to conceptualize well-being through the lens of positive psychology. Grounded in Ryff's (1989) work, and following the most agreed-upon themes of eudaimonia, identified by Huta and Waterman (2014), the strong resemblance of optimal psychological functioning with holistic tourists' motivations, sought benefits, and outcomes are illustrated in Figure 9.3.

***Personal growth – self-realization, self-actualization, maturity, and development of potentials*** (Bauer, McAdams, & Pals, 2008; Delle Fave, Brdar, Freire, Vella-Brodrick, & Wissing, 2011; Fowers, 2012; Huta, 2013; Huta & Ryan, 2010; Keyes, 2002; Ryan, Huta, & Deci, 2008; Ryff, 1989; Seligman, 2002; Steger, Shin, Shim, & Fitch-Martin, 2013; Tiberius & Hall, 2010; Vittersø et al., 2009; Waterman, 1993).

Personal growth is defined as the continuation of developing one's potential, and to grow and expand as a person. It is closely aligned with Aristotle's notion of eudaimonia, as well as the emphasis of life-span theories on continual growth and taking new challenges at different stages of life. Being open to new experiences is a feature of fully functioning people (Ryff, 1989; Ryff & Singer, 2008).

Huta and Waterman (2014) referred to personal growth as self-actualization and the realization of personal goals. Maslow (1970) noted that people who advance to a higher level of awareness and a connection to the universe are on the way to self-actualization. Retreat visitors who participated in Kelly's (2012) mix-method analysis of retreat visitors' motivations and experiences, acknowledged the contribution of their retreat experience to personal growth, enlightenment, and achievement of new insights into the self and others. Cohen, Elliott, Oates, Schembri, and Mantri's (2017) study of the impacts of a week-long wellness retreat experience showed that in addition to healthy food and environment, retreats afford various educational programmes that support personal growth, transformation, and the adoption of a healthy lifestyle. They also found that the therapeutic community in retreat centres contribute to individuals' growth.

***Purpose in life – meaning, long-term perspective, and serving a greater purpose*** (Bauer et al., 2008; Delle Fave et al., 2011; Fowers, 2012; Fowers, Mollica, & Procacci, 2010; Huta, 2013; Huta & Ryan, 2010; Keyes, 2002; Ryan et al., 2008; Ryff, 1989; Seligman, 2002; Steger et al., 2013; Vittersø, Vittersø, et al., 2009; Waterman, 1993).

Having a sense of purpose in life refers to a personal belief that one's life is meaningful and purposeful. It can be applied to individuals who have an intention and a sense of direction while functioning positively towards their goals (Ryff, 1989; Ryff & Singer, 2008). 'Meaning' was defined by Huta and Waterman (2014) as the identification of purpose in life, as well as the attainment of a 'bigger picture' of life and contributing to it. In a secular context, spiritual enlightenment as a key dimension of holistic tourists means to achieve a sense of meaning (Bone, 2011; Kelly & Smith, 2009). Groen (2016) explained how holistic tourists' craving for spirituality implies their search for a deeper understanding of their life. Gaining meaningful insights into the 'self' in the bigger frame of life was reported in several studies as a motivation or consequence of retreat experiences (Heintzman, 2013; Kelly, 2012; Kennedy, Abbott, & Rosenberg, 2002; Norman & Pokorny, 2017; Sibi & Abraham, 2017). Different activities, silence, walks in nature, and experience sharing with like-minded people on retreat, all in all, can bring retreat visitors closer to their truth with a sense of clarity about themselves and their life (Dillette, Douglas, & Andrzejewski, 2020; Dillette, Douglas,, & Andrzejewski, 2019).

***Autonomy – existential authenticity, identity, personal expressiveness, and integrity*** (Bauer et al., 2008; Huta, 2013; Huta & Ryan, 2010; Keyes, 2002; Ryan et al., 2008; Ryff, 1989; Seligman, 2002; Steger et al., 2013; Tiberius & Hall, 2010; Waterman, 1993).

Autonomy is defined as self-determination, independence, and the evaluation of behaviour based on internal personal standards (Ryff, 1989; Ryff & Singer, 2008). These concepts are linked with the notion of existential authenticity and self-expressiveness, which means to identify and develop one's best potential and to stay connected with one's true self. Maslow (1970) also describes authenticity and autonomy as behaving in ways consistent with one's potentials (Huta & Waterman, 2014; Maslow, 1970; Waterman, 1993).

"The growth in holistic tourism suggests that there is an increasing desire to focus on the 'self' rather than the 'other', and in existential rather than objective authenticity" (Rocha et al., 2016, p. 178). Holistic tourism is described as being more than a 'touristic physical movement' or a 'mere escape' (Smith & Kelly, 2006), a paradoxical journey that engenders an identity crisis, which requires one to lose his/her self to get in touch with the 'true self' rather than the 'ego-driven self', the 'social' or the 'ideal' self (Kelly & Smith, 2017). Concerning the human desire to be free of control by others, Reisinger and Steiner (2006) and Steiner and Reisinger (2006) proposed that such journeys stem from a desire to get away from everyday life stress, in a quest for unity and a sense of authenticity to revitalize personal identity.

Smith and Kelly (2006) adopted Wang's notion of 'existential authenticity' (1999) and declared that, in this quest, the object of tourist gaze is, not creating a temporary or exaggerated self, as is the case with mass tourism, but seeking an 'authentic sense of self' that can be discovered through confronting or accepting all his/her dormant aspects. Steiner and Reisinger (2006, 9) depicted this confrontation appropriately as "getting in touch with what is inside or outside us without reasoning it through or even having the words to express what is going on".

Fu et al.'s (2015) analysis of online guest reviews about retreat visitors' experience in Thailand showed that experiencing a feeling of meaninglessness, alienation, and existential crisis had motivated people to consider retreats as a way to discover the essence of self. Retreats provide individuals, who have lost touch with their identity, with supportive, nurturing, and restorative experiences to re-connect with themselves (Kelly & Smith, 2017). Bone (2013); Gill et al. (2019); Voigt, Howat, and Brown (2010) found that participants on retreats, who are detached from their ordinary lives, have time and freedom to learn about themselves and reflect and develop deep insights into their 'authentic true self'. Thal (2015) also acknowledged that such autonomous experiences can contribute to autonomous self-regulation.

***Positive relations with others – relatedness, belonging, social well-being*** (Bauer et al., 2008; Fowers, 2012; Fowers et al., 2010; Keyes, 2002; Ryan et al., 2008; Ryff, 1989; Seligman, 2011; Steger et al., 2013).

Positive relations with others can be interpreted as warm and trusting interpersonal relationships. It also means to hold strong empathy for others (Ryff, 1989; Ryff & Singer, 2008). Holistic retreats provide their visitors with a sense of belonging to a supportive, like-minded community (Smith & Kelly, 2006). in Bone's (2011, 2013) study, this sense of community was described by participants, as non-judgemental friendships, feelings of safety, positive group energy, relationship 'authenticity', and helping or caring for others. The strong sense of camaraderie in retreat was also reported by Thal (2015), Cohen, Elliott, et al. (2017) and Thal and Hudson (2019).

Wellness retreats have the power to transform individuals through fostering interpersonal relationships within a community of like-minded people. Through sharing experiences of vulnerability and learning about different cultural backgrounds in a supportive environment, these relationships normally lead to the attainment of meaningful insights and a shift in perspectives (Dillette, Douglas, & Andrzejewski, 2019; Kelly, 2012). In Gill, Packer, and Ballantyne's (2018) research, the collective atmosphere at the retreat was a significant aspect that not only resulted in new insights for individuals but also contributed to a greater sense of belonging (Gill et al., 2019). The feeling of being understood by peers and having a supportive network usually continues virtually once back home after the retreat (Gill et al., 2019; Smith, 2013).

***Environmental mastery – competence, achievement, and accomplishment*** (Bauer et al., 2008; Keyes, 2002; Ryan et al., 2008; Ryff, 1989; Vittersø, Overwien, & Martinsen, 2009; Vittersø, Vittersø, et al., 2009).

Environmental mastery can be defined as the ability to choose or manage the environment in favour of personal needs and values (Ryff, 1989; Ryff & Singer, 2008). It is connected with the notion of competence that was defined by Deci and Ryan (2000) as the tendency to affect the environment, as well as to achieve beneficial outcomes within it. Many retreat centres have their roots in environmental conservation. They tend to be grounded in nature with principles around respectful and meaningful lifestyles (Bone, 2013). An analysis of the link between retreat design and visitors' restorative experience illustrates how being mentally away from ordinary life's demands and responsibilities, and engaged in a peaceful and relaxing environment of natural beauty is crucial for reflective contemplation and restoration (Gill et al., 2019). Retreats can also contribute to visitors' overall sense of competence as reported by participants in Thal's (2015) study of retreats' contributions to psychological well-being. She realized that while staying at a retreat centre, one can attain higher levels of confidence and capability to cope, whether it be managing a healthy diet or enhanced proficiency in doing physical activities (Thal, 2015; Thal & Hudson, 2016).

***Self-acceptance*** (Bauer et al., 2008; Keyes, 2002; Ryff, 1989).

Self-acceptance is described as having positive beliefs in one's self and his/her past life (Ryff, 1989). Ryff and Singer (2008) considered this feature as a kind of long-term self-evaluation that involves awareness and acceptance of both personal strengths and weaknesses (Ryff & Singer, 2008). Retreats' environment and community of like-minded people offer individuals time and freedom to express the 'self', which has been repressed in their everyday lives. The sense of belonging to a community of non-judgmental peers tends to encourage participants to express themselves overtly. This feeling consequently contributes to the development of self-acceptance (Bone, 2011). The study of yoga retreats' contribution to transformation, conducted by Dillette et al. (2019), also supported this claim as their participants reflected on themselves with self-love and acceptance.

Given the intersections of eudaimonic well-being and holistic tourism, it can be argued that retreat centres provide their guests with personal growth and sense of achievement, meaning and purpose in life, existential authenticity and self-expressiveness, positive relationships with others, and a sense of bonding and belonging to a community of like-minded people, a sense of competency and control over their surroundings, and a feeling of self-love and -acceptance. Even though there is vigorous evidence that Ryff's (1989) self-report instrument is strong enough to conceptualize eudemonic well-being (Waterman, 2008), to comprehensively understand how a holistic experience on retreat is eudaimonic in nature, two more characteristics of eudaimonia, which are central to holistic tourism, are further explored: 1) flow, which implies a loss of self-consciousness (Csikszentmihalyi, 1975); 2) mindfulness, which conveys a state of being fully conscious of the self (Brown & Ryan, 2003).

### *9.3.2 Engagement and flow*

Engagement is defined as experiences when one can engage completely with the present moment, thus entering the state of 'flow' (Seligman, 2011). Flow is conceptualized by Csikszentmihalyi (1975) in terms of challenge and skill at the level of behaviour (Delle Fave et al., 2011; Vittersø, Overwien, & Martinsen, 2009; Vittersø, Vittersø, et al., 2009; Waterman, 1993), and a sense of deep immersion and engagement at the level of experience (Bauer et al., 2008; Delle Fave et al., 2011; Ryan & Deci, 2001; Vittersø, Overwien, et al., 2009; Vittersø, Vittersø, et al., 2009).

**Challenge and skill – effort, and engaging with the challenge**: Existential transformation requires effort, and achieving a holistic experience is not a convenient process. The self-development programmes in holistic retreats are often out of individuals' comfort zone and might be emotionally and psychologically tough, demanding, challenging, and even frustrating at the time. However, through overcoming the adversity and persevering the struggle, individuals are afforded delayed positive effects that can lead to transformation. In contrast to hedonic pleasure-seeking activities that take place in spa tourism, the transformative experience of holistic retreats is often the result of unpleasant, high-effort activities (Voigt, 2017; Voigt, Howat, & Brown, 2010).

***Deep immersion and engagement – absorption, interest, and optimal experience*:** Holistic retreats are normally located in "aesthetically pleasing, environmentally lush and culturally rich surroundings" (Smith & Kelly, 2006, p. 15). The findings of several studies revealed how these aspects of beauty can lead individuals to a sense of flow and deep absorption in their experiences. Groen's (2016) evaluation of retreats' contribution to Environmental Adult Education, acknowledged how some guidance could make guests deeply engaged with nature. Dillette et al. (2019) also realized that a transformative experience can be a result of immersion and deep absorption. Some of their participants attributed their perspective shifts to intense levels of engagement in the retreat's natural setting. Another study by Dillette et al. (2020) indicates that retreat visitors tend to experience spiritual wellness through a deep immersion into the beauty and sounds of natural surroundings, the culture of the retreat community, as well as spiritual practices. Experiencing spiritual engagement (Fu et al., 2015), immersion, and mental absorption on retreat have also been identified as predictors of a restorative experience (Thal & Hudson, 2019).

***Mindfulness – awareness, contemplation, and self-reflection*** (Bauer et al., 2008; Ryan & Deci, 2001; Steger et al., 2013).

Mindfulness refers to a "state of being attentive to and aware of what is taking place in the present" (Brown & Ryan, 2003, p. 822). A mindful experience enables one to be aware and conscious of thought (Norman & Pokorny, 2017). Holistic retreats are known as places of quiet reflection. These centres provide their visitors with a range of mindful practices that bring them back to the present and make them aware of the self (Kelly & Smith, 2017), so that they can engage in a process of reflection and contemplation away from the concerns and distractions of home (Smith & Kelly, 2006). Retreat practices and activities like journalling, mindfulness (Thal, 2015), yoga, meditation, and Tai Chi, which require one to be focused, tend to afford enhanced levels of self-awareness (Fu et al., 2015).

Mindfulness is a key precursor of holistic wellness (Dunn, 1959). Thal and Hudson (2019) found that retreat elements such as the time and space to meditate and enjoy nature, psychological workshops and seminars, catered meals, and structured programmes, altogether can facilitate the act of mindfulness and contemplative reflection. This aspect of eudaimonic well-being – mindfulness and self-reflection – has also emerged from studies in the context of meditation retreats (e.g. Cohen, Jensen, Stange, Neuburger, & Heimberg, 2017; Norman & Pokorny, 2017).

Contemplating the intersecting points of holistic tourism and eudaimonic well-being, this study affirms that retreat centres can provide Millennials with time, space and freedom for self-reflection to be able to retrieve their well-being and cope with what is going on in their everyday life. This claim is consistent with the existent literature about Millennials' psychological health. In particular, an empirical study about the sources of happiness of Generation Y, conducted by Suar et al. (2020), illustrated that prosocial behaviors, close and warm interpersonal relationships, positive self-evaluation, goal achievement, and self-fulfillment, and being engaged with nature are amongst the main factors that can reduce Millennials' stress, connect them to their roots and, consequently, enhance their happiness and well-being.

## 9.4 Conclusion

A holistic journey is a still and slow type of tourism centred on inner peace and self-reflection. Recognizing Millennials' need for such experiences to build their identity and find their authentic selves, this chapter was an attempt to understand how retreat centres can foster eudaimonic well-being. Since a eudaimonic experience involves hardship and a balance of challenge and skills, it has been mainly conceptualized with physically and socially challenging activities. However, this study, mainly grounded in Ryff's (1989) theoretical framework and the most agreed-upon themes of eudaimonia (Huta & Waterman, 2014), argues that a holistic journey on retreat is an opportunity to attain an 'authentic sense of self', which is primarily considered as a eudaimonic experience. As illustrated in Figure 9.3, the resemblance of optimal psychological functioning with holistic tourists' motivations, sought benefits and outcomes can be summarized as follows:

- **Personal growth, enlightenment, achievement**: educational programmes in retreat centres enable personal growth, enlightenment, and the adoption of a healthy lifestyle.
- **Meaning and purpose in life:** Various activities, silence, walks in nature, and experience sharing with like-minded people on retreat provide meaningful insights into the 'self' and a sense of clarity about life.
- **Existential authenticity and self-expressiveness:** Retreats provide time, freedom, and a nurturing environment to individuals, who have lost touch with their identity, to reconnect with themselves and develop deep insights into their 'authentic true self'.
- **Bonding and belonging to a community of like-minded people:** Retreat centres foster interpersonal relationships within a supportive community of like-minded people, in which people can share their experiences of vulnerability and learn about different cultural backgrounds.
- **Competency and control over the environment:** Through providing a peaceful and relaxing environment away from ordinary life's demands and responsibilities, retreats contribute to people's overall sense of competence, confidence and capability to cope.
- **Self-acceptance and self-love**: Through overtly expressing the 'oppressed self' in a supportive environment and community of like-minded people in retreat centres, people tend to develop self-acceptance and self-love
- **Overcoming the adversity and unpleasant emotions**: Retreats' self-development programmes, which are often out of individuals' comfort zone and might be psychologically tough and challenging, afford people delayed positive effects that can lead to transformation if they persevere the struggle and overcome the emotional adversity.
- **Spiritual engagement and mental absorption:** Retreats' awe-inspiring surroundings can lead individuals to a deep sense of flow and intense levels of engagement in their experiences.
- **Mindfulness and contemplative reflection:** Psychological workshops and mindful practices in retreats are designed to enhance levels of self-awareness by bringing people back to the present so that they can engage in a process of reflection and contemplation away from the concerns and distractions of home.

## 9.5 Future directions

Given that global capitalism, especially in western developed countries, is leading the post-modern human to feelings of emptiness and alienation, and, consequently, mental and behavioural disorders like stress, anxiety and depression, people are becoming increasingly concerned with identity and meaning of life. In response, retreat centres, as a special sub-sector of wellness tourism, provide time, space and activities for self-reflection, personal growth, identity-building and the development of psychological well-being. However, due to the rapid spread of COVID-19, which has led to the shutdown of many retreat centres and restrictions on border crossing, people are less willing and able to physically visit retreat centres. Regardless of such unprecedented conditions, retreat tourism is recognized as a luxury product directed to a high-profile market. The busy and hectic schedule of life and time poverty presents an additional barrier to engaging in on-site retreat holidays. Therefore, given that Millennials as digital natives are round-the-clock connected to the Internet through various tools and platforms, further research across the wellness industry might investigate how holistic retreats can redesign their offerings to make them more accessible and affordable through digital platforms. Furthermore, the effectiveness of online wellness programmes in fostering higher levels of well-being would be interesting for future studies.

A thorough analysis of the measures of eudaimonic well-being in the context of holistic tourism was beyond the scope of this conceptual study. Therefore, the way forward could be empirical qualitative research to examine and refine the captured dimensions of holistic tourists' eudemonic experience. There is also an opportunity to utilize hybrid models of psychological well-being like PERMA (Seligman, 2011) and DRAMMA (Newman, Tay, & Diener, 2014) to revise, extend or complement the proposed picture of eudaimonic experience in wellness retreat centres.

## References


Alexander, Z., & Bakir, A. (2013). The development of self through volunteer tourism. In S. Filep & P. L. Pearce (Eds.), *Tourist experience and fulfilment; Insights from positive psychology* (pp. 147–164). New York: Routledge.

American Psychological Association. (2014). Stress in AmericaTM: Are teens adopting adults' stress habits. *Stress in America Surveys* (⟨http://www.apa.org/news/press/releases/stress/2013/stress-report.pdf⟩).

Annas, J. (1993). *The morality of happiness*. New York: Oxford University Press.

Bauer, J. J., McAdams, D. P., & Pals, J. L. (2008). Narrative identity and eudaimonic well-being. *Journal of Happiness Studies*, 9(1), 81–104. doi:10.1007/s10902-006-9021-6.

Bone, K. (2011). *Spiritual Retreat Tourism in Aotearoa/New Zealand.* (Masters of Arts), The University of Auckland, New Zealand.

Bone, K. (2013). Spiritual Retreat Tourism in New Zealand. *Tourism Recreation Research*, 38(3), 295–309.

Boudreau, P., Mackenzie, S. H., & Hodge, K. (2020). Flow states in adventure recreation: A systematic review and thematic synthesis. *Psychology of Sport and Exercise*, 46, 101611. doi: 10.1016/j.psychsport.2019.101611.

Brown, K. W., & Ryan, R. M. (2003). The Benefits of being present: Mindfulness and its role in psychological well-being. *Journal of Personality and Social Psychology*, 84(4), 822–848. doi:10.1037/0022-3514.84.4.822.

Buzza, J. S. (2017). Are you living to work or working to live? What millennials want in the workplace. *Journal of Human Resources*, 5(2), 15–20.

Coghlan, A., & Weiler, B. (2018). Examining transformative processes in volunteer tourism. *Current Issues in Tourism*, 21(5), 567–582. doi:10.1080/13683500.2015.1102209.

Cohen, J., Jensen, D., Stange, J., Neuburger, M., & Heimberg, R. (2017). The Immediate and Long-Term Effects of an Intensive Meditation Retreat. *Mindfulness*, 8(4), 1064–1077. doi:10.1007/s12671-017-0682-5.

Cohen, M. M., Elliott, F., Oates, L., Schembri, A., & Mantri, N. (2017). Do wellness tourists get well? An observational study of multiple dimensions of health and well-being after a week-long retreat. *J Altern Complement Med*, 23(2), 140–148. doi:10.1089/acm.2016.0268.

Csikszentmihalyi, M. (1975). *Beyond boredom and anxiety* (1st ed.). San Francisco: Jossey-Bass Publishers.

Csikszentmihalyi, M., & Seligman, M. (2000). Positive psychology. *American Psychologist*, 55(1), 5–14.

Deci, E. L., & Ryan, R. M. (2000). The "What" and "Why" of goal pursuits: Human needs and the self-determination of behavior. *Psychological inquiry*, 11(4), 227–268. doi:10.1207/s15327965pli1104_01.

Delle Fave, A., Brdar, I., Freire, T., Vella-Brodrick, D., & Wissing, M. P. (2011). The Eudaimonic and Hedonic Components of Happiness: Qualitative and Quantitative Findings. *Social indicators research*, 100(2), 185–207. doi:10.1007/s11205-010-9632-5.

Deloitte Global. (2019). *The Deloitte Global Millennial Survey 2019*. Retrieved from https://www2.deloitte.com/content/dam/Deloitte/global/Documents/About-Deloitte/gx-2018-millennial-survey-report.pdf.

Deloitte Global. (2020). *Millennials, Gen Z and mental health: Managing mental health in the workplace*. Retrieved from https://www2.deloitte.com/content/dam/Deloitte/global/Documents/About-Deloitte/gx-millennial-survey-mental-health-whitepaper.pdf.

Diener, E. (2009). *Subjective well-being. In The science of well-being: The collected works of Ed Diener* (Vol. 37). New York: Springer.

Dillette, A. K., Douglas, A. C., & Andrzejewski, C. (2019). Yoga tourism – a catalyst for transformation? *Annals of Leisure Research*, 22(1), 22–41. doi:10.1080/11745398.2018.1459195.

Dillette, A. K., Douglas, A. C., & Andrzejewski, C. (2020). Dimensions of holistic wellness as a result of international wellness tourism experiences. *Current Issues in Tourism*, 1–17. doi:10.1080/13683500.2020.1746247.

Dunn, H. L. (1959). What High-Level Wellness Means. *Canadian Journal of Public Health / Revue Canadienne de Sante'e Publique*, 50(11), 447–457.

Eddy, S. W. N. G., Charles W. G., & Richard, W. (2016). Millennials and public service renewal: introduction on millennials and public service motivation (PSM). *Public Administration Quarterly*, 40(3), 412–428.

Fowers, B. J. (2012). An Aristotelian framework for the human good. *Journal of Theoretical and Philosophical Psychology*, 32(1), 10–23. doi:10.1037/a0025820.

Fowers, B. J., Mollica, C. O., & Procacci, E. N. (2010). Constitutive and instrumental goal orientations and their relations with eudaimonic and hedonic well-being. *The Journal of Positive Psychology*, 5(2), 139–153. doi:10.1080/17439761003630045.

Fu, X., Tanyatanaboon, M., & Lehto, X. Y. (2015). Conceptualizing transformative guest experience at retreat centers. *International Journal of Hospitality Management*, 49, 83–92. doi:10.1016/j.ijhm.2015.06.004.

Gable, S., & Haidt, J. (2005). What (and Why) Is Positive Psychology? *Review of General Psychology*, 9. doi:10.1037/1089-2680.9.2.103.

Gill, C., Packer, J., & Ballantyne, R. (2018). Exploring the restorative benefits of spiritual retreats: the case of clergy retreats in Australia. *Tourism Recreation Research*, 43(2), 235–249. doi:10.1080/02508281.2017.1410972.

Gill, C., Packer, J., & Ballantyne, R. (2019). Spiritual retreats as a restorative destination: Design factors facilitating restorative outcomes. *Annals of Tourism Research*, 79. doi:10.1016/j.annals.2019.102761.

Glouberman, D., & Cloutier, J.-A. (2017). Community as holistic healer on health holiday retreats: The case of Skyros. In M. K. S. A. L. Puczkó (Ed.), *The Routledge Handbook of Health Tourism* (pp. 152–167). New York: Routledge.

Groen, J. (2016). Rediscover, reawaken, renew: The potential role of spiritual retreat centres in environmental adult education. *The Canadian Journal for the Study of Adult Education (Online)*, 28(1), 83–96.

Hart, K. A. (2006). Generations in the workplace: finding common ground. *MLO. Medical laboratory observer*, 38(10), 26–27.

Heintzman, P. (2013). Retreat Tourism as a Form of Transformational Tourism. In Y. Reisinger (Ed.), *Transformational Tourism: Tourist Perspectives* (pp. 68–81). London, UK: CAB International.

Holm, M. R., Lugosi, P., Croes, R. R., & Torres, E. N. (2017). Risk-tourism, risk-taking and subjective well-being: A review and synthesis. *Tourism Management*, 63, 115–122. doi:10.1016/j.tourman.2017.06.004.

Houge Mackenzie, S., & Brymer, E. (2018). Conceptualizing adventurous nature sport: A positive psychology perspective. *Annals of Leisure Research*, 23(1), 79–91. doi:10.1080/11745398.2018.1483733.

Houge Mackenzie, S., & Raymond, E. (2020). A conceptual model of adventure tour guide well-being. *Annals of Tourism Research*, 84. 102977. doi:10.1016/j.annals.2020.102977.

Howe, N., & Strauss, W. (2000). *Millennials Rising: The Next Great Generation*. New York: Vintage Books.

Huta, V. (2013). Eudaimonia. In S. David, I. Boniwell, & A. C. Ayers (Eds.), *Oxford handbook of happiness* (pp. 201–213). Oxford, GB: Oxford University Press.

Huta, V., & Ryan, R. (2010). Pursuing pleasure or virtue: The differential and overlapping well-being benefits of hedonic and eudaimonic motives. *Journal of Happiness Studies*, 11(6), 735–762. doi:10.1007/s10902-009-9171-4.

Huta, V., & Waterman, A. S. (2014). Eudaimonia and its distinction from Hedonia: Developing a classification and terminology for understanding conceptual and operational definitions. *Journal of Happiness Studies*, 15(6), 1425–1456. doi:10.1007/s10902-013-9485-0.

Kelly, C. (2010). Analysing wellness tourism provision: A retreat operators' study. *Journal of Hospitality and Tourism Management*, 17(1), 108–116. doi:10.1375/jhtm.17.1.108.

Kelly, C. (2012). Wellness Tourism: Retreat Visitor Motivations and Experiences. *Tourism Recreation Research*, 37(3), 205–213. doi:10.1080/02508281.2012.11081709.

Kelly, C., & Smith, M. K. (2009). Holistic Tourism integrating body, mind, spirit. In R. Bushell & P. J. Sheldon (Eds.), *Wellness and Tourism: Mind, Body, Spirit, Place* (pp. 69–83). New York: Cognizant.

Kelly, C., & Smith, M. K. (2017). Journeys of the self the need to retreat. In L. P. M.K. Smith (Ed.), *The Routledge Handbook of Health Tourism* (pp. 138–151). London: Routledge.

Kennedy, J. E., Abbott, R. A., & Rosenberg, B. S. (2002). Changes in spirituality and well-being in a retreat program for cardiac patients. *Alternative Therapies in Health and Medicine*, 8(4), 64–70.

Keyes, C. L. M. (2002). The Mental Health Continuum: From Languishing to Flourishing in Life. *Journal of Health and Social Behavior*, 43(2), 207–222. doi:10.2307/3090197.

Kontogeorgopoulos, N. (2017). Finding oneself while discovering others: An existential perspective on volunteer tourism in Thailand. *Annals of Tourism Research*, 65, 1–12. doi:10.1016/j.annals.2017.04.006.

Lea, J. (2008). Retreating to nature: rethinking 'therapeutic landscapes'. *Area*, 40(1), 90–98.

Maslow, A. H. (1970). *Motivation and Personality* (2nd ed.). New York: Harper & Row.

Matteucci, X., & Filep, S. (2017). Eudaimonic tourist experiences: the case of flamenco. *Leisure Studies*, 36(1), 39–52. doi:10.1080/02614367.2015.1085590.

Medeiros, S. A. d., Gonçalves, M. A., Veiga, R. T., & Caraciolo, G. B. (2020). The most important trip for who I am: exploring the dimensions of the Eudaimonic Tourist Experience. *Revista Brasileira de Pesquisa em Turismo*, 14(1), 14–33. doi:10.7784/rbtur.v14i1.1599.

Mueller, H., & Kaufmann, E. L. (2001). Wellness tourism: Market analysis of a special health tourism segment and implications for the hotel industry. *Journal of Vacation Marketing*, 7(1), 5–17. doi:10.1177/135676670100700101.

Narottama, N. (2016). Spiritual tourism A case study of foreigners' participation in the Pitrayajna ceremony in the Desa Pakraman of Muncan, Selat, Karangasem. *E-Journal of Tourism*, 3(1), 33–44.

Nawijn, J., & Filep, S. (2016). Two directions for future tourist well-being research. *Annals of Tourism Research*, 61, 221–223. doi:10.1016/j.annals.2016.07.007.

Newman, D., Tay, L., & Diener, E. (2014). Leisure and Subjective Well-Being: A Model of Psychological Mechanisms as Mediating Factors. *Journal of Happiness Studies*, 15(3), 555–578. doi:10.1007/s10902-013-9435-x.

Norman, A., & Pokorny, J. J. (2017). Meditation retreats: Spiritual tourism well-being interventions. *Tourism Management Perspectives*, 24, 201–207. doi:10.1016/j.tmp.2017.07.012.

O'Malley, A. L., & Williams, D. E. (2012). Emerging Leaders: The Roles of Flourishing and Religiosity in Millennials' Leadership Development Activity. *Journal of Spirituality, Leadership and Management*, 6(1), 48–€58.

Ouellette, P., Kaplan, R., & Kaplan, S. (2005). The monastery as a restorative environment. *Journal of Environmental Psychology*, 25(2), 175–188. doi:10.1016/j.jenvp.2005.06.001.

Pew Research Center (2019). *Gen Z, Millennials and Gen X outvoted older generations in 2018 midterms*. Retrieved from: https://www.pewresearch.org/fact-tank/2019/05/29/gen-z-millennials-and-gen-x-outvoted-older-generations-in-2018-midterms/

Rahmani, Z., & Carr, A. (forthcoming). Holistic Tourism. In D. Buhalis (Ed.), *Encyclopedia of Tourism Management and Marketing*. UK: Edward Elgar Publishing Limited.

Reisinger, Y., & Steiner, C. J. (2006). Reconceptualizing object authenticity. *Annals of Tourism Research*, 33(1), 65–86. doi:10.1016/j.annals.2005.04.003.

Rocha, G., Seabra, C., Silva, C., & Abrantes, J. L. (2016). Holistic Tourism: Motivations, Self-Image and Satisfaction. *Journal of Tourism Research & Hospitality*, S2-004, 1–8. doi:10.4172/2324-8807.S2-004.

Ryan, R. M., & Deci, E. L. (2001). On happiness and human potentials a review of research on hedonic and eudaimonic well-being. *Annu. Rev. Psychol.*, 52, 141–166.

Ryan, R. M., Huta, V., & Deci, E. (2008). Living well: A self-determination theory perspective on eudaimonia. *Journal of Happiness Studies*, 9(1), 139–170. doi:10.1007/s10902-006-9023-4.

Ryff, C. D. (1989). Happiness is everything, or is it? Explorations on the meaning of psychological well-being. *Journal of Personality and Social Psychology*, 57(6), 1069–1081. doi:10.1037/0022-3514.57.6.1069.

Ryff, C. D., & Keyes, C. L. M. (1995). The structure of psychological well-being revisited.pdf. *Journal of Personality and Social Psychology*, 69(4), 719–725.

Ryff, C. D., & Singer, B. (1996). Psychological well-being: meaning, measurement, *and implications for psychotherapy research. Psychother. Psychosom.*, 65(1), 14–23. doi:10.1159/000289026.

Ryff, C. D., & Singer, B. (2008). Know thyself and become what you are: A Eudaimonic approach to psychological well-being. *Journal of Happiness Studies*, 9(1), 13–39. doi:10.1007/s10902-006-9019-0.

Seligman, M. E. P. (2002). *Authentic Happiness: Using the new positive psychology to realize your potential for lasting fulfillment*. New York: Free Press.

Seligman, M. E. P. (2011). *Flourish: a visionary new understanding of happiness and well-being* (1st Free Press hardcover ed.). New York.

Sharpley, R., & Sundaram, P. (2005). Tourism: a sacred journey? The case of ashram tourism, *India. International Journal of Tourism Research*, 7(3), 161–171. doi:10.1002/jtr.522.

Sibi, P. S. & Abraham, S. (2017). Wellness Tourism - An Analysis of Tourist Preferences and Perceptions in Puducherry.pdf. *Journal of Hospitality Application & Research*, 12(2), 61–79.

Smith, M. K. (2013). Transforming quality of life through wellness tourism. In Y. Reisinger (Ed.), *Transformational Tourism: Tourist Perspectives* (pp. 55–67). London, UK: CAB International.

Smith, M. K., & Diekmann, A. (2017). Tourism and wellbeing. *Annals of Tourism Research*, 66, 1–13. doi:10.1016/j.annals.2017.05.006.

Smith, M. K., & Kelly, C. (2006). Holistic Tourism: Journeys of the Self? *Tourism Recreation Research*, 31(1), 15–24. doi:10.1080/02508281.2006.11081243.

Smith, M. K., & Puczkó, L. (2009). *Health and Wellness Tourism*. Oxford, UK: Elsevier Ltd.

Steger, M. F., Shin, J.-Y., Shim, Y., & Fitch-Martin, A. (2013). Is meaning in life a flagship indicator of well-being? In A. Waterman (Ed.), *Eudaimonia* (pp. 159–182). Washington, DC: APA Press.

Steiner, C. J., & Reisinger, Y. (2006). Ringing the Fourfold: A Philosophical Framework for Thinking about Wellness Tourism. *Tourism Recreation Research*, 31(1), 5–14. doi:10.1080/02508281.2006.11081242.

Suar, D., Jha, A. K., Das, S. S., Alat, P., & Patnaik, P. (2020). What Do Millennials Think of Their Past, Present, and Future Happiness, and Where Does Their Happiness Reside? *Journal of Constructivist Psychology*, 1–17. doi:10.1080/10720537.2020.1805657.

Suresh, S., Ravichandran, S., & Ganesan, P. (2011). Understanding wellness center loyalty through lifestyle analysis. *Health Mark Q*, 28(1), 16–37. doi:10.1080/07359683.2011.545307.

Thal, K. (2015). *Self-Determination Theory and Wellness Tourism- How Do Wellness Facilities Contribute to Wellbeing?* (Doctor of Philosophy), University of South Carolina, United States.

Thal, K., & Hudson, S. (2016). *Fitter and healthier but happier? Assessing psychological wellbeing as the outcome of a wellness vacation using Self-Determination Theory.* Paper presented at the *Travel and Tourism Research Association: Advancing Tourism Research Globally.*

Thal, K., & Hudson, S. (2019). A conceptual model of wellness destination characteristics that contribute to psychological well-being. *Journal of Hospitality & Tourism Research*, 43(1), 41–57. doi:10.1177/1096348017704498.

Tiberius, V., & Hall, A. (2010). Normative theory and psychological research: Hedonism, eudaimonism, and why it matters. *The Journal of Positive Psychology*, 5(3), 212–225. doi:10.1080/17439761003790971.

Vittersø, J., Overwien, P., & Martinsen, E. (2009). Pleasure and interest are differentially affected by replaying versus analyzing a happy life moment. *The Journal of Positive Psychology*, 4(1), 14–20. doi:10.1080/17439760802060602.

Vittersø, J., Vittersø, J., Oelmann, H. I., Oelmann, H. I., Wang, A. L., & Wang, A. L. (2009). Life Satisfaction is not a Balanced Estimator of the Good Life: Evidence from Reaction Time Measures and Self-Reported Emotions. *Journal of Happiness Studies*, 10(1), 1–17. doi:10.1007/s10902-007-9058-1.

Voigt, C. (2010). *Understanding wellness tourism: an analysis of benefits sought, health-promoting behaviours and positive psychological well-being.* Thesis (PhD)--University of South Australia, 2010, Retrieved from https://find.library.unisa.edu.au/primo-explore/fulldisplay?vid=UNISA&search_scope=All_Resources&docid=UNISA_ALMA11146634470001831.

Voigt, C. (2017). Employing hedonia and eudaimonia to explore differences between three groups of wellness tourists on the experiential, the motivational and the global level. In S. Filep, J. Laing, & M. Csikszentmihalyi (Eds.), *Positive Tourism* (pp. 105–121). New York: Routledge.

Voigt, C., Hall, C. M., Brown, G., & Howat, G. (2011). Wellness tourists: in search of transformation. *Tourism Review*, 66(1/2), 16–30. doi:10.1108/16605371111127206.

Voigt, C., Howat, G., & Brown, G. (2010). Hedonic and Eudaimonic Experiences among Wellness Tourists: An exploratory enquiry. *Annals of Leisure Research*, 13(3), 541–562. doi:10.1080/11745398.2010.9686862.

Voigt, C., Laing, J., Wray, M., Brown, G., Howat, G., Weiler, B., & Trembath, R. (2010). *Health Tourism in Australia: Supply, Demand and Opportunities.* Retrieved from Australia.

Waterman, A. S. (1993). Two conceptions of happiness: Contrasts of personal expressiveness (eudaimonia) and hedonic enjoyment. *Journal of Personality and Social Psychology*, 64(4), 678–691. doi:10.1037/0022-3514.64.4.678.

Waterman, A. S. (2008). Reconsidering happiness: a eudaimonist's perspective. *The Journal of Positive Psychology*, 3(4), 234–252. doi:10.1080/17439760802303002.

Winn, G. L., & Dykes, A. C. (2019). Identifying toxic leadership and building worker resilience. *Professional Safety*, 64(3), 38–45.

Zainal Badri, S. K., Mun, C. T. Y., & Ramos, H. M. (2020). Work passion and psychological wellbeing among millennial employees in Malaysia: The moderating role of personality. *International Journal of Employment Studies*, 28(1), 28–47.

# 10 Millennials and social media marketing

## The case of Indian UNESCO world cultural heritage sites

*Raymond Rastegar and Zohreh (Zara) Zarezadeh*

### 10.1 Introduction

Previous research suggests that UNESCO WHS label or branding can lead to an increase in tourist flows (Landorf, 2009; Rastegar, Zarezadeh, & Gretzel, 2021), and support for sustainable destination development (Baral, Kaul, Heinen, & Ale, 2017; Kaltenborn, Thomassen, Wold, Linnell, & Skar, 2013; King & Halpenny, 2014). Many of these heritage sites are known to play an important role in attracting travellers interested in local culture, religion and spirituality. To use this opportunity, heritage tourism marketers and destination managers actively engage in different activities to promote their attractions. With the significant advancement of information technology, social media provides unique opportunities for tourism marketers to interact and build relationships with existing and potential travellers (Cabiddu, Carlo, & Piccoli, 2014). This has created many opportunities for tourism business involved in heritage tourism (Surugiu & Surugiu, 2015). Social media marketing in tourism is a powerful marketing approach for reaching a large number of travellers and greatly influencing their perceptions and travel intentions. This is particularly important as social media is one of the best ways to reach the Millennial audience. Millennials have interacted with technology since birth and are much more digitally literate than previous generations. They are natural and active users of social media (Zhang et al., 2017). Therefore, tourism marketers in many countries perceive social media marketing as an effective approach to communicate with Millennials (Bolton et al., 2013).

With 30 unique world cultural heritage sites inscribed on the UNESCO World Heritage list (UNESCO, 2020), cultural and spiritual tourism is the primary focus of India's efforts to attract travellers. At the same time, research shows that the majority of young international and domestic travellers use social media platforms to search for information to plan their trips in India (Rathore, Joshi, & Ilavarasan, 2017). Therefore, it is crucial for Indian WHSs to understand the effective social media marketing strategies to engage with Millennials. Consequently, to gauge these heritage attractions' ability to

DOI: 10.4324/9781003170631-13

engage successfully with tourists, and particularly Millennials, over social media, this study will evaluate their social media presence and engagement on Facebook and TripAdvisor, which are, in global terms, the most widely used platforms and have a particular significance within the context of tourism. The theoretical and practical implications of the study will discuss approaches for the effective use of social media marketing to promote cultural attractions and successfully attract travellers with special emphasis on Millennials.

## 10.2 Literature review

### *10.2.1 Social media marketing and millennials*

Social media provides new paths for creating awareness and promoting tourism destinations and attractions. The rapid development of social media and its use by both travellers and the tourism industry has resulted in phenomenal changes in the field of tourism (Xiang & Gretzel, 2010). The use of social media in tourism will continue to grow rapidly and social media marketing will continue to develop as a significant marketing instrument (Hays, Page, & Buhalis, 2013). It is particularly important for the Millennial Generation or digital natives as they were born into the world of technology, qualified in digital knowledge and their relationships are formed on social media sites and apps (Llopis-Amorós, Gil-Saura, Ruiz-Molina, & Fuentes-Blasco, 2019). The generation is identified as an important consumer group as they are roughly a third of world's total population (Duffett & Wakeham, 2016). Similarly, the preference, communication styles and behaviours of Millennials can greatly influence the future of tourism marketing. For example, Millennials are found to pay more attention to the information they receive from friends or other consumers rather than information from companies or traditional media (Smith, 2012). It is as a high degree of travellers' reliance on other travellers' experiences is considered as a tool to boost exchange utility and reduce uncertainty (Yoo, Lee, & Gretzel, 2007). Similarly, Duffett Rodney (2015) found that content on social media platforms such as Facebook has a positive influence on intention to purchase among Millennials. Therefore, it is not surprising to see managers interests to understand Millennials social media usage, as it can help them predict their perceptions and behaviours (Bolton et al., 2013).

According to Kaplan and Haenlein (2010), there are six types of social media based on the degree of social presence and self-presentation. Their categories of social media comprise blogs, social networking sites, virtual social worlds, collaborative projects, content communities, and virtual game worlds. The various social media types offer a unique opportunity in terms of type of content to share and ways of sharing which engage different users (Gretzel & Yoo, 2013; Zarezadeh, Rastegar, & Gretzel, 2018). Therefore, it offers various options for marketers in terms of content and the social media platform used. However, Llopis-Amorós et al. (2019), in an investigation of social media communication for Millennials, argue that managers and researchers might have limited knowledge as how to influence the perception and behaviour of consumers. Providing relevant and up-to-date content on social media platforms encourages travellers to link to the content through their social media platforms, which increases the visibility of a social media page/account. Particularly in the case of Millennials, social media should facilitate social interaction, involvement and co-creation of the experiences (Skinner, Sarpong, & White, 2018). As a result, presenting interesting, relevant and updated content is one of the more decisive strategies in social media marketing (Erdoğmuş & Çiçek, 2012).

The main difference between social media marketing and traditional marketing is the role of consumers in co-creating the marketing content (Evans, 2008). Similarly, Millennials believe social media to be a more reliable source of information compared to traditional marketing communication methods (Christodoulides, Jevons, & Bonhomme, 2012; Llopis-Amorós et al., 2019). Therefore, building relationships with travellers is a fundamental strategy in social media marketing (Barefoot & Szabo, 2010). In order to build a relationship, social media marketers need to be able to engage travellers in conversation and also to shape the conversation. As travellers are active partners who co-create value with marketers in the social media conversation (Mangold & Faulds, 2009), social media marketers must apply strategic approaches to influence and manage the conversations (Mangold & Faulds, 2009). However, it is also argued that social media marketers cannot/should not control the conversation but rather, manage it by influencing the conversation (Gretzel & Yoo, 2013).

To investigate the issue of customer engagement using social platforms, researchers have studied the applications of different theories. For instance, the Diffusion of Innovation theory was applied to show compatibility and ease of use as key factors for the adoption of technology by users (Ahuja & Shakeel, 2017; Folorunso, Vincent, Adekoya, & Ogunde, 2010). Applications of the Uses and Gratifications theory also showed social platforms fulfilling the emotional needs of the consumers (Raacke & Bonds-Raacke, 2008). It is also discussed that theories of relationship marketing outline the use of technology to build relationships with customers (Sheth & Parvatiyar, 1995). Considering the application of these theories in investigating consumer engagement, Ahuja and Shakeel (2017) argue that, in the field of digital marketing, customer engagement requires the active use of online social platforms to share meaningful information and allow individuals' participation. Customer engagement in the marketing process also requires innovative approaches to provide added value by meeting individuals' emotional needs. For example, it is suggested that effective DMO websites should be designed in such a way as to engage customers based on their varying needs (Choi, Lehto, & Oleary, 2007). It is also argued that engaging customers by meeting their needs based on local knowledge can increase website performance (So & Morrison, 2004) and consequently increase visitor satisfaction.

### *10.2.2 Cultural heritage and social media*

Despite the strong influence of social media on travellers' decision-making, there is still a limited effort, particularly in developing countries, to adopt social media marketing. Dwivedi, Yadav, and Venkatesh (2012) investigated the use of social media by National Tourism Organizations (NTOs) in 195 countries and found that almost 59% of the countries did not rely on any popular social media platforms to promote their destinations. They concluded that these destinations had provided limited information to official NTOs or the embassy websites. Their finding showed that countries such as India and Nepal were present on only one platform. The authors argued that the reason might be that the NTOs had the impression that being present on several platforms may not have any advantages. Bethapudi (2013) also identified issues such as lack of attention to social media marketing, top-level management not being aware of updates, the majority of tourism websites not being mobile-ready, and the lack of Information and Communication Technologies (ICT) skills as important challenges for the Indian tourism industry. Similarly, Rathore et al. (2017) found that in popular tourism destinations such as Rajasthan, the department of tourism was present only on Facebook and Twitter with minimum user engagement and out-of-date information. The authors maintained that one reason might be the active presence of other stakeholders on social media to promote the destination. The above finding can be supported by the outcomes of the social media marketing trends in India by Shah and Jani, Johnson, Banu, and Shah (2016), who contended that, despite the increasing interest by most brands in investing in digital marketing platforms, the greatest challenge is the customer engagement on the online platforms.

The emerging ICT has created many opportunities for tourism business involved in heritage tourism (Surugiu & Surugiu, 2015). Social media and digital marketing are playing key roles in digital cultural heritage communication (Stuedahl, 2009). It is because social media platforms can be used as an instrument to raise awareness of cultural heritage status and promote the cultural heritage (Gaitan, 2014). Digital marketing allows the DMOs to reach existing and potential consumers by creating a platform for communication. For example, when studying the representation of the intangible cultural heritage of Bangladesh, Khalid and Chowdhury (2018) discuss the roles social media can play in preserving and promoting cultural heritage by increasing awareness and creating knowledge. However, their analysis showed that, in this case, social media had failed to play its role due to the lack of strategic actions to engage customers by the responsible organizations. They suggest the strategic development of media content to build a space for communication which can result in disseminating cultural heritage to travellers. Media content and the type of message and promotion of social media is effective only if it encourages the curiosity of the social media users to visit the cultural heritage site. Sharron and Abraham (2015), in studying the role of curiosity in the use of digital content promoting cultural heritage, found that to encourage potential travellers to visit a cultural heritage site, the social media content should be created in a way that comprises the message of communicating and unveiling the inherent cultural heritage. Implementing social media marketing can help by using different platforms to send the message based on the chosen communication strategies. Sending such a message through different media will be effective in engaging social media users by encouraging them to visit the cultural heritage and to find further information.

Social media marketing is all about using social media platforms to promote a brand and/or its products (Barefoot & Szabo, 2010). In the tourism industry now, social media has provided the unique opportunity to engage travellers in creating marketing content. In the case of marketing WHSs and unique cultural heritage, this opportunity can be used to establish a two-way communication to build relationships with potential and existing visitors, create a space for conversation, raise visitor awareness, and promote and preserve the cultural attraction. This is particularly important in the case of Millennials as such approaches can facilitate social interaction, involvement and co-creation of the experiences.

## 10.3 Study context

With more than 426 million (or approximately 34% of its total population), India has one of the largest Millennial populations in the world (Outlook India, 2020). At the same time, India, as a vast and culturally diverse country, is a popular tourism destination for many travellers. According to the Ministry of Tourism, India received more than 10 million international tourists in 2017, representing a 15.6% annual growth rate from 2016 and a 1.17% share of the world's international tourist arrivals. However, in the same year, the number of domestic travellers only recorded a 2.3% annual growth rate, reaching about 1652 million visits in the same year. In 2017, Bangladesh, with 21.5%, was the first source country for international arrivals to India, followed by the United States (13.7%) and the United Kingdom (9.8%) (India Ministry of Tourism, 2018). According to a report by the World Travel and Tourism Council, tourism contributed US$234.0 billion to GDP in India in 2017 and this is forecast to rise by 6.9% p.a. to reach US$492.2 billion in 2028 (World Travel and Tourism Council, 2018).

Recognising the potential to become a modern tourism destination and using unique attractions such as cultural heritage, the Government of India has invested in tourism development in the last 20 years. For example, the government launched its Incredible India Campaign (IIC) in 2002, which is viewed as responsible for increasing the number of international arrivals to India from 2.65 million in 2000 to 5.38 million in 2008 (Kant, 2009). Acknowledging the challenges in branding a diverse nation in a fast-developing country, the IIC targeted both domestic and international travellers (Kerrigan, Shivanandan, & Hede, 2012).

With the aim to facilitate the means of spreading information in their marketing campaign in 2012, the government's department of Information and Communication Technology formulated a framework to encourage the use of social media. Recognising the great influence of social media on international visitors, the guidelines aim to enable different agencies to create strategies to use social media (Indian Institute of Tourism and Travel Management, 2017). Such efforts indicate the increasing interest and popularity of digital marketing (Shah & Jani, 2016) and the use of social media by tourists. For example, a study showed that 86% of both international and domestic travellers have used social media platforms to search for information to plan their trip to Rajasthan, India (Rathore et al., 2017). Similarly, another research on social media and its influence on international tourists visiting India found that more than 89% of tourists used social media to explore information about the destinations in the country (Indian Institute of Tourism and Travel Management, 2017). The same study reports that TripAdvisor and Facebook were the two most popular platforms used by tourists during their decision-making process to travel to India. It is argued that despite the presence of such unique resources, the heritage branding of India is still under-explored (Chhabra, 2017).

Despite acknowledging the significance of social media and digital marketing in tourism development, very little is known about social media marketing by Indian WHSs. Therefore, an investigation of the representation of Indian WHSs can provide a valuable insight regarding the international competitiveness of cultural tourism in India.

## 10.4 Methodology

To measure the ability of Indian WHSs to attract Millennials, an analysis of their presence on TripAdvisor and Facebook was conducted. TripAdvisor and Facebook platforms are considered to be the two main social media channels used by Millennials when deciding to visit cultural and heritage sites (social media as an influencer among Millennials). While a framework and guidelines to the use of social media as an interaction tool with travellers have been developed by the Department of Electronic and Information Technology at the Ministry of Communication and Information Technology in 2012, adopting social media as a marketing tool among government agencies was still novel. Therefore, it is crucial to gauge how and to what extent, Indian WHSs use social media as their marketing tool.

India with 30 cultural, seven natural, and one mixed WHSs stands as the country with the second largest number of WHSs after China in the Asia and Pacific region. The large number of UNESCO cultural WHSs in India plays a significant role in its tourism marketing. Therefore, the analysis examined data collected from 30 Indian cultural WHSs' TripAdvisor accounts and Facebook pages during October and November 2020. The cultural Heritage Sites were listed on UNESCO World Heritage between 1983 and 2019. The main management authority of the WHSs is the Indian Ministry of Culture.

The Indian Ministry of Tourism initially proposed "an adoption scheme" for WHSs in 2017, in collaboration with the Ministry of Culture and the Archaeological Survey of India (Bhargava, 2018). According to this scheme, the government invited companies and individuals to partner in order to preserve Indian Heritage by providing basic amenities including cleanliness, public conveniences, safe drinking water, and easy access for tourists (The Times of India, 2018). Under the new scheme, the ownership of the Indian WHSs has not changed as the Indian government is still the owner of WHSs. Following this scheme, only the management of some of the Indian WHSs that changed under "an adoption scheme" for a certain period of time. The scheme aims to increase visitor attractiveness of the sites and promote Indian culture and heritage (Binayak, 2018).

The analysis focused on two main objectives: 1) identifying whether Indian WHSs managed the pages and/or accounts on social media, 2) whether the pages/ accounts engaged by users could serve as marketing tools. We commenced our data analysis by looking at the India Ministry of Culture's website. The WHSs icon was in the home page of the website and both tangible and intangible World Heritage were listed. Background information and the history of WHSs was available by a name and also a map location of WHSs was provided. Indian WHSs' history was the only information provided on the website with no link to any social media platforms or official websites of the monuments being provided.

### *10.4.1 Sample description*

The Indian cultural WHSs have been inscribed by UNESCO since 1983, and are located across 14 different states. Maharashtra, with five World Cultural Heritage Sites, has the largest number of WHSs in India. There are different sizes and types of heritage sites, ranging from a temple (such as the Sun temple in Konark) to an entire city, such as the historical city of Ahmadabad. Most heritage sites are accessible for visitors through road, air or train (see Table 10.1).

## 10.5 Results

### *10.5.1 Facebook result*

The analysis of Facebook (see Table 10.2) shows that all but two of the Indian WHSs have pages. Twelve WHSs have more than one page, either with the same name such as *Mahabodhi Temple* and *Fatehpur Sikri*, or with a similar name, such as *The Taj Mahal* and *Taj Mahal Agra*. The results also show that 19 WHSs have Facebook pages with recent posts, including the Jantar Mantar, Konark Sun Temple, and Great Living Chola Temple. Except for one page which are completely in Hindi and eleven pages which are in both the Hindi and English language, the rest are in English. Twenty-three Facebook pages share visitors' ratings and almost all of these are higher than

*Table 10.1* Indian heritage sites description and location

| *Heritage Sites* | *Year of UNESCO Inscription* | *State* | *Type* |
|---|---|---|---|
| Agra Fort | 1983 | Uttar Pradesh | A fort |
| Ajanta Caves | 1983 | Maharashtra | 30 caves |
| Archaeological Site of Nalanda Mahavihara | 2016 | Bihar | A monastery |
| Buddhist Monuments | 1989 | Madhya Pradesh | A group of monuments |
| Champaner-Pavagadh Archaeological Park | 2004 | Gujarat | 11 different types of buildings |
| Chhatrapati Shivaji Terminus (formerly Victoria Terminus) | 2004 | Maharashtra | A rail station |
| Churches and Convents of Goa | 1986 | Goa | Seven churches |
| Elephanta Caves | 1987 | Maharashtra | A collection of cave temples |
| Ellora Caves | 1983 | Maharashtra | Over 100 caves at the site, 34 caves open to public |
| Fatehpur Sikri | 1986 | Uttar Pradesh | A town in the Agra district |
| Great Living Chola Temples | 1987, 2004 | Tamil Nadu | The monument includes the Brihadisvara temple at Thanjavur which was recognized as a WHS in 1987, the temple of Gangaikonda Cholapuram and the Airavatesvara temple at Darasurm were added as extension to the site in 2004 |
| Group of Monuments at Hampi | 1986 | Karnataka | A village |
| Group of Monuments at Mahabalipuram | 1984 | Tamil Nadu | Including 40 sanctuaries |
| Group of Monuments at Pattadakal | 1987 | Karnataka | Includes 9 Hindu temples and a Jain sanctuary |
| Hill Forts | 2013 | Rajasthan | Six majestic forts: Chittor fort/Kumbhalgarh fort/ Ranthambore fort/ Gagron fort/ Amer fort/ Jaisalmer fort |

(*Continued*)

*Table 10.1* (Continued)

| *Heritage Sites* | *Year of UNESCO Inscription* | *State* | *Type* |
|---|---|---|---|
| Historical City of Ahmedabad | 2017 | Gujarat | A city |
| Humayun's Tomb | 1993 | Delhi | A tomb |
| Jaipur City | 2019 | Rajasthan | A city |
| Khajuraho Group of Monuments | 1986 | Madhya Pradesh | About 20 temples remain |
| Mahabodhi Temple Complex | 2002 | Bihar | A complex |
| Mountain Railways of India | 1999, 2005, 2008 | West Bengal, Tamil Nadu, Himachal Pradesh | Darjeeling Himalayan Railway was the first to inscribe by UNESCO |
| Qutb Minar and its Monuments | 1993 | Delhi | A complex |
| Rani-ki-Vav (the Queen's Stepwell) | 2014 | Gujarat | Renowned for its stepped corridors, sculptures and stone carvings |
| Red Fort Complex | 2007 | Delhi | A complex was built as the palace |
| Rock Shelters of Bhimbetka | 2003 | Madhya Pradesh | Comprise five clusters of natural rock shelters |
| Sun Temple, Konarak | 1984 | Odisha | A temple |
| Taj Mahal | 1983 | Uttar Pradesh | |
| The Architectural Work of Le Corbusier, an Outstanding Contribution to the Modern Movement | 2016 | Chandigarh | A complex |
| The Jantar Mantar | 2010 | Rajasthan | A collection of 19 architectural astronomical instruments |
| Victorian Gothic and Art Deco Ensembles of Mumbai | 2018 | Maharashtra | A collection of 19th-century Victorian neo-Gothic |

four out of five. The ratings in the majority of pages are based on the reviews from more than 100 people who rate the monument post after their visits.

Another finding from the analysis shows that none of the heritage sites and monuments use Facebook for marketing purposes. While more than 80% of WHSs have Facebook pages, the majority of the pages are community pages rather than official pages. Even the famous monuments such as *Taj Mahal* and *Agra Fort* are managed by the community. Some monuments have unofficial pages such as *Historical city of Ahmadabad* and *Mahabodhi Temple*,

*Table 10.2* Facebook analysis of Indian WHSs

| *Heritage Site* | *Likes* | *Followers* | *Language* | *Type of Page* | *Posts* |
|---|---|---|---|---|---|
| Agra Fort | 476 | 489 | English | Community | August 192015 |
| Ajanta Caves | 5807 | 5790 | English | Community | Up-to-date |
| Archaeological Site of Nalanda Mahavihara | No Facebook page, only photos hashtag by travellers | | | | |
| Buddhist Monuments | 73 | 73 | English/ Hindi | City | Up-to-date |
| Champaner-Pavagadh Archaeological Park | 1654 | 1653 | English | City | Up-to-date |
| | Pavagadh Temple 75,270 | 78,615 | English/ Hindi | Community | Up-to-date |
| Chhatrapati Shivaji Terminus (formerly Victoria Terminus) | Chhatrapati Shivaji Terminus 14,700 | 14,650 | English/ Hindi | Community | Up-to-date |
| Churches and Convents of Goa | Basilica od Bom Jesus 7483 | 7460 | English | City | Up-to-date |
| | Churches and Convents of Goa 654 | 658 | English | | Up-to-date |
| Elephanta Caves | 1011 | 1022 | English | Community | May 5, 2018 |
| | Elephanta caves 10,074 | 10,052 | English | Community | Up-to-date |
| Ellora Caves | 389 | 394 | English | Community | August 7, 2017 |
| | The Ajantha and Ellora Caves 956 | 959 | English | Community | May 8, 2015 |
| | The Ellora and Ajanta Caves 3703 | 3723 | English | Community | March 9, 2012 |
| Fatehpur Sikri | Fatehpur Sikri Dargah 1616 | 1642 | English | Community | March 24, 2017 |
| Great Living Chola Temples | 1370 | 1368 | English | | Up-to-date |
| Group of Monuments at Hampi | 115 | 115 | English | Community | January 31, 2020 |

| | | | | | |
|---|---|---|---|---|---|
| Group of Monuments at Mahabalipuram | 233 | 232 | English | | Up-to-date |
| Group of Monuments at Pattadakal | No face book page | | | | |
| Hill Forts | Chittorgarh fort 4326 | 4420 | English | Community | Up-to-date |
| | Chittorgarh Fort 2618 | 2673 | English/ Hindi | Community | Up-to-date |
| Historical City of Ahmedabad | 41 | 41 | English/ Hindi | Unofficial page | Up-to-date |
| Humayun's Tomb | Humayun's Tomb-Nizamuddin Basti Urbane Renewal Initiative 32,426 | 32,512 | English | | Up-to-date |
| | Homayun's Tomb 20,679 | 20,608 | English | | Up-to-date |
| Jaipur City | Jaipur- Beautiful Pink City 23,563 | 23,691 | English | | December 19, 2017 |
| Khajuraho Group of Monuments | Khajuraho Temple7627 | 7724 | English/ Hindi | Community | January 23,2018 |
| | Glimpses of Khajuraho – The Temple of Love 1034 | 1049 | mostly Hindi English | Community | November 7, 2017 |
| Mahabodhi Temple Complex | Mahabodhi Temple 6200 | 6182 | English | Unofficial page | Up-to-date |
| | Mahabodhi Temple 753 | 777 | English | Community | Up-to-date |
| Mountain Railways of India | 227 | 227 | English | Unofficial page | August 12, 2016, only few personal photos |

(*Continued*)

*Table 10.2* (Continued)

| *Heritage Site* | *Likes* | *Followers* | *Language* | *Type of Page* | *Posts* |
|---|---|---|---|---|---|
| Qutb Minar and its Monuments | 33<br>Qutb Minar<br>31,987 | 33<br>31,895 | English/<br>Hindi<br>English | Community | Up-to-date<br>Up-to-date |
| Rani-ki-Vav (the Queen's Stepwell) | Rani Ki vav<br>2295<br>Rani Ki Vav<br>1348<br>Rani Ki Vav<br>6262 | 2334<br>1414<br>6259 | English/<br>Hindi<br>English/<br>Hindi<br>English/<br>Hindi | official page | Up-to-date<br>Up-to-date<br>Up-to-date |
| Red Fort Complex | Red Fort<br>31,195 | 30,952 | English | | Up-to-date |
| Rock Shelters of Bhimbetka | Bhimbetka rock shelters 1285 | 1284 | English/<br>Hindi | | Up-to-date |
| Sun Temple, Konarak | Konark Sun temple<br>5166<br>Konark Sun Temple Visit<br>5414 | 5325<br>5487 | English<br>English | Community | January 28, 2018<br>Only two posts<br>Up-to-date |
| Taj Mahal | The Taj Mahal<br>903,652<br>Taj Mahal Agra<br>14,449 | 893,590<br>14,549 | English<br>English/<br>Hindi | Community<br>Community | November 23,<br>2019<br>Up-to-date |
| The Architectural Work of Le Corbusier, an Outstanding Contribution to the Modernist Movement | No Facebook Page | | | | |
| The Jantar Mantar | Jantar Mantar, Jaipur 13,716 | 13,451 | English | Unofficial page | Up-to-date |
| Victorian Gothic and Art Deco Ensembles of Mumbai | Victorian Gothic and Art Deco Ensembles of Mumbai-UNESCO<br>2 | 2 | English | Unofficial page | July 8, 2020 |

which means the pages are created because people on Facebook have shown interest in such places. Therefore, it is not affiliated with or endorsed by anyone associated with the government, marketing department or management of the WHSs.

The number of likes and followers of Facebook pages show how much the page has been successful in attracting and engaging users. The number of likes and followers for most of WHSs' Facebook pages are low. Twenty-six WHSs' Facebook pages have fewer than 5000 likes and followers, such as *Hills Forts* and *Ellora Caves*, which means only a small number of people see the posts and photos. Nine pages, including *Humayun's Tomb* and *Taj Mahal*, have more than 10,000 likes and followers, but most of the posts in these pages are 'selfie' photos or posts from other travellers, or introductory information about the monuments.

### *10.5.2 TripAdvisor analysis*

The analysis of TripAdvisor (see Table 10.3) shows that except for three monuments, *Fatehpur Sikri, Jaipur city*, and the *Historical City of Ahmedabad*, the rest of the WHSs have reviews in TripAdvisor. Although *Jaipur city* and *Historical City of Ahmedabad* had no reviews in TripAdvisor as a UNESCO-listed cultural sites, they have reviews as a city which travellers shared their experience about accommodation, transportation, attraction and activities. However, the lack of management presence is evident as none of the questions or reviews have been answered or commented on by authorized people from WHSs' management. *Taj Mahal* (with 31,602 reviews) has the highest number of reviews among WHSS on TripAdvisor, while the *Victorian Gothic* and *Art Deco Ensembles of Mumbai*, with only three reviews, has the lowest number of reviews. Except for *Great Living Chola Temples* and *Victorian Gothic* and *Art Deco Ensembles of Mumbai*— their last reviews being June 2018 and December 2019, respectively—the rest of the monuments had recent reviews.

All monuments and heritage sites have photos posted by visitors, with the exception of the *Mountain Railways of India*, which had no photos. *Taj Mahal*, with more than 29,050 photos, had the highest number of photos and the *Victorian Gothic* and *Art Deco Ensembles of Mumbai*, with just 3 photos, had the fewest. The photos had been posted by travel agents or visitors.

The quality rating for the Indian WHSs on TripAdvisor are quite high, ranging from 4 to 5 stars. *Taj Mahal, Konark Sun Temple, Mahabodhi Temple, Khajuraho Temple, Group of Monuments at Pattadakal, Group of Monuments at Hampi*, and the *Great Living Chola Temples* were the most popular places, and these received 5-star ratings. Most of the WHSs have an address, contact number, opening hours and information about the monument. Some of the Indian WHSs have more than one page in TripAdvisor, including *Mountain Railways of India*, which has three different pages based on its three stations, and *Sun Temple* and *Konark*, which has two different pages, both pages being active.

*Table 10.3* Trip Advisor analysis of Indian WHSs

| *Heritage Site* | *Reviews* | *Rating* | *Last Review* | *photos* | *Related Information* |
|---|---|---|---|---|---|
| Agra Fort | 11,894 | 4.5 | Up-to-date | 12,092 | Address/map and opening hours<br>Link to website |
| Ajanta Caves | 2419 | 4.5 | Up-to-date | 3031 | Address/map and opening hours |
| Archaeological Site of Nalanda Mahavihara | Nalanda University Ruins<br>157 | 4.5 | Up-to-date | 491 | Address/map |
| Buddhist Monuments | 822 | 4.5 | Up-to-date | 960 | Address/map/phone/link to email and website |
| Champaner-Pavagadh Archaeological Park | 100 | 4.5 | Up-to-date | 361 | Address/map and opening hours/link to the website/contact number |
| Chhatrapati Shivaji Terminus (formerly Victoria Terminus) | 4502 | 4.5 | Up-to-date | 1885 | Opening hour, address/ map/link to the website, contact number, Email |
| Churches and Convents of Goa | Old Goa<br>1260 | 4.5 | Up-to-date | 1507 | Map |
| Elephanta Caves | 1941 | 4 | Up-to-date | 3525 | Opening hours/Address/Some information about distance and transport /Map/Link to the website/ contact number |
| Ellora Caves | 2648 | 4.5 | Up-to-date | 4085 | Opening hour/ address/map |
| Fatehpur Sikri | 244 | 4.5 | Up-to-date | 739 | Map/link to the website |
| Great Living Chola Temples | 12 | 5 | Last post June 2018 | 28 | -- |
| Group of Monuments at Hampi | 592 | 5 | Up-to-date | 2771 | Address/map |
| Group of Monuments at Mahabalipuram | 1680 | 4.5 | Up-to-date | 1956 | Address/opening hour/ map |
| Group of Monuments at Pattadakal | 199 | 5 | Up-to-date | 973 | Opening hours/address/ map/Link to the website/ email/contact number |

| | | | | | |
|---|---|---|---|---|---|
| Hill Forts | Amber fort 18,101 | 4.5 | Up-to-date | 18,919 | Map/Address |
| | Kumbhalgarh fort 1910 | 4.5 | Up-to-date | 2767 | Link to the website/map/address |
| Historical City of Ahmedabad | No page | | | | |
| Humayun's Tomb | 10,436 | 4.5 | Up-to-date | 8060 | Opening hours/Address/ map/ link to the website/contact number/email |
| Jaipur City | No page | | | | |
| Khajuraho Group of Monuments | Khajuraho temples 2684 | 5 | Up-to-date | 3780 | Address/map |
| Mahabodhi Temple Complex | 996 | 5 | Up-to-date | 963 | Opening hours/ Address/ map |
| Mountain Railways of India | Nilgiri Mountain Railway 1584<br>Kalka-shimla Railway 1787<br>Darjeeling Himalayan railway 864 | 4.5<br>4.5<br>4 | Up-to-date<br>Up-to-date<br>Up-to-date | No Photo<br>No Photo<br>No Photo | Open hour/ link to the website/contact number/Email<br>Address/map/link to the website<br>Address/ map/ link to the website/ contact number/Email |
| Qutb Minar and its Monuments | 13,155 | 4.5 | Up-to-date | 11,270 | Address/ map |
| Rani-ki-Vav (the Queen's Stepwell) | 346 | 4.5 | Up-to-date | 947 | Address/map/link to the website/ contact number |
| Red Fort Complex | 5491 | 4 | Up-to-date | 5211 | Opening hours/address/map/link to the website |
| Rock Shelters of Bhimbetka | 639 | 4.5 | Up-to-date | 931 | Address/map/link to the website/ contact number |
| Sun Temple, Konarak | Konark sun temple 1169 | 4.5 | Up-to-date | 2735 | Opening hours/Address/map/link to the website/contact number/Email |
| | Konark sun temple 42 | 5 | Up-to-date | 209 | Map |

(*Continued*)

*Table 10.3* (Continued)

| *Heritage Site* | *Reviews* | *Rating* | *Last Review* | *photos* | *Related Information* |
|---|---|---|---|---|---|
| Taj Mahal | 31,602 | 5 | Up-to-date | 29,050 | Address/Map/Link to the website/ contact number/Email |
| The Architectural Work of Le Corbusier, an Outstanding Contribution to the Modern Movement | No page… | | | | |
| The Jantar Mantar | 7511 | 4.5 | Up-to-date | 5020 | Address/map/contact number/Email |
| Victorian Gothic and Art Deco Ensembles of Mumbai | Victorian Gothic and Art Deco Ensembles of Mumbai – UNESCO 3 | 4 | December 2019 | 3 | Address/ website link of UNESCO |

## 10.6 Discussion and conclusion

The findings of this study show that the adoption of social media marketing is often a challenge for smaller tourism venues in developing countries such as India. Developing countries usually face limited resources, little infrastructure, low education levels, poor levels of technological literacy, and a lack of proactive government policies to adopt new technologies. India is a country with unique heritage and cultural attractions. However, our findings show that WHS marketing in India faces challenges such as limited use of social media by tourism bodies, minimum user engagement, lack of skills and out-of-date information.

Indian WHSs are managed by the government and, in some cases, have been adopted by individuals or corporations. In both cases, the heritage sites' management focus mostly on preservation and maintenance of the monuments rather than marketing to attract more tourists. This circumstance is reflected on social media platforms where there is a notable absence of government and also managers of WHSs. Cases where individuals have created, and managed unprofessional social media pages have resulted in drawing a rather gloomy picture of Indian WHSs for travellers searching for information online. The issue appeared in both Indian WHSs; those managed by the government and those adopted by individual or corporate bodies. Considering that travellers come across social media pages during their information search (Xiang & Gretzel, 2010), and their strong influence on their decision making (Osei, Mensah and Amenumey, 2018; Rathore et al., 2017), the unprofessional and confusing information on these pages can be highly problematic. This is even more significant in the case of Millennials, as they mainly use social media platforms for their information search. This fact is supported by the literature showing the majority of international travellers to India use social media platforms to search for information and decision-making (Indian Institute of Tourism and Travel Management, 2017). Therefore, unprofessional pages, outdated information and a lack of engagement can negatively affect the travellers' decision especially Millennials to visit the destination.

The finding of this study showed that most of the WHSs in India are present on one or more social media platforms. However, while the government and managers of the WHSs are responsible for marketing and managing the monuments, other stakeholders, such as local people, travel agencies and tourists, create pages and share information on social media. Similarly, Rathore et al. (2017) found that the lack of government and management presence on social media in the case of WHSs in Rajasthan, India was due to the lack of marketing strategy, planning and skilled resources, which allowed other stakeholders to provide information on their behalf. Sending different messages composed by different parties can create confusion for tourists looking for information online. It can also create an image of the WHS in the traveller's mind which can be undesirable and different to the desirable image perceived by DMOs or the management system. No doubt the marketing campaigns such as IIC and initiatives by the government aim to brand the destination and promote attractions in India; however, to effectively market the attractions, there is a need for implementing integrated marketing communication (IMC) to be consistent in sending the same message to travellers to India. Such a strategy can help not only to promote the WHSs but also to create a desirable image of the whole country as a destination and also the cultural attractions.

It can also be suggested that to avoid social media users' confusion when coming across different pages as seen for some of the WHSs, the official page must be identifiable. Using a logo and words such as *Ministry* or *Official* can help the users to ensure they have reached the official page (Dwivedi et al., 2012). The social media pages should also be linked to the official website, to provide assurance for the users and also to provide ease of accessibility from one to the other. It could also be argued that the involvement of different stakeholders in the process, such as government bodies, tourism sectors and local communities, as custodians of the heritage can enhance its effectiveness (Rastegar, 2019; Rastegar et al., 2021). A better collaboration between government bodies responsible for preservation and promoting the WHSs is also critical to the process (Rastegar, 2017; Rastegar, 2020).

It is therefore for the WHSs managers to understand that the use of social media in tourism is growing quickly and information provided online is strongly affecting Millennials' perceptions and travel intentions to visit a destination or attraction. Social media as a conversation space can be monitored to ensure the effectiveness of marketing campaigns. In this conversation space, travellers are active participants who co-create value and marketing content by contributing to the discussion (Vargo & Lusch, 2004). Similarly, social media marketing is related to relationship marketing, whereby marketers need to shift from "trying to sell" to "making connections" with the travellers when promoting destinations (Gordhamer, 2009).

Therefore, for an effective communication with Millennials and other potential travellers, consumer engagement on social media platforms is a must. However, WHSs managers need to realize that Millennials' engagement goes beyond just 'liking' or 'favouriting' a post on social media platforms. Millennials are interested in the information they receive from other consumers on online platforms (Llopis-Amorós et al., 2019). It therefore requires active presence and involvement of the management in the conversation and providing information in more interactive ways (Narangajavana, Callarisa Fiol, Moliner Tena, Rodríguez Artola, & Sánchez García, 2017). For example, the travellers must feel valued by receiving replies to their comments. Similarly, seeing such conversation by other travellers can positively affect their perception. It also sends a positive message to the travellers that the management system is happy to hear and make the necessary changes to accommodate their needs. Findings in the literature support this argument that content of social media platforms significantly influence Millennials intentions to purchase a product (Duffett Rodney, 2015).

Finally, information is shared, conversation is happening, and words spread rapidly on social media. Therefore, it is critical for the manager and marketer to ensure that they try to influence, rather than control, the conversation (Gretzel & Yoo, 2014). This suggests that managers can take an even more active role not only to observe the conversation but also to manage it. This can be done by ensuring all communication channels such as different social media platforms are used to promote the WHSs. To achieve the above goal, an effective two-way communication is critical. Therefore, in the first step, communication strategies have to be developed to deliver the opportunity for the dialogue to take place so that up-to-date information can be provided to travellers and feedback can be received.

### *10.6.1 Limitations and future directions*

Despite the significant findings related to social media marketing and Millennials, the results of this study should be considered in the light of some limitations which constitute potential lines of research for the future. This study is based on the available data collected from 30 Indian cultural WHSs on two popular social media platforms, TripAdvisor accounts and Facebook pages, during October and November 2020. However, we believe that this study opens new avenues for future research. Future research is suggested to expand the scope of investigation by alternative data collection methods and to study the presence of Indian WHSs on other social media platforms and their visitor engagement. Longitudinal methods are also suggested for future research to examine heritage marketing and the approaches in engaging Millennials in different contexts which will benefit academic and practitioners interested in heritage site marketing.

## References

Ahuja, V., & Shakeel, M. (2017). Twitter presence of jet airways-deriving customer insights using netnography and wordclouds. *Procedia Computer Science*, 122, 17–24.

Baral, N., Kaul, S., Heinen, J. T., & Ale, S. B. (2017). Estimating the value of the world heritage site designation: A case study from Sagarmatha (Mount Everest) National Park, Nepal. *Journal of Sustainable Tourism*, 25(12), 1776–1791.

Barefoot, D., & Szabo, J. (2010). *Friends with benefits: A social media marketing handbook*. San Francisco, CA: No Starch Press.

Bethapudi, A. (2013). The Role of ICT in Tourism Industry. *Journal of Applied Economics and Business*, 1(4), 67–79.

Bhargava, Y. (2018). What is 'Adopt a Heritage' scheme? *The Hindu*. Retrieved from https://www.thehindu.com/news/national/what-is-adopt-a-heritage-scheme/article23866697.ece.

Binayak, P. (2018). Why India's World Heritage Sites Are Up For 'Adoption'. *Culture Trip*. Retrieved from https://theculturetrip.com/asia/india/articles/why-indias-world-heritage-sites-are-up-for-adoption/.

Bolton, R. N., Parasuraman, A., Hoefnagels, A., Migchels, N., Kabadayi, S., Gruber, T., Loureiro, Y.K., & Solnet, D. (2013), "Understanding Generation Y and their use of social media: a review and research agenda", *Journal of Service Management*, 24(3), 245–267.

Brito, M. (2011), 3 Reasons Why Relevant Content Matters. Retrieved from http://www.socialmediaexplorer.com/social-mediamarketing/3-reasons-why-relevant-content-matters/.

Cabiddu, F., Carlo, M. D., & Piccoli, G. (2014). Social media affordances: Enabling customer engagement. *Annals of Tourism Research*, 48, 175–192.

Chhabra, D. (2017). Heritage branding of India: a Gandhi tourism perspective. *Annals of Tourism Research*, 62, 110–112.

Choi, S., Lehto, X. Y., & Oleary, J. T. (2007). What does the consumer want from a DMO website? A study of US and Canadian tourists' perspectives. *International Journal of Tourism Research*, 9(2), 59–72.

Christodoulides, G., Jevons, C., & Bonhomme, J. (2012). Memo to marketers: Quantitative evidence for change: How user-generated content really affects brands? *Journal of Advertising Research*, 52(1), 53–64.

Duffett, R. G., & Wakeham, M. (2016). Social media marketing communications effect on attitudes among millennials in South Africa. *The African Journal of Information Systems*, 8(3), 2.

Duffett Rodney, G. (2015). Facebook advertising's influence on intention-to-purchase and purchase amongst Millennials. *Internet Research*, 25(4), 498–526.

Dwivedi M, Yadav A, & Venkatesh, U. (2012). Use of social media by national tourism organizations: a preliminary analysis. *Information Technology and Tourism*, 13(2), 93–103.

Erdoğmuş, I., & Çiçek, M. (2012). The Impact of Social Media Marketing on Brand Loyalty. *Procedia – Social and Behavioral Sciences*, 58, 1353–1360.

Evans, D. (2008). *Social media marketing: An hour a day*. Indianapolis, IN: John Wiley & Sons.

Folorunso, O., Vincent, R. O., Adekoya, A. F., & Ogunde, A. O. (2010). Diffusion of innovation in social networking sites among university students. *International journal of computer science and security*, 4(3), 361–372.

Gaitan, M. (2014). Cultural heritage and social media. *e-dialogos*, 4, 38–45.

Gordhamer, S. (2009), 4 Ways Social Media is Changing Business. retrieved from http://Mashable.Com/2009/09/22/Social-Media-Business.

Gretzel, U., & Yoo, K. (2014). Premises and promises of social media marketing in tourism. *Routledge Handbook of Tourism Marketing* (pp. 491–504). New York: Taylor & Francis.

Gretzel, U., & Yoo, K. H. (2008). Use and impact of online travel reviews. In P. O'Connor, W. Höpken, & U. Gretzel (Eds.), *Information and communication technologies in tourism* (pp. 35–46). New York: Springer

Gretzel, U., & Yoo, K.-H. (2013). Premises and promises of social media marketing in tourism. In McCabe, S. (Ed.), *The Routledge handbook of tourism marketing* (pp. 491–504). New York: Routledge.

Hays, S., Page. S., & Buhalis, D. (2013). Social media as a destination marketing tool: its use by national tourism organisations. *Current Issues in Tourism*, 16(3), 211–239.

India Ministry of Tourism (2018). *India tourism statistics at a glance*. Retrieved from http://tourism.gov.in/sites/default/files/Other/ITS_Glance_2018_Eng_Version_for_Mail.pdf.

Indian Institute of Tourism and Travel Management. (2017). *Social media as an influencer among foreign tourists visiting India*. New Delhi, India.

Jani, S., Johnson, R. S., Banu, S., & Shah, A. (2016). Cross-cultural bias in the diagnosis of borderline personality disorder. *Bulletin of the Menninger Clinic*, 80(2), 146–165.

Kaltenborn, B. P., Thomassen, J., Wold, L. C., Linnell, J. D., & Skar, B. (2013). World Heritage status as a foundation for building local futures? A case study from Vega in Central Norway. *Journal of Sustainable Tourism*, 21(1), 99–116.

Kant, A. (2009), *Branding India: An Incredible Story*. Uttar Pradesh, India: HarperCollins.

Kaplan, A. M., & Haenlein, M. (2010). Users of the world, unite! The challenges and opportunities of social media. *Business Horizons*, 52(1): 59–68.

Kerrigan, F., Shivanandan, J., & Hede, A. (2012). Nation branding: A critical appraisal of incredible India. *Journal of Macromarketing*, 32 (3), 319–327.

Khalid, S., & Chowdhury, SA. (2018). Representation of intangible cultural heritage of Bangladesh through social media, *Anatolia*, 29(2), 194–203.

King, L. M., & Halpenny, E. A. (2014). Communicating the World Heritage brand: visitor awareness of UNESCO's World Heritage symbol and the implications for sites, stakeholders and sustainable management. *Journal of Sustainable Tourism*, 22(5), 768–786.

Landorf, C. (2009). Managing for sustainable tourism: a review of six cultural World Heritage Sites. *Journal of Sustainable Tourism*, 17(1), 53–70.

Llopis-Amorós, M.-P., Gil-Saura, I., Ruiz-Molina, M. E., & Fuentes-Blasco, M. (2019). Social media communications and festival brand equity: Millennials vs Centennials. *Journal of Hospitality and Tourism Management*, 40, 134–144.

Mangold, W. G., & Faulds, D. J. (2009). Social Media: the new hybrid element of the promotion mix. *Business Horizons*, 52(3), 357–365.

Monuments adopted under 'Adopt A Heritage' project: Culture minister (2018, Dec 17). *The Times of India*. Retrieved from https://timesofindia.indiatimes.com/india/10-monuments-adopted-under-adopt-a-heritage-project-culture-minister/articleshow/67132866.cms

Narangajavana, Y., Callarisa Fiol, L. J., Moliner Tena, M. Á., Rodríguez Artola, R. M., & Sánchez García, J. (2017). The influence of social media in creating expectations. An empirical study for a tourist destination. *Annals of Tourism Research*, 65, 60–70.

Osei, B., Mensah, I., & Amenumey, E. (2018). Utilisation of social media by international tourists to Ghana. *Anatolia*, 29 (3), 411–421.

Outlook India (2020). Life insurance opportunities for millennials. Retrieved from https://www.outlookindia.com/outlookmoney/insurance/life-insurance-opportunities-for-millennials-4206.

Raacke, J., & Bonds-Raacke, J. (2008). My space and Facebook: Applying the uses and gratifications theory to exploring friend-networking sites. *Cyberpsychology & Behavior*, 11(2), 169–174.

Rastegar, R. (2017). Environmental protection and local resident attitudes at early stages of tourism development (Doctoral dissertation). The University of Queensland, UQ Business School.

Rastegar, R. (2019). Tourism development and conservation, do local resident attitudes matter?. *International Journal of Tourism Science*, 19 (3), 181–191.

Rastegar, R. (2020). Tourism and justice: Rethinking the role of governments. *Annals of Tourism research*. 85, 102884.

Rastegar, R., Zarezadeh, Z., & Gretzel, U. (2021). World heritage and social justice: Insights from the inscription of Yazd, Iran. *Journal of Sustainable Tourism*, 29(2–3), 520–539.

Rathore, A. K., Joshi, U., & Ilavarasan, P. (2017). Social media usage for tourism: A case of Rajasthan tourism. *Procedia Computer Science*, 122, 751–758.

Sharron, S., & Abraham, J. (2015). The role of curiosity in making up digital content promoting cultural heritage. *Procedia – Social and Behavioral Sciences*, 184, 259–265.

Sheth, J.N., & Parvatiyar, A. (1995). The evolution of relationship marketing. *International business review*, 4(4), 397–418.

Skinner, H., Sarpong, D., & White, G. R. (2018). Meeting the needs of the millennials and generation Z: Gamification in tourism through geocaching. *Journal of Tourism Futures*, 4(1), 93–104.

Smith, K. T. (2012). Longitudinal study of digital marketing strategies targeting Millennials. *Journal of Consumer Marketing*, 29(2), 86–92.

So, S. A., & Morrison, A. M. (2004). Internet marketing in tourism in Asia: An evaluation of the performance of East Asian national tourism organization websites. *Journal of Hospitality Marketing & Management*, 11(4), 93–118.

Stuedahl, D. (2009). Digital cultural heritage engagement – A new research field for ethnology. *Ethnologia Scandinavica*, 39, 67–81.

Surugiu, M.-R., & Surugiu, C. (2015). Heritage Tourism Entrepreneurship and Social Media: Opportunities and Challenges. *Procedia - Social and Behavioral Sciences*, 188, 74–81.

Times of India (2018). 10 monuments adopted under 'Adopt A Heritage' project: Culture minister. Retrieved from https://timesofindia.indiatimes.com/india/10-monuments-adopted-under-adopt-a-heritage-project-culture-minister/articleshow/67132866.cms

UNESCO (2020). *World heritage sites in India*. Retrieved from http://whc.unesco.org/en/statesparties/IN

Vargo, S. L., & Lusch, R. F. (2004). Evolving to a new dominant logic for marketing. *Journal of Marketing*, 68, 1–17.

Verma, R. K., Kumar, S., & Ilavarasan, P. V. (2017). Government portals, social media platforms and citizen engagement in India: Some insights. *Procedia Computer Science*, 122, 842–849.

World Travel and Tourism Council (2018). *Travel and tourism economic impact 2018: India*. London, UK.

Xiang, Z., & Gretzel, U. (2010). Role of social media in online travel information search. *Tourism Management*, 31(2), 179–188.

Yoo, K.H., Lee, KS, & Gretzel, U. (2007). *The role of source characteristics in eWOM: What makes online travel reviewers credible and likeable. Proceedings of the 14th international ENTER conference*. Ljubljana, Slovenia.

Zarezadeh, Z. Z., Rastegar, H. R., & Gretzel, U. (2018). Reviewing the past to inform the future: A literature review of social media in tourism. *Czech Journal of Tourism*, 7(2), 115–131.

Zhu, L., Anagondahalli, D., & Zhang, A. (2017). Social media and culture in crisis communication: McDonald's and KFC crises management in China. *Public Relations Review*, 43(3), 487–492.

# 11 Millennials in the search for spiritual ecstasy

## Touristic consumption of shamanic mushroom rituals in Huautla de Jimenez, Mexico

*Diego Hannon Ovies*

### 11.1 Introduction

Also known as the net-generation (Sandars & Morrison, 2007), Gen Me (Twenge, 2006), the Google generation (Rowlands et al., 2008), the digital generation (Morris, 2009) and the festival generation (Benelli, 2019), Millennials are an innovative force reshaping the global travel industry by demanding ever-more vivid, intense, and personally meaningful travel experiences that allow them to discover exotic cultural traditions and explore remote corners of the world (Cavagnaro, Staffieri, & Postma, 2018). Compared with their parents and grandparents, Millennials are waiting longer to get married and have children, choosing education, travel and spirituality instead (Pond, Smith, & Clement, 2010). Indeed, Millennials place a higher value on knowledge and personally meaningful experiences rather than material products and are thus more likely than previous generations to try something new, go on an adventure, and travel to collect memories, learn, and improve well-being (Future Cast, 2016; Gelfeld, 2017).

For the Millennial generation – those born between 1980 and 2000 – (Rainer & Rainer, 2011), travel is an essential element of identity, "a vital experience that helps them understand, grow and continuously reinvent their sense of self" (Future Cast, 2016, p. 3). Consequently, alternative forms of travel have expanded in recent decades, and the offer of touristic services has diversified globally to include a wide range of cultural practices from different ethnic traditions and spiritual philosophies. This chapter examines an understudied practice carried out by spiritual Millennial travellers: the use of indigenous sacred plants with psychedelic properties in shamanic contexts. Predominantly, the literature on the subject of spirituality and tourism focuses on Asian and European destinations and spiritual practices adhered to Eastern philosophies such as meditation and yoga while overlooking the vast history of shamanic tourism that, since the 1960s, has taken place in Latin America.

The Mazatec community of Huautla de Jimenez (HDJ), in the southern Mexican state of Oaxaca, was selected as the field research location for the

DOI: 10.4324/9781003170631-14

study due to its long history as a shamanic tourism destination. The study relies on field-based research to understand the motivations of Millennial travellers to participate in a shamanic mushroom ritual in HDJ, as well as the effects and outcomes of their experience. Although the primary research goal was to understand the local perspectives about the impacts of sacred heritage commodification, sufficient interviews with Millennial tourists were carried out to delineate a broad perspective on their motivations to participate in a sacred mushroom ceremony, as well as the outcomes and effects of their experience.

The findings are based on semi-structured interviews with eight subjects that visited HDJ in the last five years and actively participated in a Mazatec mushroom ceremony. These indicate that Millennial tourists have a wide range of motives to participate in a mushroom ceremony. However, they share common themes such as the possibility of living a non-ordinary experience of deep personal significance, searching for spiritual wellness, and healing physical, spiritual and mental imbalances.

In addition to addressing the lack of literature on an understudied phenomenon such as the touristic use of shamanic mushroom rituals, the chapter aims to encourage academic discussion on the touristic appropriation of indigenous spiritual practices and the lack of inclusion of indigenous communities in the process of scientific validation of psychedelic plant medicines. The final reflection urges further research on the potential of shamanic tourism to enhance spiritual well-being in light of globally experienced challenges such as the COVID-19 pandemic.

## 11.2 Millennials and travel

In the post-war era, economic growth and technological development allowed citizens worldwide to access new forms of transportation and telecommunication for the first time. Consequently, the 1960s generation became the first generation in history to globally experience the cultural shifts of the era (Edmunds & Turner, 2005). In the 1990s, the expansion of the Internet further increased the capacity of individuals to connect, access information, and exchange goods, services, and ideas across national boundaries. Raised amidst the digital revolution, Millennials became the most globalized, well-informed, and interconnected generation in human history, which was accompanied by a rapid expansion of international travel and tourism. As noted by Edmunds and Turner (2005): "virtual contact goes hand in hand with physical contact, so although the information revolution has introduced new virtual ways of travelling, physical mobility has simultaneously increased" (p. 573).

As airfares' prices continued to cheapen towards the end of the 20th century, tourism became a widespread social practice of leisure, relaxation and entertainment among middle-class families, primarily in economically developed nations. Nowadays, travel has such relevance in the modern world that over 60 per cent of Millennials in the United States regard it as a very important aspect of their lives and make 4 or 5 trips a year (Cavagnaro et al., 2018). As global citizens and natural travellers, Millennials are inclined to learn new languages, spend money on meaningful and culturally enriching events, and explore different cultural realities and non-touristy areas for long periods, often blending with locals for weeks or months (Garikapati, Pendyala, Morris, Mokhtarian, & McDonald, 2016).

Rather than tourists or vacationers, Millennial travellers see themselves as cultural explorers and knowledge seekers in genuine destinations. For them, "travel means to them novelty: the possibility to evade the quotidian, to try a different lifestyle, to live new experiences, to visit new places and to acquire new knowledge" (Cavagnaro et al., 2018, p. 5). Likewise, Millennials care about creating a better world, so they place a higher value on social and environmental justice and sustainability than previous generations, often making travel decisions based on ideological convictions (Deloitte, 2020; Kaplan, 2020).

Nonetheless, as middle- and upper-class Millennials have an unprecedented capacity to access travel deals and cheap airfares at the touch of a button, most low-earning Millennials live in a context of growing economic disparity, financial uncertainty and socio-political instability in which travel as a leisure practice is still far from reach (Counted & Arawole, 2015; International Monetary Fund, 2017). Moreover, the 2020 COVID-19 pandemic has decreased the capacity of Millennials to travel by putting one in five out of work (Deloitte, 2020).

## 11.3 Millennials and new-age spirituality

Although religion and spirituality are strongly correlated and commonly deemed overlapping concepts, recent studies have identified an increase in the population that rejects traditional religiosity and instead self-identifies as spiritual but not religious (Wixwat & Saucier, 2020). Accordingly, a set of fundamental differences between both conceptual notions has been acknowledged. On the one hand, religiousness is characterized as inherently collective and tied to institutionalized religious organizations and structured sets of beliefs; on the other hand, spirituality is characterized as inherently individual and tied to a process of personal growth and search for connection with the transcendental or the sacred through subjective practices of self-improvement (Dragan, 2018; Gay & Lynxwiler, 2013; Norman, 2011). Hence, religious individuals tend to self-describe as religiously affiliated, whereas spiritual individuals often self-describe as religiously unaffiliated.

The origins of Western secular spirituality can be traced to cultural shifts incited by the 1960s generation across the world which highlighted "values of pluralism, equality, diversity, and individual freedom" (Bahan, 2015, p. 64). At the time, a countercultural, secular and globalized spirituality emerged in the main urban centres of the Western world. The New Age movement rejected the materialist, consumerist and environmentally detrimental nature of capitalist societies, advocating for harmonic development of human society in unity with nature through non-Western spiritual values and practices. Consequently, cultural practices and alternative therapies such as yoga, meditation, drum-circles, crystal healing, reiki, indigenous and native American dances, and sweat-lodge ceremonies, became popular among middle-class and well-educated individuals in Western cities. Simultaneously, the use of "mind-manifesting" (Osmond, 1957) or psychedelic substances, plants and fungi, such as LSD, mescaline, ayahuasca, and hallucinogenic mushrooms, became a popular act of consciousness expansion and countercultural rebellion during the 1960s revolution (Pollan, 2018).

As a result of the adoption of progressive values among Baby Boomer parents and the subsequent implementation of secular educational systems in Western countries, many Millennials grew up without religion at home or school (Bahan, 2015). In the US, for example:

> One-in-four members of the Millennial generation are unaffiliated with any particular faith. Indeed, Millennials are significantly more unaffiliated than members of Generation X were at a comparable point in their life cycle (20% in the late 1990s) and twice as unaffiliated as Baby Boomers were as young adults (13% in the late 1970s). Young adults also attend religious services less often than older Americans today. And compared with their elders today, fewer young people say that religion is very important in their lives.
>
> (Pond et al., 2010, p. 1)

Nonetheless, "many of the unaffiliated hold some religious or spiritual beliefs, such as belief in God or a universal spirit" (Hackett & Grim, 2012, p. 9). In larger numbers than previous generations, Millennials have moved away from structured forms of organized religion towards inner guidance through alternative forms of spirituality (Benelli, 2019). The secular, globalized, and technological society in which they were raised has granted them access at an unprecedented scale to a widely diverse repertoire of spiritual practices from multiple religious traditions. The mechanisms of self-improvement to connect with their higher consciousness are thus as diverse as the spiritual traditions themselves, and many Millennials are choosing a mix of values, ideas and practices from different cultural traditions to build a personalized spiritual identity.

## 11.4 New Age spiritual tourism and neo-shamanism

For spiritual seekers, travelling is more than a vacation or a leisure experience; it is an opportunity to push themselves beyond their comfort zone and challenge their preconceived beliefs and attitudes; in other words, a spiritual quest or sacred journey of self-discovery and inner growth away from home (Buzinde, 2020; Cheer, Belhassen, & Kujawa, 2017). According to scholars, a common goal to the practice of spiritual tourism is to attain physical, mental and spiritual well-being through personally meaningful experiences of spiritual significance such as the visit to sacred places (Basset, 2012; Olsen, 2013), the attendance to yoga and meditation retreats (Buzinde, 2020; Gill, Packer, & Ballantyne, 2019), or the participation in sacred pilgrimages and healing rituals (Laure & Hannon, 2018; Norman, 2011).

> Spiritual tourism is thus a reflexive well-being intervention driven by the sense that some aspect of everyday life needs fixing or improving, and oriented towards the space of non-work and leisure where such problems can be given full attention.
>
> (Norman & Pokorny, 2017, p. 3)

Away from the hustle of everyday life, the setting provided by the touristic holiday allows spiritual seekers to focus on solving existential questions and finding solutions to inner problems affecting their lives. Under conditions of growing discontent with contemporary culture (Olsen, 2013), many of them have resorted to the spiritual practices of indigenous traditions to construct a new sense of identity and satisfy their inner needs. Consequently, Amerindian shamanism and ethnomedicine have become a valuable structure of spiritual support for citizens in the western world, prompting the emergence of centres, clinics, workshops and retreats where indigenous knowledge is applied to provide spiritual wellness. Scholars define the process of cultural adoption of indigenous shamanic knowledge and practices by members of urban societies as neo-shamanism (Bouse, 2019; Scuro & Rodd, 2015; Wolff, 2020).

According to Bouse (2019), "shamanism is a global practice and belief system that has transcended time, geography, and culture" (p. 146). The term shaman was introduced to Western literature from Siberia in the early twentieth century. For Eliade (2020), the shamans of ancient and indigenous societies are "the manipulators of the sacred" (p. 5) and the intermediaries between the divine and the terrestrial worlds. The physiological capacity of humans to achieve shamanic states of consciousness, also known as "soul flights" or "shamanic journeys", through practices such as meditating, fasting, chanting, praying, drumming, dancing, breathing, ingesting psychoactive plants or isolating the senses from external stimuli, has incited the development of the techniques and procedures of spiritual trance that allow the human spirit to travel through higher and inferior planes of reality and communicate with mythical figures, superhuman deities, power animals, deceased ancestors, or spiritual entities to attain wisdom and bring knowledge to their tribe (Harner, 1980).

Commonly, shamanistic societies believe the terrestrial and the spiritual worlds are intrinsically linked, and any relevant outcome or event, including decease, is deemed the result of spiritual, rather than physical forces. Within this context, rituals and ceremonies are the symbolic vehicles enabling the shaman to interface with the spiritual world "outside of sensory cognition" (Bouse, 2019, p. 146) to find the origins of disease, bring wisdom to the community, or to foresee the unfolding of upcoming events (George, Michaels, Sevelius, & Williams, 2019).

Across the American continent, shamans have played a multifunctional role in the spiritual life of indigenous communities as medics, priests, mystics, artists, therapists, healers, poets and spiritual leaders. For thousands of years, they have relied on sacred plants with psychedelic properties both as medicines and communication devices with the spiritual world. During the Spanish conquest of America, the use of hallucinogenic plants was forbidden and severely punished by ecclesiastic authorities, which caused many spiritual practices to relocate from the public arena to the intimate domain. In the early twentieth century, the remaining shamanic traditions became objects of scientific interest. The resulting ethnographic and anthropological studies contributed widely to destigmatize indigenous spiritual practices within the Western world. During the 1960s cultural revolution, psychedelic sacred-plants used in shamanic ceremonies were popularized through books such as *The Teachings of Don Juan* by Carlos Castaneda, *The Doors of Perception* by Aldous Huxley, and *The Yage Letters by* Allen Ginsberg and William Burroughs. Inspired by these embellished accounts, New Age travellers and spiritual seekers decided to visit indigenous communities to live the shamanic experience for themselves and a new form of spiritual tourism was born as a result (Laure & Hannon, 2018).

With the expansion of digital media and international tourism, a revitalization process of indigenous shamanism and ethnomedicine is currently taking place. Through the Internet, Millennials are discovering the potential of indigenous shamanic practices to enhance self-improvement, and, by travelling, they are approaching the communities where the spiritual resources they so much cherish are still produced and applied. For Prayag, Mura, Hall, and Fontaine (2016): "The revival of shamanism is part of the worldwide trend of individuals seeking renewal of their spirituality through connectedness with nature" (p. 4). In the particular case of psychedelic plant medicines, the dissemination of scientific studies validating their therapeutic potential is contributing to the popularization of sacred plant rituals and the subsequent expansion of shamanic tourism in countries like Mexico, Peru, Colombia, Brazil and Ecuador, where the offer of ethno-therapeutic and shamanistic health services has diversified widely since the early 2000s (George et al., 2019; Laure & Hannon, 2018).

## 11.5 Study methods and data analysis

The study presented in the chapter's second half includes a description of the Mazatec mushroom rituals and the touristic history of the indigenous city of Huautla de Jimenez (HDJ). The study's findings rely on qualitative data obtained through face-to-face, in-depth semi-structured interviews (DiCicco-Bloom & Crabtree, 2006) conducted in HDJ with eight millennial travellers to understand their motivations and outcomes concerning their participation in a shamanic mushroom ritual. The subjects were initially met in person as the author of this chapter undertook ethnographic field research in HDJ for four months between 2016 and 2019.

Given the sensitivity of the topic and the intimate nature of undertaking an experience involving the use of psychoactive mushrooms, only eight subjects agreed to share their testimony with the researcher in a formal interview. The selection of the subjects was based on three criteria: 1) their age, so as to ensure the involvement of millennial travellers exclusively, 2) their active participation in a Mazatec mushroom ritual in HDJ, which included having ingested a dose of psilocybin mushrooms under the supervision of a Mazatec guide, and 3) their openness and willingness to share a personal experience with the researcher.

During the face-to-face, in-depth semi-structured interviews, the selected millennial subjects were inquired about their reasons for taking part in a Mazatec mushroom ceremony and the degree of satisfaction based on their expectations. By asking questions such as 'How did you learn about Mazatec mushroom rituals?', 'Why did you decided to participate in one?', 'What were the effects during the experience?', 'How did you felt afterwards?', and 'Would you qualify your experience as spiritual?', the interviewees provided a rich description of their perceptions, implications and meanings concerning the recently lived experience. All interviews were digitally recorded and transcribed. Follow-up conversations were held with the subjects via email and chat to ensure their testimonies were genuinely presented. Moreover, the author of this chapter – who also happens to be a Millennial – considered it appropriate to personally participate in a Mazatec mushroom ritual to facilitate understanding of the meanings and concepts in the subjects' testimonies.

Following the thematic analysis approach proposed by Braun and Clarke (2012), the coding of the collected qualitative data was done at the semantic level, from a bottom-up inductive approach that derives from the content of the data themselves to represent the meanings and concepts provided by the subjects. As the data were transcribed and the codes were analysed and categorized, discursive patterns and common themes tied with spirituality were identified across data sets. These revealed the subjects' motives and outcomes of their shamanic experience under a conceptual framework that allows the case study of HDJ to be compared with similar case studies in the field of spiritual tourism. For the purposes of this chapter, the terms shamanic experience, mushroom ritual, and Mazatec ceremony will be used interchangeably throughout the text.

## 11.6 Huautla de Jimenez

Huautla de Jimenez (HDJ) is the largest and most inhabited urban centre in the indigenous Mazatec region of the Mexican state of Oaxaca (see Figure 11.1). In 1955, amateur mycologist and vice-president of J.P. Morgan Bank, Gordon Robert Wasson, and her paediatrician wife, Valentina Pavlona, travelled from New York to HDJ to experience the sacred mushroom ritual for themselves. Two years later, *Life Magazine* published a detailed account of Wasson's experience of "holy communion" through the "astonishing

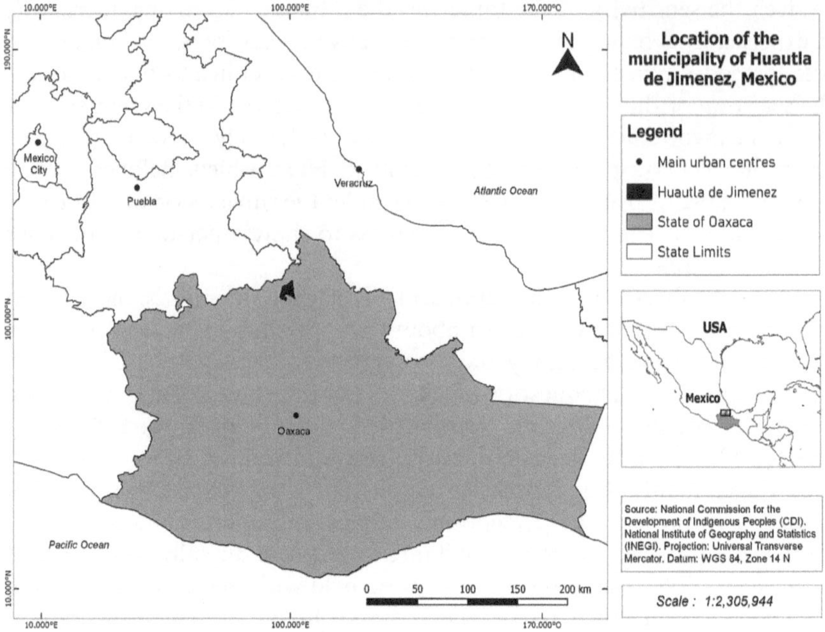


*Figure 11.1* Map of Huatla de Jimenez.

effects of the mushroom" (Wasson, 1957). The article inspired journalists and academics to visit HDJ to discover the millenary mushroom ritual and experience the shamanic state of consciousness described by Wasson. As the word spread in universities and intellectual circles, writers and scientists tied to the emerging New Age and hippie movements, such as Albert Hoffmann, Richard Evan Schultes, and Stanislav Grof, travelled to HDJ in the 1960s. Finally, crowds of hippies descended upon the Mazatec village for purposes of drug-induced spiritual exploration, fostering the development of a small infrastructure of hotels, restaurants, transportation companies and other tourism services.

As a result, HDJ was popularized through romanticized accounts, articles, books and documentaries, and ultimately transformed into a spiritual hub, "the city of magic mushrooms" (Feinberg, 2018) and the ideal destination for shamanic tourism. Eventually, Mazatec shamans such as Maria Sabina, Julieta Casimiro and Ines Cortes became internationally recognised figures and tourist attractions themselves. Nowadays, spiritual seekers recognise HDJ as a place that played a pivotal role in the countercultural revolution of the 1960s; some will say, the birthplace of the hippie movement. According to a millennial American tourist:

> This is the place where it all began. Without Wasson's publications, the psychedelic revolution, the hippie movement, it would never have

> happened. Some say Gordon Wasson was a CIA agent working to spread a cultural revolution in the US; it was all a government-run experiment that got out of hand.
>
> (Participant 8, American War Veteran)

Six decades of uninterrupted tourism activities in HDJ have brought positive and negative outcomes to the Mazatec community. On the one hand, previously stigmatized cultural practices such as the *velada* have been revitalized, and the sense of Mazatec ethnic pride has been reinforced in light of international recognition. Similarly, the creation of jobs and the establishment of family-owned businesses have empowered the community while increasing the quality of life of many local families by providing them with an additional monetary income source. For example, *Casa Cejota* and *Hotel Casas Enpi*, both managed by Millennial Mazatecs, offer accommodation, tours, and sacred mushroom experiences through digital platforms like Booking.com and Facebook.

On the other hand, an intimate ritual practice such as the mushroom ceremony has been trivialized and transformed into a publicly available touristic product, creating tensions among residents with contrasting views. Under these circumstances, the economic opportunities provided by the commodification of mushroom rituals have prompted the emergence of plastic shamans that present themselves to outsiders as authentic Mazatec healers but are deemed illegitimate by the community. These contemporary shamans offer a simulated version of the ritual that differs drastically from the traditional Mazatec format; for instance, by allowing large groups of up to 10 or more subjects with no previous preparation to participate. Although Mazatecs condemn the transformation and commodification of their ancient tradition for tourism, some families have willingly diminished their relations within the community in favour of more profitable relations with outsiders as a means to achieve socio-economic development. The profile of visitors coming to HDJ was drafted by Piña (2019) from a sample of 60 travellers, revealing that "these are men and women between the ages of 16 and 50, from middle and urban socio-economic levels" (p. 54), a common characteristic was the inclination towards spirituality, shamanism, and New Age beliefs.

## 11.7 Spirituality and psilocybin mushrooms

The strain of mushrooms in the genus *Psilocybe* belong to the *Basidiomycota* division and are characterized by the production of a tryptamine with psychedelic or psychoactive properties called psilocybin, which molecularly resembles to the neurotransmitter serotonin. Although humans have used these mushrooms for sacramental and medicinal purposes since at least 7,000 years ago (Stamets, 1996), the scientific recognition of the therapeutic properties of psilocybin is a consequence of very recent clinical trials that reassert the capacity of psilocybin to interact with the central nervous system and induce mystical and transcendental experiences of deep personal significance and spiritual meaning (Griffiths, Richards, Johnson, McCann, & Jesse, 2008). By increasing neuronal connectivity, these experiences have the potential to improve the well-being of subjects suffering from anxiety (Goldberg, Pace, Nicholas, Raison, & Hutson, 2020), depression (Carhart-Harris et al., 2018), addiction (Johnson, Garcia-Romeu, Cosimano, & Griffiths, 2014) and post-traumatic stress disorder or PTSD (Krediet, Bostoen, Breeksema, van Schagen & Vermetten, 2020). The subjects under the effects of psilocybin often describe a transformative psychological experience of "awe and connectedness to a superior power (...) paired with an intense sense of unity, insightfulness and bliss" (Kometer, Pokorny, Seifritz, & Volleinweider, 2015, p. 3664).

As the psilocybin-induced spiritual journey unfolds, subjects often confront a range of emotional challenges and engage in a process of deep self-reflection. Analogous to a religious pilgrimage, this inner spiritual odyssey allows subjects to reconceptualize maladaptive behavioural patterns and obtain a renovated desire to make the necessary changes to improve their lives. Therefore, it is common for similar outcomes such as "a feeling that certain behavioural or social structures have been broken" (Norman, 2011, p. 10) to be ascribed to both religious pilgrimages and psychedelic-induced spiritual journeys. According to a 14-month follow-up study, psilocybin-occasioned experiences were regarded by subjects as one of the most personally meaningful and spiritually significant events of their lives (Griffiths et al., 2008). In light of the 2020 COVID-19 global pandemic, Kelly et al. (2020) have suggested that "psilocybin therapy has the potential to play an important therapeutic role for various psychiatric disorders in post-COVID-19 clinical psychiatry" (p. 1).

The scientific recognition of psilocybin's therapeutic potential and low toxicity has led to its legalization for therapy in Oregon along with the decriminalization of psychedelic plants and fungi in Washington, both in the US. Another culturally relevant outcome of the ongoing psychedelic and neo-shamanic boom is the involvement of transnational pharmaceutical companies such as ATALI Life Sciences and Compass Pathways in the development of psilocybin-based treatments for mental health disorders, for which they have invested millions of dollars in preliminary research (Nutt, Erritzoe, & Carhart-Harris, 2020; Shead, 2020). In countries where psilocybin mushrooms are decriminalized, such as Jamaica and the Netherlands, healing centres and clinics providing spiritual retreats based on psilocybin therapy have become flourishing businesses in the last decade (Rucker et al., 2020).

In the indigenous Mazatec town Huautla de Jimenez, our case study, psilocybin ritual experiences have been part of the local offer of touristic services since the 1960s under the protection of the Indigenous and Tribal Peoples Convention of 1989, to which Mexico adheres since 1990. Until before the pandemic of COVID-19, many researchers and scientists from international institutes had a presence in the community with the goal of tracking the endemic species of mushrooms to map their DNA and isolate their chemical compounds (Garcia, Acosta, & Piña, 2020; Piña, 2019). Moreover, the author of this chapter held a conversation in the municipal market of HDJ with two researchers from the Institute of Transpersonal Psychology of Barcelona, whose goal in the community was to study the therapeutic structure of the Mazatec mushroom ritual hoping to offer similar healing services and "ethnotherapeutic journeys" in Europe, for which they established an alliance with Doña Ines Cortes, a popular Mazatec shaman that supposedly introduced John Lennon to psilocybin mushrooms in the 1960s.

The recent presence of international scientists from different research institutes has sparked the debate among the local population of HDJ around the scientific appropriation of the Mazatec biocultural heritage. From this perspective, Mazatecs are being excluded from the process of scientific validation and public recognition of their system of knowledge and reduced to silent spectators as international pharmaceutical companies and research institutes compete to control the means of exploitation of psilocybin (Piña, 2019). The failure to integrate Mazatec constructs and procedures into the discussion suggests that these are neither recognised, nor well understood by the psychedelic scientific community, or that researchers are unwilling to share the spotlight with non-scientific actors.

## 11.8 Mazatec shamanism and veladas

The Mazatec ethnic group is one of the 17 different indigenous traditions that inhabit the multilingual and pluricultural Mexican state of Oaxaca. Since the early 1900s, the Mazatec tradition has drawn the attention of anthropologists and ethnobotanists (Johnson, 1939; Reko, 1919; Schultes, 1939). Throughout the centuries, Mazatecs have developed a complex cultural system consisting of linguistic variations, agricultural and artisanal techniques, plant-based medicinal practices, and orally transmitted religious myths and beliefs. According to the Mazatec religious worldview, the caves, canyons, mountains and rivers of the earth are inhabited by evil and benevolent supernatural lords or *chikones* that require cyclical tribute and religious offerings to maintain the harmonic balance of the universe. Simultaneously, Mazatecs celebrate Catholic holidays and pay tribute to Jesus Christ, the saints, and Mary, mother of Jesus, thus constituting a syncretic system of religious beliefs and practices that is unique to their tradition.

For Mazatecs, one of the most sacred healing resources gifted to them by *chikones* is the *ndi xijtho* or "the little ones that sprout", a name that is given to the endemic species of psilocybin mushrooms (Faudree, 2015). The Mazatec men and women of knowledge or *chjota chjine* are the shamans that specialize in the medicinal and ceremonial use of endemic plants and fungi for healing purposes. In the Mazatec region, the most commonly used species of psilocybin mushrooms are *Psilocybe caerulescens, Stropharia cubensis and P. Mexicana Heim*. Often called "saint children" or "little things", the mushrooms are regarded as sacred sentient entities with their own voice, soul and personality. When consumed in a ceremonial context, mushrooms allow the shaman to diagnose and cure disease. The ceremonial act in which they are orally ingested for healing purposes is known as *velada* or *vijnachoan*, which means "we live awake" or "we stay awake" (Minero, 2015).

The *velada* takes places at nightfall before a Catholic altar composed of candles, images of saints, cacao seeds, flowers and copal incense. The two main participants of the ritual ceremony are the shaman and the patient, although close family members of any of both might also participate if deemed necessary by the shaman. During the ritual, the shaman's chants and prayers work as a linguistic vehicle between the divine and the terrestrial worlds. The effects of psilocybin, along with the subtle candlelight, the chants and prayers of the healer, and the smell of herbs and incense in the dark room, create the perfect conditions for the patient to transit from a normal state of consciousness to a state of ecstatic trance or spiritual ecstasy. As the experience intensifies, the shaman takes care of the patient's feelings and makes sure suppressed emotions, traumas, and physical distress are gently released. After the climax, the shaman's chants and prayers come to an end, the candles are blown out, and the room is absorbed by silence and darkness, allowing the patient to quietly self-reflect as the effects of psilocybin gradually decline. Depending on each patient and the size of the ingested amount of mushrooms, a *velada* might take between four to eight hours.

## 11.9 Findings

The findings in the chapter's last section display the motivations of eight Millennial subjects to participate in a shamanic mushroom ritual in HDJ, as well as the outcomes and effects of their experiences. The testimonies were obtained through face-to-face, in-depth interviews with each of the subjects. Follow-up conversations were also held via chat and email to ensure the subjects' views and testimonies were accurately presented.

## 11.10 Motivations

From a total of eight Millennial subjects, six of them reported having travelled to HDJ specifically to experience the mushroom ritual and admitted having learned beforehand about psilocybin mushrooms and Mazatec shamanism, either from a person they know or on the internet. Only a couple of German tourists were made aware of the existence of mushroom rituals by their Airbnb host in Mexico City and decided to visit HDJ to satisfy their newly-acquired curiosity. Concerning the reasons to participate in a mushroom ritual, the subjects' answers were widely diverse, but they shared similar purposes of curiosity satisfaction, individual development, self-improvement, and spiritual, emotional or physical healing (Table 11.1).

A female massage therapist from the United States (S1) declared she travelled to HDJ to "*get away from the US's toxic social environment*". By taking the mushrooms in a ceremonial context, she hoped to "*find the source*" of her

*Table 11.1* Summary of subjects' motives for experiencing a mushroom ceremony

| Subject | Age | Gender | Nationality | Religious affiliation | No. of previous shamanic journeys | Motives for experiencing a mushroom ceremony |
|---|---|---|---|---|---|---|
| S1 | 24 | Female | US | SBNR (Spiritual but no religious) | 0 | Mental health improvement (anxiety). |
| S2 | 25 | Female | Mexico | SBNR | 1 | Spiritual healing and emotional forgiveness. |
| S3 | 33 | Male | Chile | SBNR | 4 | Ego understanding/ psychological development. |
| S4 | 35 | Male | Mexico | SBNR | 10 or more | Consciousness exploration, creative/ artistic inspiration. |
| S5 | 24 | Female | Costa Rica | SBNR | 0 | Curiosity satisfaction. Desire to learn about indigenous shamanism and medicine. |
| S6 | 28 | Male | Germany | Atheist, non-spiritual | 0 | Curiosity satisfaction/search for non-ordinary experiences. |
| S7 | 31 | Female | Germany | SBNR | 0 | Curiosity satisfaction/search for non-ordinary experiences. |
| S8 | 39 | Male | US | Catholic | 0 | Mental health improvement (depression, PTSD). |

Based on data collected by the authors

anxiety and "*improve mentally*". The subject emphasized having researched extensively on the Internet about Mazatec shamanism and the therapeutic properties of psilocybin before travelling to HDJ with a friend to try the mushrooms for the first time. Likewise, a female lawyer from Mexico City (S2) mentioned she wanted to participate in a mushroom ceremony to "*reflect on my family relationships*", "*become more tolerant*", and "*forgive wounds of the past*". She claimed this was her second experience under the same Mazatec healer's guidance, so she hoped to have a "*deeper spiritual journey*" this time.

For a male Chilean chef (S3), his primary motivation to "*work with the mushrooms*" was to "*understand his ego*". The subject emphasized this was his fourth experience in a shamanic ritual. Correspondingly, a male painter and sculptor from Oaxaca City (S4) decided to visit HDJ and take part in a mushroom ritual to "*explore his consciousness*" and "*gain creativity to inspire his artistic work*". The subject reported having a long experience using psychedelic plant medicines for purposes of creative and artistic inspiration, claiming to have been in at least ten shamanic rituals since he was 16.

For a female anthropology student from Costa Rica (S5), her motive to travel to HDJ was "*to work on a research project for a thesis about Mazatec traditional medicine and shamanism*". She mentioned she learned about Maria Sabina through a book and became fascinated by her story, hence decided to do a thesis on the topic that she had read so much about. However, she claimed she felt discouraged by the commodification of the sacred practices in HDJ. Hence, she travelled to a more isolated community in the Mazatec region with the hope of finding a "*more authentic shaman*". By taking the mushrooms, she hoped to obtain first-hand knowledge about indigenous spirituality and gain a deeper understanding of traditional healing procedures.

Similarly, a male German web designer (S6) stated he was recommended by his accommodation host in Mexico City to visit HDJ and participate in a mushroom ritual. Following his recommendation, the subject programmed a mushroom ceremony with a healer to "*live an authentic indigenous ritual*" and "*experience something out of the ordinary*". A female German journalist (S7) accompanying him reported she also wanted to experience the mushroom ritual to satisfy her curiosity by "*learning how it feels to have them [the mushrooms] with an experienced guide taking care of you*".

Contrastingly, a male war veteran from the US (S8) reported his primary motivation to use the mushrooms in a shamanic context was to overcome his post-traumatic stress disorder (PTSD) and "*reconnect with himself, with nature and the universe*". The subject revealed he was undertaking a "*continental healing journey*" that would include multiple shamanic experiences from different indigenous traditions throughout the Americas. The source of his emotional distress was a near-death experience during an armed conflict in the Middle East. After trying medications commonly used to treat PTSD with very little success, the subject claimed he learned about the therapeutic potential of psychedelics from a fellow veteran and decided to try them under the framework of an indigenous-shamanic tradition, "*where the knowledge on these medicines is more authentic and developed*".

## 11.11 Effects and outcomes

From a total of eight Millennial subjects, seven regarded the shamanic mushroom ritual as personally meaningful, and six described it as spiritual; only subject four (S4) revealed feeling no effects after ingesting the mushrooms in a shamanic ceremony. Nonetheless, after ingesting them for a second time outside the ritual context, the subject reported having a profound spiritual experience. In his own words:

> I didn't feel a thing in the ceremony, surely because the amount of mushrooms was too small for a big guy like me (laughs). But afterwards, in the Hill of Worship [a sacred site in the mountains surrounding HDJ], I had a stronger journey. I felt Pacha-Mama [Mother Earth] taking care of me and us humans, like the little children that we really are.
>
> (S4)

Similarly, for a female massage therapist from the US (S1), the mushroom ritual was "*a powerful experience (…) they [the mushrooms] showed me my shortcomings and forced me to face my fears and process my dark emotions instead of always running away from them*". Likewise, a female lawyer from Mexico City (S2) reported having a spiritual experience during the ritual and feeling "*a rush of energy*" before having an encounter with her deceased grandmother in which she played and sang with her like when she was a child. After the ceremony, she reported feeling "*relief and inner peace*". The subject also mentioned she planned to continue with her spiritual journey by retaking the mushrooms in San Jose del Pacifico, another town in the state of Oaxaca that has recently popularized among spiritual seekers due to the availability of psilocybin mushrooms.

A male Chilean chef (S3) claimed he felt disappointed after not fulfilling his expectations during the ritual experience. Nonetheless, he admitted feeling satisfied and having learned a lot from it. In his own words, "*the mushrooms will never give you what you want, only what you need*". By contrast, after finally finding what she deemed as a "*more authentic*" (S5) shaman outside HDJ, a female anthropology student from Costa Rica reported her experience was physically intense:

> I wanted to pay attention to the shaman's chants and prayers, but it was impossible because I could not stop focusing on my body. I really got in touch with my body in a way I never had before. I learned that Mother Nature can provide us with the most beautiful experiences.
>
> (S5)

For a male German web designer (S6), the emotional intensity of the experience amazed him. In his words, "*I was expecting to see colourful visions, but I was actually very emotional. I hadn't cried in years, and I was suddenly unable to control myself.*" Similarly, a German female journalist (S7) travelling with him reported:

> It was very challenging and uncomfortable at the beginning. Eventually, I was able to understand why this is seen as spiritual for many people. It does get you in touch with something sacred. But you have to overcome that initial stage of fear. I guess that applies to many aspects of life as well.
> (S7)

Finally, a male war veteran from the US (S8) described his experience as "*really hard to put into words. At one point I felt like a kid again. I was laughing, and Eugenia [his healer] scolded me for that (…) I think I needed to allow myself to be that vulnerable again.*" After finishing his continental shamanic journey throughout the continent, the subject reported in a follow-up conversation: "*I finally began to find the peace that I've been looking for.*"

## 11.12 Discussion

Reasserting the individualist nature of contemporary spirituality (Cheer et al., 2017; Gay & Lynxwiler, 2013), all eight subjects' motives to participate in a shamanic mushroom ritual revolved around subjective personal goals such as the satisfaction of curiosity, the search for knowledge, the healing of mental and spiritual imbalances, or the desire for self-improvement and inner wellness. Concerning the subjects' religious affiliation, six of them self-described as spiritual but not religious, whereas one self-identified as a Catholic, and one as a non-spiritual atheist. The predominance of spiritual, non-religious individuals can be interpreted as representative of the preference of millennial travellers for culturally exotic spiritual practices and life-affirming travel experiences as an alternative to institutionalized religious practices and traditional vacation experiences.

Moreover, references to the mushroom shamanic experience as part of a larger individual process of spiritual development emerged three times in the testimonies, portraying how spiritual Millennials regard travel not as a means for leisure but as a tool of long-term personal development and inner growth through personally meaningful experiences. Correspondingly, the search for authenticity was mentioned by three subjects as a motive to participate in a shamanic ritual, which reflects the continuous search of spiritual millennials for experiential intensity and cultural authenticity (Dragan, 2018; Gay & Lynxwiler, 2013; Norman, 2011).

Likewise, the emergence of common themes, such as the desire to temporarily leave behind the rhythm of urban life or getting away from a "*toxic social environment*" (S1), reassert Olsen's (2013) view on the growing discontent among spiritual seekers with an excessively technified, materialist and consumerist Western culture and the subsequent search for a new core of spiritual values in non-Western cultural traditions. Concerning the effects of the shamanic mushroom experience, two subjects reported feeling a sense of peace and relief after the ritual, this reaffirms the capability of psilocybin-induced spiritual experiences to provide inner well-being and peacefulness (Griffiths et al., 2008). Correspondingly, the subjects regarded psilocybin mushrooms as safe and effective for treating and even curing mental, physical, emotional, and spiritual imbalances.

Additionally, the multiple references to mother nature (or 'Pacha Mama') mirrors multiple testimonies of ayahuasca ritual participants (Prayag et al., 2016; Wolff, 2020) and illustrates the deep sense of connection to nature often provided by psychedelic-induced spiritual experiences. Concerning the role of the shaman, seven subjects regard it as essential for the development of a beneficial experience. They also described the shaman as a "spiritual guide" (S4; S8), a "caregiver" (S1) and an "ancient therapist" (S7). The same seven subjects admitted having established a trust relationship with their shaman and expressed their intentions to come back to HDJ to have another experience with the same guide.

According to Bahan (2015), the spiritual experiences sought by millennials "are defined by a sense of connectedness between the individual and something beyond the self that does not pertain to concepts of religion or divinity" (p. 73). Indeed, shamanic mushrooms experiences in HDJ are regarded as spiritual by Millennial spiritual seekers because these allow them to transcend the ordinary reality and reflect on one-self from a new perspective, providing them with valuable insights on how to live a better life after returning home. As part of an identity-construction and meaning-seeking process, shamanic experiences enable Millennials to perceive a profound sense of "awe stimulated by nature and beauty" while making them "feel bigger than themselves" (Wixwat & Saucier, 2020, p. 4), thus gaining a new appreciation for humankind and nature.

The subjects' descriptions of their shamanic mushroom experiences in HDJ mirror those given by participants in clinical psilocybin studies (Griffiths et al., 2008) and ayahuasca ritual studies (Kavenská & Simonová, 2015; Prayag et al., 2016; Wolff, 2020). Likewise, conceptually equivalent descriptions have been previously collected from spiritual seekers in destinations like India and Spain (Norman, 2011). Although the eight subjects considered in this study reported having generally positive experiences, accounts of negative experiences such as panic attacks and nausea have been reported in other studies involving the use of psilocybin (Rucker et al., 2020). A larger sample of travellers is thus required to articulate a broader perspective on the motivations, outcomes, risks and benefits tied to the touristic use of psilocybin mushrooms.

## 11.13 Final reflection

Each generation is shaped by the unique historical events and cultural changes taking place around them (Kurz, Li, & Vine, 2019). In the case of Millennials, the globally interconnected society in which they grew up has enabled them to collective experience traumatic events, such as the terrorist attacks of September 11, the Middle East armed conflicts and population displacements, and the earthquakes and tsunamis of 2004 in the Indian

Ocean and 2011 in Japan, to name a few. More recently, the global COVID-19 pandemic has transformed the lives of societies around the world. The travel sector, in particular, has been significantly affected and might take years to recover fully. According to the World Tourism Organization, international tourist arrivals declined 70 per cent in 2020, putting an end to 10 years of sustained growth (UNWTO, 2020).

Nonetheless, globally experienced traumatic events present an opportunity to create a collective awareness to positively transform the world (Galvani, Lew, & Perez, 2020). From this perspective, the COVID-19 pandemic has sparked a public conversation on the vital relevance of mental health and the most effective mechanisms to enhance psycho-social wellness, a topic that is increasingly attached to spirituality. As the gap between science and spirituality narrows and social awareness on mental health increases, the demand for spiritual tourism experiences that enhance mental wellness will likely grow in the post-pandemic recovery years. As part of the process, innovative ethno-therapeutic and shamanistic services and procedures will integrate into the global marketplace of wellness and spirituality.

> Neo-shamanic practice could have value as an adjunct to contemporary mental health practices because of its use in treating the spiritual aspect of mental distress and aiding contemporary therapists in standard clinical and humanistic practices to integrate the patient's spiritual aspect with their mental and physical states.
>
> (Bouse, 2019, p. 143)

Given the rising popularity of non-Westerner spiritual practices among Millennials and younger generations, it is crucial to acknowledge the contributions of indigenous groups by including them in the processes of scientific validation and public legitimization of their spiritual practices in order to avoid a sort of spiritual neo-colonialism in which the holders of such valuable knowledge are left behind and reduced to living historical artefacts to be studied by researchers and admired by tourists. Likewise, greater efforts from governments, public institutions, academics, community leaders and activists in destination countries are necessary to devise and implement mechanisms of biocultural heritage conservation and sustainable development that guide the shamanic tourism sector towards a favourable outcome for local indigenous communities. The promotion of responsible behaviour among tourists is also of paramount importance to reduce the risk of ingesting psychoactive plants and fungi while avoiding the degradation of indigenous spiritual heritage.

Luckily for our case, Millennials happen to be the most socially aware generation in history, and ethnic inclusion, social equality, and sustainability are familiar conceptual notions and essential issues to many of them. Hopefully, as they come to occupy leadership positions in the upcoming decades, they will lead the urgently needed change towards the creation of a more harmonic and humane world.

## References


Bahan, S. A. (2015). The spirituality of atheist and no-religion individuals in the millennial generation: developing new research approaches for a new form of spirituality. *The Arbutus Review*, 6, 63–75. doi: 10.18357/ar.bahans.612015.

Basset, V. (2012). Del turismo al neochamanismo: ejemplo de la reserva natural sagrada de Wirikuta en México [From tourism to neo-shamanism: the example of the natural sacred reserve of Wirikuta in Mexico]. *Cuicuilco*, 19, 245–266. https://www.redalyc.org/pdf/351/35128270007.pdf.

Benelli, R. (2019). Reiki and millennials. *Reiki News Magazine*. Fall, 2019. https://reikilifestyle.com/reiki-and-millenials/.

Bouse, K. J. (2019). *Neo-shamanism and Mental Health*. Cham: Springer Nature.

Braun, V., & Clarke, V. (2012). *Thematic analysis*. In H. Cooper, P. M. Camic, D. L. Long, A. T. Panter, D. Rindskopf, & K. J. Sher (Eds.), *APA handbook of research methods in psychology, Vol.* 2. (pp. 57–71). American Psychological Association. doi: 10.1037/13620-004.

Buzinde, C. N. (2020). Theoretical linkages between well-being and tourism: The case of self-determination theory and spiritual tourism. *Annals of Tourism Research*, 83, 102920. doi: 10.1016/j.annals.2020.102920.

Carhart-Harris, R. L., Bolstridge, M., Day, C. M. J., Watts, R., Erritzoe, D. E., Giribaldi, B., Bloomfield M. Pilling, & Rickard, J. A. (2018). Psilocybin with psychological support for treatment-resistant depression: six-month follow-up. *Psychopharmacology*, 235(2), 399–408. doi: 10.1007/s00213-017-4771-x.

Cavagnaro, E., Staffieri, S., & Postma, A. (2018). Understanding millennials' tourism experience: values and meaning to travel as a key for identifying target clusters for youth sustainable tourism. *Journal of Tourism Futures*, 4(1), 31–42. doi: 10.1108/JTF-12-2017-0058.

Cheer, J. M., Belhassen, Y., & Kujawa, J. (2017). The search for spirituality in tourism: Toward a conceptual framework for spiritual tourism. *Tourism Management Perspectives*, 24, 252–256. doi: 10.1016/j.tmp.2017.07.018.

Counted, A. V., & Arawole, J. O. (2015). "We are connected, but constrained": internet inequality and the challenges of millennials in Africa as actors in innovation. *Journal of Innovation and Entrepreneurship*, 5(3), 1–21. doi: 10.1186/s13731-015-0029-1.

Deloitte. (2020). The Deloitte Global Millennial Survey 2020. Resilient generations hold the key to creating a better normal. https://www2.deloitte.com/global/en/pages/about-deloitte/articles/millennialsurvey.html.

DiCicco-Bloom, B., & Crabtree, B. F. (2006). The qualitative research interview. *Medical Education*, 40(4), 314–321. doi: 10.1111/j.1365-2929.2006.02418.x.

Dragan, D. (2018). *Religion and spirituality conceptualisations across the lifespan: a mixed methods approach*. (Doctoral dissertation, University of Alabama Libraries). http://ir.ua.edu/handle/123456789/5277.

Edmunds, J., & Turner, B. S. (2005). Global generations: social change in the twentieth century. *British Journal of Sociology* 56(4), 559–577. doi: 10.1111/j.1468-4446.2005.00083.x.

Eliade, M. (2020). *Shamanism: Archaic techniques of ecstasy* (Vol. 76). Princeton, NJ: Princeton University Press.

Faudree, P. (2015). Tales from the land of magic plants: textual ideologies and fetishes of indigeneity in Mexico's Sierra Mazateca. *Comparative Studies in Society and History*, 57(3), 838–869. doi: 10.1017/S0010417515000304.

Feinberg, B. (2018). Undiscovering the pueblo magico: Lessons from Huautla for the psychedelic renaissance. In B. C. Labate & C. Cavnar (Eds.), *Plant Medicines, Healing and Psychedelic Science* (pp. 37–54). Springer Publishing AG.

Feinberg, B. (2018). Undiscovering the pueblo magico: Lessons from Huautla for the psychedelic renaissance. In *Plant medicines, healing and psychedelic science* (pp. 37–54). Cham: Springer.

Future Cast. (2016). The millennial brief on travel and lodging. Barkley Inc. and Future Cast LLC. https://cutt.ly/SkZuQzJ.

Galvani, A., Lew, A. A., & Perez, M. S. (2020). COVID-19 is expanding global consciousness and the sustainability of travel and tourism. *Tourism Geographies*, 22(3), 567–576. doi: 10.1080/14616688.2020.1760924.

Garcia, I., Acosta, R., & Piña, S. (2020). Niños Santos, psilocybin mushrooms and the psychedelic renaissance. *Chacruna Institute for Psychedelic Plant Medicines*. https://bit.ly/2JfRwFg.

Garikapati, V., Pendyala, R., Morris, E., Mokhtarian, P., & McDonald, N. (2016). Activity patterns, time use, and travel of millennials: a generation in transition?, *Transport Reviews*, 36(5), 558–584. doi: 10.1080/01441647.2016.1197337.

Gay, D. A., & Lynxwiler, J. (2013). Cohort, spirituality, and religiosity: A cross-sectional comparison. *Journal of Religion & Society*, 15, 1–17. http://hdl.handle.net/10504/64325.

Gelfeld, V. (2017). Travel Research: 2018 Travel Trends. *AARP*. https://www.aarp.org/content/dam/aarp/research/surveys_statistics/life-leisure/2017/2018-travel-trends.doi.10.26419%252Fres.00179.001.pdf.

George, J., Michaels, T., Sevelius, J., & Williams, M. (2019). The psychedelic renaissance and the limitations of a white-dominant medical framework: A call for indigenous and ethnic minority inclusion. *Journal of Psychedelic Studies*, 1–12. doi: 10.1556/2054.2019.015.

Gill, C., Packer, J., & Ballantyne, R. (2019). Spiritual retreats as a restorative destination: Design factors facilitating restorative outcomes. *Annals of Tourism Research*, 79, 102761. doi: 10.1016/j.annals.2019.102761.

Goldberg, S. B., Pace, B. T., Nicholas, C. R., Raison, C. L., & Hutson, P. R. (2020). The experimental effects of psilocybin on symptoms of anxiety and depression: A meta-analysis. *Psychiatry Research*, 112749. doi: 10.1016/j.psychres.2020.112749.

Griffiths, R., Richards, W., Johnson, M., McCann, U., & Jesse, R. (2008). Mystical-type experiences occasioned by psilocybin mediate the attribution of personal meaning and spiritual significance 14 months later. *Journal of Psychopharmacology*, 22(6), 621–632. doi: 10.1177/0269881108094300.

Hackett, C., & Grim, B. J. (2012). The global religious landscape: a report on the size and distribution of the world's major religious groups as of 2010. *Pew Research Center*. https://assets.pewresearch.org/wp-content/uploads/sites/11/2014/01/global-religion-full.pdf.

Harner, M. (1980). *The way of the shaman: a guide to power and healing*. San Francisco: Harper and Row.

International Monetary Fund. (2017). Millennial and the future of work. *Finance and Development* 54(2). https://www.imf.org/external/pubs/ft/fandd/2017/06/pdf/fd0617.pdf.

Johnson, B. (1939). *Some notes on the Mazatec*. Austin: University of Texas Press.

Johnson, M. W., Garcia-Romeu, A., Cosimano, M. P., & Griffiths, R. R. (2014). Pilot study of the 5-HT2AR agonist psilocybin in the treatment of tobacco addiction. *Journal of Psychopharmacology*, 28(11), 983–992. doi: 10.1177%2F0269881114548296.

Kaplan, E. B. (2020). The Millennial/Gen Z leftists are emerging: Are sociologists ready for them? *Sociological Perspectives*, 63(3), 408–427. doi: 10.1177/0731121420915868.

Kavenská, V., & Simonová, H. (2015). Ayahuasca Tourism: Participants in Shamanic Rituals and their Personality Styles, Motivation, *Benefits and Risks. Journal of Psychoactive Drugs*, 47(5), 351–359. doi: 10.1080/02791072.2015.1094590.

Kelly, J. R., Crockett, M. T., Alexander, L., Haran, M., Baker, A., Burke, L., Brennan, C., & O'Keane, V. (2020). Psychedelic science in post-COVID-19 psychiatry. *Irish Journal of Psychological Medicine*, 1–6. Advance online publication. doi: 10.1017/ipm.2020.94.

Kometer, M., Pokorny, T., Seifritz, E., & Volleinweider, F. X. (2015). Psilocybin-induced spiritual experiences and insightfulness are associated with synchronisation of neuronal oscillations. *Psychopharmacology*, 232(19), 3663–3676. doi: 10.1007/s00213-015-4026-7.

Krediet, E., Bostoen, T., Breeksema, J., van Schagen, A., Passie, T., & Vermetten, E. (2020). Reviewing the potential of psychedelics for the treatment of PTSD. *International Journal of Neuropsychopharmacology*, 23(6), 385–400.

Kurz, C. J., Li, G., & Vine, D. J. (2019). Are millennials different? *Handbook of US Consumer Economics*, (pp. 193–232). doi: 10.1016/B978-0-12-813524-2.00008-1.

Kurz, C. J., Li, G., & Vine, D. J. (2019). Are millennials different?. In *Handbook of US consumer economics* (pp. 193–232). Academic Press.

Laure, A. & Hannon, D. (2018). Psychedelic tourism in Mexico. *A thriving trend. PASOS: Revista de Turismo y Patrimonio Cultural*, 16(4), 1037–1050. http://www.pasosonline.org/es/articulos/1209-x-86.

Minero, F. (2015). Viaje nocturno a lo sagrado [Nocturnal journey into the sacred]. En W. Jacorzynski y M.T. Rodríguez (Eds.), *El encanto discreto de la modernidad* (pp. 161–179). Publicaciones de la Casa Chata.

Morris, N. (2009). Understanding digital marketing: marketing strategies for engaging the digital generation. *Journal of Direct, Data and Digital Marketing Practice*, 10(4), 384–387. doi: 10.1057/dddmp.2009.7.

Norman, A. (2011). *Spiritual Tourism: Travel and Religious Practice in Western Society*, London and New York: Bloomsbury Academic.

Norman, A., & Pokorny, J. J. (2017). Meditation retreats: Spiritual tourism well-being interventions. *Tourism Management Perspectives*, 24, 201–207. doi: 10.1016/j.tmp.2017.07.012.

Nutt, D., Erritzoe, D., & Carhart-Harris, R. (2020). Psychedelic Psychiatry's Brave New World. *Cell*, 181(1), 24–28. doi: 10.1016/j.cell.2020.03.020.

Olsen, D. (2013). *Definitions, motivations and sustainability: The case of spiritual tourism*. In *First UNWTO International Conference on Spiritual Tourism for Sustainable Development*, Ninh Binh Province, Viet Nam (pp. 21–22). https://www.e-unwto.org/doi/pdf/10.18111/9789284416738.

Osmond, H. (1957). A review of the clinical effects of psychotomimetic agents. *Annals of the New York Academy of Sciences*, 66(3), 418–434. doi: 10.1111/j.1749-6632.1957.tb40738.x.

Piña, S. (2019). Turismo y chamanismo, dos mundos imbricados: el caso de Huautla de Jiménez, Oaxaca [Tourism and shamanism, two intertwined worlds: the case of Huautla de Jimenez]. *Cuicuilco*, 26(75) 43–66. https://www.revistas.inah.gob.mx/index.php/cuicuilco/article/view/14716.

Pollan, M. (2018). *How to change your mind*. New York: Penguin Press.

Pond, A., Smith, G., & Clement, S. (2010). Religion among the millennials. Less religiously active than older Americans, but fairly traditional in other ways.

*Pew Research Center*. https://assets.pewresearch.org/wp-content/uploads/sites/11/2010/02/millennials-report.pdf.

Prayag, G., Mura, P., Hall, C. M., & Fontaine, J. (2016). Spirituality, drugs, and tourism: tourists' and shamans' experiences of ayahuasca in Iquitos, Peru. *Tourism Recreation Research*, 41(3), 314–325. doi: 10.1080/02508281.2016.1192237.

Rainer, T. S., & Rainer, J. (2011). *The Millennials: Connecting to America's Largest Generation*. Nashville, TN: B & H Books.

Reko, B. P. (1919). On Aztec botanical names. *El México Antiguo*, 1(5), 113–157.

Rowlands, I., Nicholas, D., Williams, P., Huntington, P., Fieldhouse, M., Gunter, B., Withey, R., Jamali, H. R., Dobrowolski, T., & Tenopir, C. (2008). The Google generation: the information behaviour of the researcher of the future, *Aslib Proceedings*, 60(4), 290–310. doi: 10.1108/00012530810887953.

Rucker, J., Schnall, J., D'Hotman, D., King, D., Davis, T., & Neill, J. (2020). *Medicinal use of psilocybin*. The Adam Smith Institute. https://bit.ly/3cJK9mc.

Sandars, J., & Morrison, C. (2007). What is the Net Generation? The challenge for future medical education. *Medical Teacher*, 29(2–3), 85–88. doi: 10.1080/01421590601176380.

Schultes, R. E. (1939). Plantae Mexicanae II. The identification of Teonanacatl, a narcotic basidiomycete of the Aztecs. *Botanical Museum Leaflets*, 7(3), 37–54. doi: 10.2307/275845.

Scuro J., & Rodd R. (2015) Neo-Shamanism. In: Gooren H. (Eds.). *Encyclopedia of Latin American Religions*. Cham: Springer. doi: 10.1007/978-3-319-08956-0_49-1.

Shead, S. (2020, November 23). Peter Thiel backs Berlin start-up making psychedelics in $125 million round. *CNBC*. https://www.cnbc.com/2020/11/23/peter-thiel-backs-psychedelics-startup-atai.html.

Stamets, P. (1996). *Psilocybin Mushrooms of the World: An Identification Guide*. Berkeley, CA: Ten Speed Press.

Twenge, J. (2006). *Generation me: Why today's young Americans are more confident, assertive, entitled—and more miserable—than ever before*. New York: Free Press.

UNWTO. (2020, October). Impact assessment of the COVID-19 outbreak on international tourism. *World Tourism Organization*. https://www.unwto.org/impact-assessment-of-the-covid-19-outbreak-on-international-tourism.

Wasson, R. G. (1957, May 13). Seeking the magic mushroom. *Life*, 49(19), 100–102. Retrieved from: https://rb.gy/bbdy3p.

Wixwat, M., & Saucier, G. (2020). Being spiritual but not religious. *Current Opinion in Psychology*. doi: 10.1016/j.copsyc.2020.09.003.

Wolff, T. J. (2020). Neo-shamanism. In *The Touristic Use of Ayahuasca in Peru* (pp. 25–28). Wiesbaden: Springer VS.

# 12 From gen Xers to millennials

## Transformation of self through tourism

*Reni Polus and Neil Carr*

### 12.1 Introduction

Spirituality is a highly problematic topic of study, both within its links to religion and faith, and in its transcendent form, as well as across all three. For decades, the study of the sacred in the tourism experience was situated exclusively within the religious point of view as for many the term *sacred* is linked to religion (Willson, 2012). Yet there have been increasing calls to embrace spirituality in tourism studies (Sharpley & Jepson, 2011; Willson, 2012). In contemporary times, travelling has been noted as an essential element of spiritual practices. Young people, currently often referred to as 'Millennials' (Strauss & Howe, 1991; Best & Kellner, 2003), have become the largest age cohort to embrace this form of travel (McIntosh & Mansfeld, 2006; Wilson & Harris, 2006; Benckendorff, Moscardo, & Pendergast, 2010; Sharpley & Jepson, 2011). They view the sacred in the tourism experience as not merely about religion, but as a label for a different category of experience, in this case spiritually transformative experiences, conceptualized as a journey towards transformation of self (Sharpley & Jepson, 2011; Voigt, Brown, & Howat, 2011; Reisinger, 2013; Kirillova, Lehto, & Cai, 2017). The rise of this conceptualization of travel was presaged by Brunton (1934) in his book, *A Search in Secret India,* where he predicted that the concept of spirituality would become the main driver for tourism.

The term *Millennials* encompasses a heterogeneous subculture, exhibiting critical distinctions between previous youth generations and those to come (Strauss & Howe, 1991; Pew Research Center, 2014). Understanding the present tourism experiences and desires of Millennials is critical as they are noted as the tourists of the future (Richards, 2011; Monaco, 2018), with different needs, wants and travel behaviour than the contemporary middle-aged tourist (Cavagnaro, Staffieri, & Postma, 2018). Accordingly, it is paramount to understand the generational change in order to recognise changes in tourism consumption and draw possible future scenarios. Robinson and Schänzel (2019, p. 127) have highlighted this issue, maintaining that "generational change is one such occurrence, rending both opportunities and challenges for tourism destinations". Thus, the purpose of this chapter is to provide a conceptual clarification of the differences in how Millennials, compared with

DOI: 10.4324/9781003170631-15

their Generation X siblings, identify and define spiritual transformative experiences through tourism. The comparative assessment of spiritually transformative experiences of the different generations of youth is one way of exploring meaningful differences in this market. Most importantly, the comparison also allows the travel industry to anticipate any changes of emphasis among future youth generations to enable them to better serve the upcoming generations of youth travellers driven by spirituality.

The chapter begins by defining generational cohorts, highlighting how differences exist across the youth generations. This focuses on the transition from being 'spiritual and religious' to 'spiritual but not religious'. The chapter then discusses the influence of generational shifts on tourist behaviour, linking it to the notion of the transition from modern to postmodern tourism. It is followed by a discussion of the two significant trends in the conceptualization of spiritually transformative experience in tourism: from the quest for object-related authenticity to the quest for activity-related authenticity, and from transformative experience as an *outcome* to transformative experience as a *process*.

## 12.2 Defining generations

Before understanding generational differences, it is critical to define the generations discussed in this chapter. A generation has multiple meanings which may include families/groups with similar origins, cohorts, life stages and historical periods (Kertzer, 1983). This chapter recognises that *generation* refers to a cohort group. Strauss and Howe (1991, p. 34) defined a generation as "a cohort-group whose length approximates the span of a phase of life and whose boundaries are fixed by peer personality". They then defined peer personality as "generational persona recognised and determined by (1) common age location; (2) common beliefs and behaviour; and (3) perceived membership in a common generation" (Strauss & Howe, 1991, p. 64). While members from a certain generation cohort may share similar values and habits, they are likely to have completely different attitudes, behaviours and lifestyle preferences to other generations. While generations represent distinct cohorts, they are not internally homogeneous. Instead, generations encompass heterogeneous subcohorts distinct from, though interrelated with, each other.

Two generational cohorts that represent most of the tourism market in the contemporary world are Generation X and Millennials (Benckendorff et al., 2010; Huang & Petrick, 2010). Generation X (also known as Gen X or Xers) incorporates those who were born between 1965 and 1980 (ages 40 to 55 in 2020) and Millennials (also known as Generation Y) who were born between 1981 and 1997 (ages 23 to 39 in 2020) (Pew Research Center, 2014). Although the oldest Millennials are adults now, the majority of them are still considered as young people. Using the chronological definitions of 'young' posited by Carr (1998), the 'young people' population or 'youth' can be said to be those between 15 and 35 years of age. The Millennials age group thus largely matches the chronological definition of youth tourists. Yet it is important to note that the definition of young people is more than a matter of simple chronological age. Rather, it is also a product of socio-cultural conditions and indicators (Carr, 1998). This implies that there is a certain fuzziness of the definition of young people, something that leaves its edges open to debate.

The life of members of Generation X has not been easy as they are the generation that grew up around high divorce rates, poor economies that included significant global shocks (e.g., 1987 stock market crash), and high crime rates (Cahill & Sedrak, 2012). On the other hand, the Millennials have been noted as the "trophy kids" (Cahill & Sedrak, 2012, p. 7). This generation is closely protected by their "helicopter parents" as they are considered as the "special" generation (Cahill & Sedrak, 2012, pp. 7–8). Additionally, the Millennials are growing up in the digital era, with their lives consequently being deeply influenced by the use of technologies, especially the Internet (Andert, 2011; Cahill & Sedrak, 2012). Most importantly, Millennials observe and live within more non-traditional families and associated values than previous generations (Andert, 2011). Included within this context, Millennials are living and growing up in a time of unprecedented questioning of religion and religious affiliations and ways of life (Cox & Thomson-DeVeaux, 2019; Percy, 2019).

Scholars have noted that there is a paradigm shift taking place in religious affiliation from Generation Xers to Millennials (e.g., Crockett & Voas, 2006; Pew Research Center, 2014; Percy, 2016; Cox & Thomson-DeVeaux, 2019; Percy, 2019). Generation Xers was more religious than the younger generation (Crockett & Voas, 2006; Norris & Inglehart, 2011). However, as the Generation Xers came of age, they tended to leave religious life and search for the *sacred* within a non-traditional, novel or new contexts (Crockett & Voas, 2006; Norris & Inglehart, 2011). This gels with the term *spirituality,* which is described as a process of discovering the *sacred* beyond the traditional concept of God through a limitless number of pathways involving systems of belief, incorporating traditional religious practices, newer spirituality movements, and more individualized worldviews (Zinnbauer & Pargament, 2005). When asked what religion they follow, Millennials typically say 'none' (Percy, 2019). Indeed, Percy (2019: 166) states the average Millennial "believes in many good things, and also in God: but does not join a faith-organisation to express this". In other words, Millennials place a greater emphasis on individualism and spirituality (Pew Research Center, 2014). Accordingly, a shift has occurred from being the 'spiritual and religious' of Generation X during their youth to the 'spiritual but not religious' of Millennials during their youth. The leaving behind, or indeed never exchanging with, religion of Millennials at an earlier age than their Generation X predecessors may be seen as an evolving pattern, showing the continuity between generations, rather than a radical shift and disjuncture between them. It speaks to the continually evolving nature of society (Crockett & Voas, 2006; Cox & Thomson-DeVeaux, 2019; Percy, 2019).

The 'spiritual and religious' group are associated with people applying religion as a guiding point in everyday life, engaging not only in traditional forms of worship such as prayer, church attendance and religious orthodoxy but also favouring spirituality elements such as spiritual growth and monism (Zinnbauer et al., 1997; Roof, 1999; Berghuijs, Pieper, & Bakker, 2013). The 'spiritual but not religious' group, on the contrary, is characterized as being less likely to engage in traditional forms of worship but holding non-traditional 'New Age' beliefs or 'New Age Spirituality' and being more individualistic (Zinnbauer et al., 1997; Roof, 1999). Hill et al. (2000) reason that the rise of spirituality indicates increased respect for the inner, contemplative practices of traditional religious systems. Taken together, these different lifestyles and values that exist in the different generations lead to differences in consumption patterns in tourism.

## 12.3 Tourism consumption patterns: from modernity to postmodernity

From a cross-generational perspective, several studies demonstrate that generational shifts in the demographic profile of society are exerting a major influence on tourism behaviour (Nedelea, 2008; Gardiner, King, & Grace, 2013; Gardiner, Grace, & King, 2015). This suggests that tourism is being anticipated and experienced in different ways than previously. As predicted by Pine and Gilmore (1999), we are currently in the experience economy, where tourists seek a series of experiences that lead to learning, growth and the achievement of aspirations and goals. They seek special, extraordinary events that can trigger highly emotional responses which, in turn, lead to self-realization and self-exploration (Brown, 2013; Kirillova et al., 2017). As such, these experiences serve as a vehicle for personal transformation or profound intra-personal changes, and are conducive to optimal human functioning (Brown, 2013; Kirillova et al., 2017). In this way, the tourist not only enjoys the experience but is also transformed by it (Pine & Gilmore, 1999). The desire for transformative experience has been a major driver of tourism in modern society, but, for the postmodern society, it is preeminent (Hall, 1992; Pine & Gilmore, 1999).

The concept of postmodernism is described as "everywhere and nowhere" and is characterized by its compromising nature that is not "either-or", but "both-and" (Denzin, 1991, p. 151). In other words, postmodern consumers do not fixate on experiencing only one thing at a time, but seek to experience diversity. In the tourism context, the postmodern tourist or post-tourist is often characterized as a traveller who enjoys combining different types of experience in a single trip (Feifer, 1985). Post-tourists habitually switch from one mode of experience to another. In this way, their desires and behaviour are fluid and open to change (Maoz & Bekerman, 2010). They are afraid of "being bound and fixed" and are driven by the constant need to seek out authentic experiences as "the joys of the familiar wear off and cease to allure" (Baumann, 1996, p. 29). Hence, it can be argued that how Millennials, living and growing up in a postmodern reality, compared with their Generation X siblings who grew up in modernity, identify and define spiritually transformative experiences through tourism is also experiencing significant changes.

## 12.4 Transformative experiences and authenticity

The search for a transformative experience in tourism is often linked with a search for authenticity. Historically, it was MacCannell (1973) who introduced the concept of authenticity in the studies of tourist motivations and experiences. He argued that the tourist was searching for authenticity in travel, which is similar to the pilgrim's search, thus making tourism a modern surrogate for religion. Steiner and Reisinger (2006, p. 299) defined authenticity as "genuineness or realness of artifacts or events, and also as a human attribute signifying being one's true self or being true to one's essential nature". In tourism, authenticity is often related to tourism objects, sites, attractions and experiences (Rickly-Boyd, 2012). While a large body of literature exists that is focused on notions of authenticity in tourism, this chapter utilizes Wang's (1999) two dimensions of authenticity: authenticity of object-related (objective and constructive); and authenticity of activity-related, which is also referred to as existential authenticity.

The conception of authenticity has undergone a series of conceptual shifts, with objectivist framings giving way to a social construction perspective and, more recently, existentialist conceptions (Wang, 1999; Kelner, 2001; Steiner & Reisinger, 2006; Kim & Jamal, 2007). Within this context, it may also be claimed that authenticity has different meanings for different generations. Indeed, Pine and Gilmore (1999) claim that the Millennials focus less on quantity and more on quality; less on material acquisition and more on personal attainment. Furthermore, Cohen (2008, p. 332) states that "the veracity of the 'objective' authenticity of attractions has been replaced by a growing concern with the tourists' experiences of 'subjective' authenticity on their trip... resembling that found at the climax of intense erotic or religious experiences". This implies that post-tourists are seeking 'profound' experiences which are similar to the so-called spiritual, sacred or mystical experiences of the pilgrim. Consequently, spiritual search has been noted as a dominant feature of late-twentieth-century life where travelling is a "...source of spiritual meaning or refreshment" (Sharpley & Jepson, 2011, p. 52). It can thus be suggested that the transition from Generation Xers to Millennials has been marked by a change in the discourse of authenticity from the tourist quests for object-related authenticity of attractions to tourists' experiences of existential authenticity on their trip.

### *12.4.1 From the quest for object-related authenticity to the quest for existential authenticity*

The search for spiritually transformative experiences through objective authenticity can be linked to how Generation Xers have and are travelling. Wang (1999, pp. 351–352) defined objective authenticity as that which involves a "museum linked usage of the authenticity of the originals that are also the toured objects to be perceived by tourists". In this way, objective authenticity refers to the "authenticity of originals". As noted earlier,

Generation Xers tend to be more religious than the younger generation. It is therefore likely that connections exist between objective authenticity and religious Generation Xers travelling for spiritual transformation, usually to recognised religious sites. In this context, the search for spiritual transformative experiences primarily focuses on the genuineness of the site, its structures and social context (Barthel, 1996; Wang, 1999). For example, Muslims go to Mecca (the birthplace of Muhammad) on *Hajj*, Christians go to the Vatican City in Rome (the home of the Pope) and Buddhists go to Bodh Gaya in India (the site where Buddha attained Supreme Enlightenment). Therefore, the originality of a site could be a major factor, if not the only one, causing pilgrims to travel to the sites. As such, the pilgrims recognise the sites that they encounter as authentic due to the fact that all these sites are synonymous with where their religions started. Hence, being able to experience the social context at these holy sites can foster spiritually transformative experiences.

As the Generation Xers move into and mature through adulthood, they carry with them another form of authenticity. This reflects the notion that as the Generation Xers came of age, they tended to drop out of religious life and sought spiritual experiences beyond the traditional concept of religion (Crockett & Voas, 2006; Norris & Inglehart, 2011). This process is reflected in the postmodern consumers proactively seeking constructivist authenticity. Unlike objective authenticity, constructive authenticity refers to tourists seeking symbolically authentic tourism experiences or tourism objects and what is authentic depends on the context and ideology of the tourist (Wang, 1999). In other words, authenticity or inauthenticity is the result of how an individual perceives things, based on "points of view, beliefs, perspectives, or powers" (Wang, 1999, p. 351). This form of authenticity is linked to the concept of tourism as a 'sacred journey' (Graburn, 1977). For Graburn (1977), the journey is about self-transformation, and the acquisition of knowledge and status through contact with the extraordinary. Scholars have noted these forms of travel as a secular pilgrimage, which is incidentally related to the quest for constructive authenticity (Digance, 2006; Timothy & Conover, 2006). Timothy and Conover (2006) described the secular pilgrims as those who may or may not possess the same religious or spiritual intent or religious devotion that is common to traditional religious pilgrims. Digance (2006, p. 37) noted two distinct types of secular pilgrims: those who still observe traditional religion in their journey, and those who embrace inner spirituality. The former refers to pilgrims who still recognise traditional religions as being integral to fulfilling their spiritual needs. Collins-Kreiner (2010, p. 153) identified them as "modern secular" pilgrims. Marine-Roig (2015) identified these pilgrims and all they represent with religious tourism.

In contrast to the modern secular pilgrims, those emphasizing the experiential and embracing inner spirituality are labelled as the "New Age" pilgrims (Ivakhiv, 2003, p. 94; Digance, 2006, p. 37). This concept of travel is clearly linked to the Millennials that are more likely than any other generation to favour spirituality both in itself and also over religion. In other words, Millennials are far more likely to look for experiences that focus on the existential self rather than external objects (Smith & Kelly, 2006; Bulencer & Egger, 2015). In the context of authenticity, this form of experience is linked to the concept of existential authenticity, which relates to "personal or intersubjective feelings activated by the liminal process of tourist activities" (Wang, 1999, p. 351). Existential authenticity, according to Wang (1999, p. 352), "can have nothing to do with the authenticity of toured objects", but refers to a "potential existential state of Being that is to be activated by tourist activities".

For the Millennials, travelling is a means for an individual to obtain self-understanding and self-renewal. Even though the hedonistic aspects of travel continue to exist, for them, travel is no longer solely a tool to escape from the ordinary or mundane, or something to undertake during hedonic leisure time (Monaco, 2018). Rather, Millennials see the travel experience as an important tool to create an impact in their lives and to construct their identities (Larsen & Urry, 2016; Monaco, 2018). Some have noted that Millennials are the 'deep' tourists (Timothy & Olsen, 2006), who seek extraordinary, pure, peak or cathartic experiences. This clearly relates to Cohen's (1996) 'existential mode' of the tourist experience. Cohen claims that tourists travelling in the 'existential mode' are similar to pilgrims as both are fully committed to an 'elective' spiritual centre. Simply put, they have found a new 'elective centre' through tourism, that has a deep spiritual meaning, and the potential to provide life-changing experiences (Cohen, 1979).

The shift towards a focus on life-changing experiences has led scholars to argue that post-tourists are seeking an experience that is not merely about sightseeing and material consumption. Rather, they seek experiences focused on spiritual growth or enlightenment (Timothy & Olsen, 2006; Sharpley & Jepson, 2011; Dillette, Douglas, & Andrzejewski, 2019). The quest for spiritual horizons evokes the ideas of *the journey* in search of life meaning, fulfilment, connectedness and transcendence through touristic pursuit (Sharpley & Sundaram, 2005; Olsen & Timothy, 2006; Laing & Crouch, 2009, 2011; Matteucci, 2013). Ironically, this refers back to the deepest meaning of leisure as a search for self, which predates the modernist consumerization of leisure and tourism. Leisure as a search for self is focused on the idea of *freedom to be*, which is linked to the idea of personal enlightenment (Bammel & Burrus-Bammel, 1996; Clarke & Critcher, 2016; Carr, 2017).

### 12.4.2 *From transformative experience as an outcome to transformative experience as a process*

The transformative experience is classified into two different trends regarding how spiritual transformation is defined. Transformative experience has been described as an *outcome* as well as a *process*. Christie and Mason (2003, p. 9) define the transformative experience in terms of outcomes as "the practice of organised [travel] that leads to a positive change in attitudes and values among those who participate in the [travel] experience". This form of travel can be linked to the concept of religious pilgrimage, which Vukonić (1996) describes as a journey undertaken by a person in search of holiness, truth, and the sacred. As noted earlier, Generation Xers are significantly more religious than Millennials. Thus, the idea of religious pilgrimage reflects the Generation Xers travelling to sacred sites for spiritual transformation, where the focus is on the 'arriving at a destination'. For example, in the case of Islam, a religious pilgrimage is a ritual act of worship (Haq & Jackson, 2009), which is the fifth of the Five Pillars of Islam (Gothoni, 1993). Every Muslim must go to Mecca on *Hajj* at least once in their lifetime to submit to God's will, making *Hajj* the very essence in Islam. Meditation, prayer and performing religious rites at the worship centre are often associated with religious pilgrimage (Jackowski & Smith, 1992). Notable in this regard, it can be argued that pilgrims review their completed goals through the lens of an 'outcome' (destination) rather than a 'process' (journey).

Contrary to Generation Xers, Millennials emphasize the *processes* involved in spiritually transformative experiences. Instead of thinking about achieving a goal as 'arriving at a destination', Millennials view it as 'completing a journey'. For them, the quest is not for religion *per se*, but what is *sacred* is the significance of the purpose with which the journey is undertaken and the anticipated lasting impact of its completion (Leite & Graburn, 2009). Thus, for the Millennials, the transformation of self is a process and not an end result as transformation can occur over and over again in a person's life (Mezirow, 1991). This aspect of transformation is focused on 'meaning-making'(Gall et al., 2005; Lancaster & Palframan, 2009) and it is an individualized process (Wade, 1998; Coghlan & Weiler, 2018). Indeed, Coghlan and Weiler (2018) noted that in order to experience transformation, an individual needs to conduct a radical self-examination. This is a reflection on "what they know, how they know it, the way they think and the way they relate to others" (Coghlan & Weiler, 2018, p. 571). In other words, in reaching a state of self-enlightenment the individuals do not only understand the meaning of the travel experience, but also identify the elements that contributed to the process of transformation (Coghlan & Weiler, 2018).

Perhaps the most powerful metaphors for the spiritually transformative experience as a process among Millennial tourists are the *inner journey, quest* and *rites of passage*. All these metaphors use the concept of a 'journey'. This is a helpful metaphor to highlight the process of getting 'there', rather than simply reaching the destination. In other words, a *journey* is more than the patterns of the overall movement. In this way, a *journey* is also a metaphor relating to the 'inner journey'. It simply means the outer journey (actual physical movement through space over time) is the vehicle for the inner journey (psychological development or growth) (Morgan, 2010). When both are taken simultaneously, any *journey* can be a profound holistic body, mind and soul experience. As Neumann (1992, p. 182) aptly states: "journeys provide the opportunity to acquire experiences that become the basis for discovering and transforming one's self". Hence the term 'sacred journey' or 'spiritual journey' has been linked to tourism to denote tourism as a vehicle for an inner spiritual journey (e.g., Sharpley & Sundaram, 2005; Timothy & Olsen, 2006).

Several studies explicitly address the transformative potential of tourism as a process. The idea that a travel experience could result in a deep and enduring spiritual transformation in oneself has been discussed in various forms within contemporary tourism experiences. For example, experiential themes such as personal growth and self-renewal have been evidenced in adventure tourism (Arnould & Price, 1993). In addition, backpacking tourism has been identified as a powerful experience of self-change (Noy, 2004) and connectedness with self (Cohen, 2010). Furthermore, volunteer tourism has been identified as having the potential to provide a stimulating cathartic experience (Zahra & McIntosh, 2007) and an experience that offers the opportunity for self-exploration and relating (McIntosh & Zahra, 2008). Another notable example is Laing and Crouch's (2011) study that explores the spiritual transformation in frontier travel, focusing on the catalysts of self-transformation. They identify characteristics such as sacrifice, danger and hardship, personal transformation, enrichment of body and soul, and communitas that may all enhance the spiritual significance of frontier tourism. Conclusively, irrespective of travelling away from home and back again, the *journey* itself represents a spiritual experience that transforms the self (Sharpley & Jepson, 2011). An inner journey has taken place through the maturation process of the experience. Yet such an inner experience must be consciously engaged with in order to be experienced. This is what the Millennials are doing, aided by their inner-directed focus on the search for spirituality rather than the focus on the religion of the Generation Xers. This explains how types of tourism, like those noted above, which predate Millennials can be repurposed by tourists. In this way, the physical journey becomes a vehicle, an entity utilized by the individual for their own purpose. This speaks to the social agency of the individual to repurpose existing physical routes and vehicles to create journeys fit for their internally directed purpose, the individualized search for the spiritual.

## 12.5 Conclusion

A comparative assessment of spiritually transformative experiences of different generations of youth is one way of exploring meaningful differences in this market, particularly those seeking spiritual transformation through tourism. This chapter provides a conceptual clarification of the differences between Millennials and Generation Xers in relation to spiritually transformative experiences through tourism. The comparison allows the travel industry to anticipate changes of emphasis among future youth generations to enable them to better serve the upcoming generations of youth travellers (e.g., Generation Z) driven by spirituality. It also questions whether the industry can truly cater to the experiences that generations like the Millennials are seeking. Is it possible for an industry to construct experiences for people who are on an internal journey. Such constructions have been at the heart of the consumerist society that have led to questions about the extent to which people have abandoned the quest *to be* in favour of the desire *to have* (Rojek,

2000). This question is mirrored in Blackshaw's (2010, p. 17) suggestion that "today we appear to be prepared to surrender our hard fought freedoms to the vast decentred power-knowledge of consumer capitalism... What this means is that leisure has by and large been turned into consuming." In focusing on a spiritual search for self, Millennials are arguably turning their back on this consumerist society. Whether the tourism industry can or should provide experiences to *help* them in this quest is debatable. Rather, it is about Millennials critically and actively using pathways constructed by the tourism industry as vehicles of their own inner journey.

It is important to recognise that discussions of youth travel and spirituality must be framed within the context of global meta-trends and local diversity. This recognises the cultural heterogeneity of youth at global and local levels. While this chapter focuses on Millennials as a general concept, it is one that encapsulates different cohorts across different countries and cultures. In other words, what it means to be a Millennial, even in an era of unprecedented globalization, is not homogenous across different places and cultures. Even within a space/culture the experiences and desires of Millennials differ as they are shaped and driven by socio-cultural norms and values linked to socio-demographic characteristics. Through this, it is recognised that the engagement by Millennials with spirituality, religion and the self may differ within and across cultures and places. However, while this may alter the specific nature of how such engagements change across generations and the nuances of the focus of Millennials it is clear that such changes occur irrespective of local differences and that the direction of this change is as noted in this chapter, even if the extent of the change is specific to the physical and social locality of the Millennial.

## 12.6 Future implication

Future research should explore the nuanced differences in Millennials across cultures and places. This entails not only looking in non-Western cultures but also at the multicultural reality of Western countries, especially in this globalised 21st century. Furthermore, future research should investigate the emerging new cohort, specifically Generation Z, as they will be the driving force of tourism in the future. Born in the digital era, highly educated, creative and innovative, and notably non-religious, this young generation is likely to influence and transform the future of the tourism industry (Corbisiero & Ruspini, 2018; Robinson & Schänzel, 2019). To determine what motivates a cohort, it is critical to understand the world's events (e.g., COVID-19 pandemic) that will impact their lives and shape their worldview. Thus, several fundamental questions emerge: will the non-religious Millennials recognise the value of religion as they age? Can we assume that future young generations will be more likely to disaffiliate themselves with religion altogether? As the population ages and the Millennials will soon be adults, it will not come as a surprise if this generation completely reshapes the travel landscape of the mature travel market. Most importantly, a shifting conception of religion and spirituality amongst these different cohorts in future will transform the landscape of religious tourism and pilgrimages. Thus, keeping abreast of their changing needs, behaviours and preferences could be a leading edge of competitive advantage in tourism.

## References


Andert, D. (2011). Alternating leadership as a proactive organizational intervention: Addressing the needs of the Baby Boomers, Generation Xers and Millennials. *Journal of Leadership, Accountability, and Ethics*, 8(4), 67–83.

Arnould, E. J., & Price, L. L. (1993). River magic: Extraordinary experience and the extended service encounter. *Journal of consumer Research*, 20(1), 24–45. doi: 10.1086/209331.

Bammel, G., & Burrus-Bammel, L. L. (1996). Leisure and human behavior. *Journal of Leisure Research*, 28(2), 135.

Barthel, D. (1996). *Historic preservation*. New Brunswick, NJ: Rutgers University Press.

Baumann, Z. (1996). From pilgrim to tourist–or a short history of identity. In S. Hall & P. Du Gay (Eds.), *Questions of Cultural Identity* (pp 18–36). Sage Publications.

Benckendorff, P., Moscardo, G., & Pendergast, D. (2010). *Tourism and generation Y*. UK: CABI.

Berghuijs, J., Pieper, J., & Bakker, C. (2013). Being 'spiritual' and being 'religious' in Europe: Diverging life orientations. *Journal of Contemporary Religion*, 28(1), 15–32. doi: 10.1080/13537903.2013.750829.

Best, S., & Kellner, D. (2003). Contemporary youth and the postmodern adventure. *The Review of Education, Pedagogy & Cultural Studies*, 25(2), 75–93. doi: 10.1080/10714410390198949.

Blackshaw, T. (2010). *Leisure*. Abingdon: Routledge.

Brown, L. (2013). Tourism: A catalyst for existential authenticity. *Annals of Tourism Research*, 40, 176–190. doi: 10.1016/j.annals.2012.08.004.

Brunton, P. (1934). *A search in secret India: The classic work on seeking a guru*. London: Random House.

Bulencer, P., & Egger, R. (2015). *Gamification in tourism: Designing memorable experiences*. Norderstedt: Books on Demand.

Cahill, T. F., & Sedrak, M. (2012). Leading a Multigenerational Workforce: Strategies for Attracting and Retaining Millennials. *Frontiers of Health Services Management*, 29(1), 3–15. Retrieved from https://journals.lww.com/frontiersonline/Fulltext/2012/07000/Leading_a_Multigenerational_Workforce__Strategies.2.aspx.

Carr, N. (1998). The young tourist: a case of neglected research. *Progress in Tourism and Hospitality Research*, 4(4), 307–318. doi: 10.1002/(SICI)1099-1603(199812)4:4<307::AID-PTH133>3.0.CO;2-4.

Carr, N. (2017). Re-thinking the relation between leisure and freedom. *Annals of Leisure Research*, 20(2), 137–151. doi: 10.1080/11745398.2016.1206723.

Cavagnaro, E., Staffieri, S., & Postma, A. (2018). Understanding millennials' tourism experience: values and meaning to travel as a key for identifying target clusters for youth (sustainable) tourism. *Journal of Tourism Futures*, 4(1), 31–42. doi: 10.1108/JTF-12-2017-0058.

Christie, M. F., & Mason, P. A. (2003). Transformative Tour Guiding: Training Tour Guides to be Critically Reflective Practitioners. *Journal of Ecotourism*, 2(1), 1–16. doi: 10.1080/14724040308668130.

Clarke, J., & Critcher, C. (2016). *The devil makes work: Leisure in capitalist Britain*. New York: Palgrave.

Coghlan, A., & Weiler, B. (2018). Examining transformative processes in volunteer tourism. *Current Issues in Tourism*, 21(5), 567–582. doi: 10.1080/13683500.2015.1102209

Cohen, E. (1979). A phenomenology of tourist experiences. *Sociology*, 13(2), 179–201. doi: 10.1177/003803857901300203.

Cohen, E. (1996). A phenomenology of tourist experiences. *The sociology of tourism: Theoretical and empirical investigations*, 90–111.

Cohen, E. (2008). The changing faces of contemporary tourism. *Society*, 45(4), 330–333. doi: 10.1007/s12115-008-9108-2.

Cohen, S. A. (2010). Chasing a myth? Searching for 'Self' through lifestyle travel. *Tourist Studies*, 10(2), 117–133. doi: 10.1177/1468797611403040.

Collins-Kreiner, N. (2010). Researching pilgrimage: Continuity and transformations. *Annals of Tourism Research*, 37(2), 440–456. doi: 10.1016/j.annals.2009.10.016.

Corbisiero, F., & Ruspini, E. (2018). Millennials and generation Z: challenges and future perspectives for international tourism. *Journal of Tourism Futures*, 4(1), 3–104. doi: 10.1108/JTF-03-2018-069.

Cox, D., & Thomson-DeVeaux, A. (2019). Millennials are leaving religion and not coming back. *FiveThirtyEight–ABC News*. December 12. https://fivethirtyeight.com/features/millennials-are-leaving-religion-and-notcoming-back/.

Crockett, A., & Voas, D. (2006). Generations of Decline: Religious Change in 20th-Century Britain. *Journal for the Scientific Study of Religion*, 45(4), 567–584. doi: 10.1111/j.1468-5906.2006.00328.x.

Denzin, N. K. (1991). *Images of Postmodern Society: Social Theory and Contemporary Cinema*. London: Sage.

Digance, J. (2006). Religious and secular pilgrimage: Journeys redolent with meaning. In D. Timothy & D. Olsen (Eds.), *Tourism, Religion and Spiritual Journeys* (pp. 52–64). London and New York: Routledge.

Dillette, A. K., Douglas, A. C., & Andrzejewski, C. (2019). Yoga tourism–a catalyst for transformation? *Annals of Leisure Research*, 22(1), 22–41. doi: 10.1080/11745398.2018.1459195.

Feifer, M. (1985). *Going places: The ways of the tourist from Imperial Rome to the present day*. London: Macmillan London Limited.

Gall, T. L., Charbonneau, C., Clarke, N. H., Grant, K., Joseph, A., & Shouldice, L. (2005). Understanding the nature and role of spirituality in relation to coping and health: A conceptual framework. *Canadian Psychology/Psychologie canadienne*, 46(2), 88–104. doi: 10.1037/h0087008.

Gardiner, S., Grace, D., & King, C. (2015). Is the Australian domestic holiday a thing of the past? Understanding baby boomer, Generation X and Generation Y perceptions and attitude to domestic and international holidays. *Journal of Vacation Marketing*, 21(4), 336–350. doi: 10.1177/1356766715580188.

Gardiner, S., King, C., & Grace, D. (2013). Travel Decision Making: An Empirical Examination of Generational Values, Attitudes, and Intentions. *Journal of Travel Research*, 52(3), 310–324. doi: 10.1177/0047287512467699.

Gothoni, R. (1993). Pilgrimage= transformation journey. *Scripta Instituti Donneriani Aboensis*, 15, 101–116. doi: 10.30674/scripta.67208.

Graburn, N. (1977). Tourism: Sacred Journey. In V. L. Smith (Ed.), *Hosts and guests: The anthropology of tourism* (2nd ed., pp. 21–52). Philadelphia: University of Pennsylvania Press.

Hall, C. M. (1992). *Hallmark tourist events: impacts, management and planning*. London: Belhaven Press.

Haq, F., & Jackson, J. (2009). Spiritual journey to Hajj: Australian and Pakistani experience and expectations. *Journal of Management, Spirituality and Religion*, 6(2), 141–156. doi: 10.1080/14766080902815155.

Hill, P. C., Pargament, K. I., Hood, R. W., McCullough, J., Michael E, Swyers, J. P., Larson, D. B., & Zinnbauer, B. J. (2000). Conceptualizing religion and spirituality: Points of commonality, points of departure. *Journal for the Theory of Social Behaviour*, 30(1), 51–77. doi: 10.1111/1468-5914.00119.

Huang, Y.-C., & Petrick, J. F. (2010). Generation Y's travel behaviours: a comparison with Baby Boomers and generation X. In P. Benckendorff, G. Moscardo, & D. Pendergast (Eds.), *Tourism and generation Y* (Vol. 1, pp. 27–37). UK: CABI.

Ivakhiv, A. (2003). Nature and self in New Age pilgrimage. *Culture and Religion*, 4(1), 93–118. doi: 10.1080/01438300302812.

Jackowski, A., & Smith, V. L. (1992). Polish pilgrim-tourists. *Annals of Tourism Research*, 19(1), 92–106. doi: 10.1016/0160-7383(92)90109-3.

Kelner, S. (2001). Narrative construction of authenticity in pilgrimage touring. *Taglit-Birthright Israel.*

Kertzer, D. I. (1983). Generation as a sociological problem. *Annual Review of Sociology*, 9(1), 125–149.

Kim, H., & Jamal, T. (2007). Touristic quest for existential authenticity. *Annals of Tourism Research*, 34(1), 181–201. doi: 10.1016/j.annals.2006.07.009.

Kirillova, K., Lehto, X., & Cai, L. (2017). Tourism and Existential Transformation: An Empirical Investigation. *Journal of Travel Research*, 56(5), 638–650. doi: 10.1177/0047287516650277.

Laing, J. H., & Crouch, G. I. (2009). Lone wolves? Isolation and solitude within the frontier travel experience. *Geografiska Annaler: Series B, Human Geography*, 91(4), 325–342. doi: 10.1111/j.1468-0467.2009.00325.x.

Laing, J. H., & Crouch, G. I. (2011). Frontier tourism: Retracing mythic journeys. *Annals of Tourism Research*, 38(4), 1516–1534. doi: 10.1016/j.annals.2011.02.003.

Lancaster, B. L., & Palframan, J. T. (2009). Coping with major life events: the role of spirituality and self-transformation. *Mental Health, Religion & Culture*, 12(3), 257–276. doi: 10.1080/13674670802500684.

Larsen, J., & Urry, J. (2016). *Mobilities, networks, geographies*. New York: Routledge.

Leite, N., & Graburn, N. (2009). Anthropological interventions in tourism studies. In T. Jamal & M. Robinson (Eds.), *The SAGE handbook of tourism studies* (pp. 35–64). London: Sage.

MacCannell, D. (1973). Staged authenticity: Arrangements of social space in tourist settings. *American Journal of Sociology*, 79(3), 589–603. doi: 10.1086/225585.

Maoz, D., & Bekerman, Z. (2010). Searching for Jewish answers in Indian resorts: The postmodern traveler. *Annals of Tourism Research*, 37(2), 423–439. doi: 10.1016/j.annals.2009.10.015.

Marine-Roig, E. (2015). Religious tourism versus secular pilgrimage: The basilica of La Sagrada Família. *The International Journal of Religious Tourism and Pilgrimage*, 3(1), 25–37. doi: 10.21427/D7KM6Z.

Matteucci, X. (2013). Experiencing flamenco: an examination of a spiritual journey. In S. Filep & P. Pearce (Eds.), *Tourist experience and fulfilment: Insights from positive psychology*. New York: Routledge.

McIntosh, A. J., & Mansfeld, Y. (2006). Spiritual hosting: An exploration of the interplay between spiritual identities and tourism. *Turizam: međunarodni znanstveno-stručni časopis*, 54(2), 117–126. Retrieved from https://hrcak.srce.hr/161462.

McIntosh, A. J., & Zahra, A. (2008). Journeys for experience: the experiences of volunteer tourists in an indigenous community in a developed nation-a case study of New Zealand. *Journeys of discovery in volunteer tourism: International case study perspectives*, 166–181. doi: 10.1079/9781845933807.0166.

Mezirow, J. (1991). *Transformative dimensions of adult learning*. San Francisco: ERIC.

Monaco, S. (2018). Tourism and the new generations: emerging trends and social implications in Italy. *Journal of Tourism Futures*, 4(1), 7–15. doi: 10.1108/JTF-12-2017-0053.

Morgan, A. D. (2010). Journeys into transformation: Travel to an "other" place as a vehicle for transformative learning. *Journal of Transformative Education*, 8(4), 246–268. doi: :10.1177/1541344611421491.

Nedelea, A. (2008). Tourism Market and Demographic Changes. *Contemporary Economics*, 2(4), 61–67.

Neumann, M. (1992). The trail through experience. *Investigating Subjectivity: Research Online and Experience*, 176–201.

Norris, P., & Inglehart, R. (2011). *Sacred and secular: Religion and politics worldwide* (2 ed.). New York: Cambridge University Press.

Noy, C. (2004). This trip really changed me: Backpackers' narratives of self-change. *Annals of Tourism Research*, 31(1), 78–102. doi: 10.1016/j.annals.2003.08.004.

Olsen, D. H., & Timothy, D. J. (2006). Tourism and religious journeys. In D. Timothy & D. H. Olsen (Eds.), *Tourism, religion and spiritual journeys* (pp. 1–21). Oxon: Routledge.

Percy, M. (2016). *The Salt of the Earth: Religious Resilience in a Secular Age*. London: Bloomsbury Publishing.

Percy, M. (2019). Sketching a shifting landscape: Reflections on emerging patterns of religion and spirituality among Millennials. *Journal for the Study of Spirituality*, 9(2), 163–172. doi: 10.1080/20440243.2019.1658268.

Pew Research Center. (2014). *Comparing Millennials to other generations*. Retrieved from https://www.pewresearch.org/fact-tank/2018/03/16/how-millennials-compare-with-their-grandparents/

Pine, B. J., & Gilmore, J. H. (1999). *The experience economy: work is theatre & every business a stage*. Massachusetts: Harvard Business Press.

Reisinger, Y. (2013). Connection between travel, tourism and transformation. In Y. Reisinger (Ed.), *Transformational tourism: tourist perspectives* (pp. 27–32). UK: CABI.

Richards, G. (2011). *An economic contribution that matters. The power of youth travel*. Madrid: UNWTO, 7–8.

Rickly-Boyd, J. M. (2012). Authenticity & aura: A Benjaminian Approach to Tourism. *Annals of Tourism Research*, 39(1), 269–289. doi: 10.1016/j.annals.2011.05.003.

Robinson, V. M., & Schänzel, H. A. (2019). A tourism inflex: Generation Z travel experiences. *Journal of Tourism Futures*, 5(2), 127–141. doi: 10.1108/JTF-01-2019-0014

Rojek, C. (2000). *Leisure and Culture*. New York: St Martin's Press.

Roof, W. (1999). *Spiritual marketplace: Baby Boomers and the remaking of American religion*. Princeton, NJ: Princeton University Press.

Sharpley, R., & Jepson, D. (2011). Rural tourism: A spiritual experience? *Annals of Tourism Research*, 38(1), 52–71. doi: 10.1016/j.annals.2010.05.002.

Sharpley, R., & Sundaram, P. (2005). Tourism: A sacred journey? The case of ashram tourism, India. *International Journal of Tourism Research*, 7(3), 161–171. doi: 10.1002/jtr.522.

Smith, M., & Kelly, C. (2006). Holistic tourism: Journeys of the self? *Tourism Recreation Research*, 31(1), 15–24. doi: 10.1080/02508281.2006.11081243.

Steiner, C. J., & Reisinger, Y. (2006). Understanding existential authenticity. *Annals of Tourism Research*, 33(2), 299–318. doi: 10.1016/j.annals.2005.08.002.

Strauss, W., & Howe, N. (1991). *Generations*. New York: William Morrow & Co.

Timothy, D., & Olsen, D. (2006). *Tourism, religion and spiritual journeys*. Oxon: Routledge.

Timothy, D. J., & Conover, P. J. (2006). Nature religion, self-spirituality and New Age tourism. In D. Timothy & D. H. Olsen (Eds.), *Tourism, religion and spiritual journeys* (pp. 139–155). Oxon: Routledge.

Voigt, C., Brown, G., & Howat, G. (2011). Wellness tourists: in search of transformation. *Tourism Review*, 66(1/2), 16–30. doi: 10.1108/16605371111127206.

Vukonić, B. (1996). *Tourism and religion*. Pergamon, Oxford: Emerald Group Publishing Limited.

Wade, G. H. (1998). A concept analysis of personal transformation. *Journal of Advanced Nursing*, 28(4), 713–719. doi: 10.1046/j.1365-2648.1998.00729.x.

Wang, N. (1999). Rethinking authenticity in tourism experience. *Annals of Tourism Research*, 26(2), 349–370. doi: 10.1016/S0160-7383(98)00103-0.

Willson, G. (2012). *Finding peace in challenging times: Exploring the growing spiritual movement in tourism*. Paper presented at the *CAUTHE 2012: The new golden age of tourism and hospitality; Book 2; Proceedings of the 22nd Annual Conference*.

Wilson, E., & Harris, C. (2006). Meaningful travel: Women, independent travel and the search for self and meaning. *Turizam: međunarodni znanstveno-stručni časopis*, 54(2), 161–172. Retrieved from https://hrcak.srce.hr/161466.

Zahra, A., & McIntosh, A. J. (2007). Volunteer Tourism: Evidence of Cathartic Tourist Experiences. *Tourism Recreation Research*, 32(1), 115–119. doi: 10.1080/02508281.2007.11081530.

Zinnbauer, B. J., & Pargament, K. I. (2005). Religiousness and spirituality. *Handbook of the Psychology of Religion and Spirituality*, 21–42.

Zinnbauer, B. J., Pargament, K. I., Cole, B., Rye, M. S., Butfer, E. M., Belavich, T. G., . . . Kadar, J. L. (1997). Religion and spirituality: Unfuzzying the fuzzy. *Journal for the Scientific Study of Religion*, 36, 549–564. doi: 10.2307/1387689.

# Part IV

# The new age tourism

## Tourism's new face

The final part of this book discusses the new age of tourism. It focuses primarily on the new emerging trends, issues, threats and prospects of spiritual tourism in the future. It further analyses policies and practices for spiritual tourism development and strategic approaches to spiritual destination branding among the Millennial generation. Modern tourism research talks intensively of young people (Generation Y or Millennials) as a noteworthy section of the international tourism and travel industry. This part will help to understand Millennials and how the new generation is embracing new ways of finding peace and solace in this fast-moving world. Further, this part confirms that spiritual tourism definitely owns a place in the future of the tourism industry.

Chapter 13, "Issues, threats and prospects of Spiritual tourism in the emerging era" by Lina Munirah Binti Kamarudin, examines the experiences of tour operators in Malaysia who arrange tour packages that are related to spiritual tourism through a series of face-to-face interviews. This study reveals that spiritual tourism has the potential to be one of sustainable tourism development by reason of involving features that are not harmful to the community, economy, environment and preservation of cultural authenticity and norms.

Next, Chapter 14 "Strategic approach to spiritual destination branding among Millennials" by Ann Ashton, explained the Millennial tourist belongs to one of the big spending groups, has unique characteristics, and likes to explore new things to gain great experience. There are three components for a strategic approach towards spiritual tourism destination branding development: firstly, spiritual functional attributes; secondly, spiritual psychology attributes; and, finally, spiritual unique experience.

Chapter 15, "Millennials Want More! Proclivities and Potential of Indigenous Cultural Experiences in the Caribbean", by Johnnel Smith, highlights the motivations and latent interests of Millennial travellers towards new modes of spirituality to which they may not be exposed in their home country, through examining the indigenous Rastafarian villages in Jamaica which are interactive and offer authentic cultural experiences with a strong spiritual component.

DOI: 10.4324/9781003170631-16

Finally, Chapter 16, “Imminent Future of Millennial Generations: A Tourism Management Perspective” by Sandeep Kumar Walia and Aruditya Jasrotia, attempts to shed light on the features and travel behaviours of Millennials, and how their travel trends are going to shape the future of the tourism industry across the globe. In addition, the chapter also attempts to outline young tourists that can be approached by destination management organizations in the future. The present study is constructed to analyze contemporary academic research and various reports from the tourism industry on Millennials.

# 13 Issues, threats and prospects of spiritual tourism in the emerging era

*Lina Munirah Kamarudin*

## 13.1 Introduction

For a long time, spiritual tourism was well known as an integral part of the tourism industry. Spiritual tourism is a term that can be applied to the oldest modes of tourism, which includes a visit to historical and cultural sites along with religious elements (Zamani-Farahani & Eid, 2016). Written records and archaeological finds throughout history showed that individuals who travel for spiritual activities were usually influenced by spiritual experience (Suri & Rao, 2014). The United Nations World Tourism Organization (UNWTO) recently classified spiritual tourism as one of the fastest-growing segments (Rao, 2020). This is aligned with Kuralbayev and Abishev (2017), who declared that spiritual tourism can be identified as one of special interest tourism and has the tendency to expand.

The term spirituality is originated from the Latin word "spiritus", which means "breath of life" (Kale, 2004; Suri & Rao, 2014). Mitroff and Denton (1999) defined spirituality as "the basic feeling of being connected with one's complete self, others, and the entire universe". In Kale (2004), the psychiatrist Andrew Sims (1994) proposed five domains to spirituality: meaning in life (what a person lives for), interrelatedness, morality, wholeness and awareness of God. However, Kale (2004) mentioned that spirituality could be described as the engagement to explore deeply and meaningfully connect one's inner self, to the known world and beyond which includes four elements:

i. Inner self: Referring to the essential to spiritual experience. It was defined as the "invisible realm", where the roots of our deeper understandings are located. The essence from which our most emotions derive and valuable thoughts originate.
ii. Meaning: This element acts as the most significant aspect of spirituality, focusing on purpose and mission in life, fulfilling relationships and dimensions of mankind.
iii. Interconnectedness: conceptualizing spirituality as the fundamental desire to find ultimate meaning and purpose in one's life and to live an

DOI: 10.4324/9781003170631-17

integrated life. Instead of separation, it was interpreted that spirituality is an emphasis on integration and interaction

iv. The Beyond: The Beyond may mean many things: greater force, universal spirit or consciousness, or life as a whole either seen or unseen. A genuine spiritual experience has also been portrayed as one that involves a sense of overwhelming, fascination, amazement, wonder and fulfillment. This theme has been regarded as one of the key pillars of both spirituality as well as religion.

Drawing on these ideas, Rao (2020) defined spiritual tourism as travelling on pilgrimage to attain enlightenment, education and entertainment through the expression of gratitude to God. The trend of spiritual tourism is seen to be increasing due to destinations do not require extensive facilities and infrastructure. Therefore, spiritual tourism can be used to enhance the development of tourism and how it impacts the intent of the tourist revisit intentions (Sukaatmadja, 2020). The aims of spiritual tourism are related to reverence and association with a higher authority or deity and underlined by religious practice (Cheer, Belhassen, & Kujawa, 2017). While Choe (2020) argued that spiritual tourism aims to cultivate mental well-being, which involves the psychological concentration of the mind and social interaction with a group of believers.

In many countries, India is among the most popular with spiritual tourism whereby almost 60 percent of domestic tourism is religion-based. Various popular religious cities in India are not only popular in the tourism sector, but also get a lot of attention in academic discussions and debates. This includes such as Varanasi, Puri, Rajasthan, Vaishno Devi Shrine and others. Not only that, but India is also known as the centre of spiritual tourism. It is projected that the contribution of the travel and tourism sector to the Indian economy is to grow from US$234.03 billion in 2017 to US$492.21 billion in 2028 (Agrawal, Choudhary, & Tripathi, 2010; Jauhari & Sanjeev, 2010; Suri & Rao, 2014; WARC, 2018).

## 13.2 Literature review

The rapid shift in people's commitment to spirituality has had a significant impact on a variety of industries, including tourism (Rawal & Sah, 2017). UNWTO hosted an international seminar and conference in 2013 in which, for the first time the theme, the focus was on spiritual tourism (UNWTO, 2013). The conference discussed ways to enhance the positive impacts of spiritual tourism on the economic and social advancement of communities. This shows that spiritual tourism is gaining attention its development is valued. Cheer et al. (2017) indicate that spiritual tourism can act as a laboratory to help individuals consider, examine and practice spirituality activities which these are not always applicable in daily life.

According to Heidari, Yazdani, Saghafi, and Jalilvand (2018), spiritual tourism seeks to explore aspects of life that go beyond the self and lead to a healthy balance between body, mind, and spirit. This is in line with Rao (2020), who suggested that spiritual tourism is not related to any single religion and is distinct from religious tourism. It is about linking the body, mind and soul, while religious tourism is about finding God's blessings and gaining redemption based on your religious values and beliefs. Spiritual tourism has a broader perspective than religious tourism where religious tourism is found to be a subset of spiritual tourism. Nonetheless, research by Haq and Wong (2010) indicates that spiritual tourism can be a strategy or approach to market certain religions. As an example, the promotion of an Islamic pilgrimage journey such as Mecca, Medina and Jeddah can be seen as a part of spiritual tourism where people will visit holy places in the search for tranquility.

Both religious and spiritual tourism are practiced in parallel (Norman, 2012). However, religious tourism consists of visitors who are entirely influenced or inspired by some religion. As religion has an enormous influence in the everyday life, the reason for visiting is strictly restricted to religion alone and closely linked to faith (Cochrane, 2009; Heidari et al., 2018). By contrast, spiritual tourism is a broader concept, as it has religion as a mere part of it and spirituality enables an individual to explore more than just one religion. A spiritual tourist relatively has their own personal interest with a motive of spiritual growth. This can be associated with his/her faith, and there may or may not be any other religious or non-religious compulsion (Suri & Rao, 2014).

A study by Roof (1993) viewed spirituality engaging emotional connection to God and involving life's deeper motivations. Religion is seen to have an institutional connotation that translated into ritual practice, dogma obedience and service participation. On the other hand, spiritual tourism can be grouped under three categories, consisting of spirituality, wellbeing, or self-discovery (Norman, 2012). In Malaysia, Islamic tourism has been implemented based on three concepts: religious tourism, cultural tourism and spiritual tourism. These three concepts are combined in order to provide attractions to tourists. This is reflected in the purpose of tourists coming to Malaysia, whose aim is to explore the cultural diversity as well as to know the religion of Islam practiced by the locals where it involves spiritual elements (Laderlah, Rahman, Awang, & Man, 2011; Mapjabil, Razak, Marzuki, & Zainol, 2015).

Millennials, who were previously engrossed in their everyday work routines, have begun to take time for themselves peculiarly in spiritual tourism. Tech channels are evolving that enable Millennials to seek inner peace through various related topics and assist them in connecting with one another, resulting in the formation of a spiritually awakened group (Cavagnaro, Staffieri, & Postma, 2018; Patra, 2019). Mya's (2020) aforementioned Millennials tend to seek authentic and specific experience where they are embarking on their journeys at a time when self-discovery is important which related to spirituality. A study from Lituanas (2017) agreed that a personal and experiential relationship with God has been included in the spiritual identities of the Millennial group.

Wellness and spirituality is related to each other by which spirituality is seen as a promoter of many wellness tourism businesses (Choe, 2020; Wellness Tourism Worldwide, 2020). They believed that tourists who are motivated to travel for wellness will participate and engage in spiritual activities. The COVID-19 pandemic has increased public awareness of the multiple dimensions of wellness. The impact of lockdown, lost friends and family, jobless and too many restrictions led to the unstable emotions of human well-being. Mental, physical, social and spiritual well-being is essential to provide a dynamic and strong wellness, where people could focus more on the spiritual side of their lives.

## 13.3 Research Methodology

A qualitative study has been conducted for the purpose of identifying the issues, threats and prospects of spiritual tourism among the tour operators. The selection of qualitative methods, especially in-depth interviews, allows researchers to understand the phenomenon of trends in the organization and implementation of spiritual tourism, where it is found to be a suitable method in understanding the phenomena of spiritual tourism holistically (Cowling & Lawson, 2016). Thus, the interview method is seen as very suitable to be used to conduct research with religious and spiritual elements provided that researchers can understand more clearly the information provided by the respondents.

Tour operators in peninsular Malaysia were chosen because they are capable agencies that assembles all spiritual tourism services under one roof, something known as tour packages. They also directly interact with tourists, suppliers and authorities. A purposive sampling technique was adopted where the researcher seeks out elements that meet specific criteria (Blackstone, 2012). The criteria include tour operators who have registered their business with the Malaysian Association of Tour and Travel Agents (MATTA), actively selling online spiritual tourism packages and operating Muslim travel packages to more than four tourist destinations. There are ten operators that have been identified and that fulfil the criteria for this study.

As mentioned above, the research examines the experiences of tour operators in peninsular Malaysia who arrange tour packages that are related to spiritual and religious tourism through in-depth interviews. Data collection was carried out using audio recording and note-taking methods. The interviews took approximately 30 to 40 minutes. For the purpose of data analysis, the researcher transcribed the recorded audio and coding the transcription according to pre-determined themes. Taylor, Bogdan, and DeVault (2016) stated that the coding process involves data analysis according to key themes, ideas, concepts, interpretations and recommendations. Along these lines, researcher coding the data according to suitable emerging themes where these themes are divided into three parts: issues, threats and prospects of spiritual tourism.

## 13.4 Research Findings

### *13.4.1 Issues in Spiritual Tourism*

In recent years, both "religious" and "spiritual" tourism have increased, leading both academics and practitioners to engage in research pertinent to segmenting the characteristics and motives and controlling the impacts of visitors to these places. Kale (2004) stated there are confusions between the concepts of religion and spirituality. To date, there has still been some ambiguity into how religious tourism can be distinguished from spiritual tourism (UNWTO, 2013). The results showed that even a services provider cannot figure out the different meaning of religious tourism and spiritual tourism terms.

> Actually, I cannot differentiate the meaning of religious tourism and spiritual tourism in detail.
>
> (Respondents 3 and 7)

> In my opinion, basically these terms are the same. It's just that they use different words.
>
> (Respondent 6)

Since there are ambiguities between religious tourism and spiritual tourism, spiritual destinations received lower attention of marketing and publicity (Heidari et al., 2018; Suri & Rao, 2014). Plenty of tourism packages and debates in academics tend to assimilate the word of 'religious tourism' and 'spiritual tourism'. Therefore, a systematic market research programme should be initiated to design spiritual tourism products in order to meet the needs of the target market (Banerjee, 2015). Currently, there are no effective methods for branding and commercializing spiritual tourism (Rao, 2020). It is essential to identify the convincing strategies to brand and market spiritual tourism globally.

One of the ways is to have a collaboration among tourism experts, practitioners, scholars and academicians to draw a blueprint on how to brand and market this tourism segment. There is also a need to enhance spiritual tourism brand image by clearly differentiating the attractions of tourist destinations. Internet usage has allowed spiritual tourism products to be promoted (Sukaatmadja, 2020). In an effort to attract more spiritual tourists, the promotion of destinations should clearly mention spiritual activities to be done, along with its uniqueness, attractive destinations and facilities provided.

> To advertise the tour packages in media, we need to highlight the features of spiritual tourism such as providing visiting historical places, religious sacred places, involve with community events and others.
>
> (Respondent 4)

> We emphasize travel itineraries and places of interest to visit when operating this package promotion. This helps to encourage clients to visit the destination.
>
> (Respondent 5)

> We also provide facilities for disabled tourists to join this spiritual tourism packages. We understand that they also want to be appreciated and being appropriately recognized as other people too.
>
> (Respondent 1)

### *13.4.2 Threats in spiritual tourism*

In this globalization era, Millennials and Generation Z use the Internet as a main source to get the desired services. In the tourism industry, tourists can not only buy flight tickets and accommodation online; they can also get all the information for free. They prefer to travel on their own and be independent without the help of a travel agent. This development will be a big challenge to the survival of tour operators. When tour operators are being asked about the biggest challenge in promoting spiritual tourism, they mentioned the Internet as their main 'enemy', followed by price manipulation from other tour operators.

> Internet booking is the biggest challenge. That is our main enemy. Second, the competition from the travel agencies itself, which is what kind of price manipulation.
>
> (Respondent 2)

> Most tourists are now very smart. They can find their own tourist information on the Internet. For example, to review accommodation they can look at search engine such as Booking.com, Expedia, Trivago and others. They are no longer dependent on travel agencies like us.
>
> (Respondent 3)

Ayorekire, Joseph, Mugizi, and Byaruhanga (2020) agreed that religious tourism should be developed and supported by providing more visitor facilities and services for worshippers, tourists and the community. Since the concept of spiritual tourism is to improve the quality of life, including body, mind, spirit and emotional energies, the facilities are most likely involve the beautiful landscape and scenery. To encourage the objectives, tour operators put an effort to provide such services compared to secular tourism. As Malaysia is currently a developing country, various constructions of man-made attractions are being developed. Therefore, this will be one of the challenges in choosing the right tourist destination that align with the objective of spiritual tourism.

> We are more likely to use highways with forest and scenery views rather than use highways in the middle of the city which are quite busy with

vehicles only. So, it will allow tourists to feel calm, appreciate the surrounding and feeling thankful to God for what they have.

(Respondent 8)

### *13.4.3 Prospects in Spiritual Tourism*

Rao (2020) mentioned that the motivation to travel in spiritual tourism is to find meaning and purpose in their lives. Spiritual tourism purposes are seen as one of the practices carried out purposefully to understand the beauty of the existence of God by making discoveries of signs, history, and man-made attractions that are His greatest gifts (Bhuiyan, Siwar, Ismai, Islam, & Ehsan, 2011; Din, 1989). Based on the respondents' views, Muslims spiritual tourists who enjoyed the beauty of nature tend to relate their feeling by appreciating the blessing of God's creations. This shows that spiritual tourism and religious tourism have the tendency to influence each other and go hand in hand.

> We don't just see the beauty of flora and fauna if we take visitors to the scenery. Actually, as needed in the Al-Qur'an, we encourage them to see the presence of a miracle that Allah has made for all human beings.
>
> (Respondent 10)

> Tourists are taken to visit sacred places such as mosques, churches and temples to gain knowledge about the religion practiced by the local community. We bring them to visit there not for worship or glorification, but simply to add knowledge and respect popular travel destinations.
>
> (Respondents 3 and 7)

> We took tourist to visit Indonesia's Borobudur temple, which was once used as a place of religious worship. It has, however, been awarded by UNESCO as one of its historic building and become one of the attraction for all tourists.
>
> (Respondent 9)

A spiritual journey is associated with culture, tradition, festivals, fairs and sacred places. It is influenced by many attractions, such as historical sites, architecture design and even cultural norms. All of these attractions are able to attract thousands of pilgrims, devotees and spiritual tourists. In order to tap the maximum of the market, there is an imperative to build a segmentation strategy and this can only be achieved by recognizing the different motivations that make people visit these destinations (Sukaatmadja, 2020; Suri & Rao, 2014).

> The aim of spiritual tourism is not for leisure or luxury. We don't provide alcohol drink, no entertainments. It is to enable tourists to see and appreciate the cultural and architecture designs at tourist destinations.
>
> (Respondents 8 and 9)

> Spiritual tourism also involves cultural elements where tourists can get to know the culture of the locals at the destination of the visit.
>
> (Respondent 4)

> Cultural differences allow them to explore tourism more deeply where tourists will be exposed to get to know the local community better.
>
> (Respondent 1)

> While visiting villages, we provide opportunities for tourists and locals to exchange opinions and stories. Through this process, tourists can learn something new about the culture and life of the local people there. This is a good opportunity to involve the local community and tourists so they can understand each other.
>
> (Respondent 2)

Spiritual tourism can be a tool for sustainable development due to its concept which bring positive impacts not only for the community, but also to the environment, economic and socio-culture (Barnerjee, 2015; UNWTO, 2013). Rao (2020) and Sukaatmadja (2020) agreed that the increasing demand of spiritual tourism will protect the local culture, nature, environment, native places, traditions and arts. In order to link spiritual tourism and community, it encompasses several elements, such as community cooperation, community potentiality and attraction potentiality (Piewdanga, Mekkamola, & Untachai, 2013). In its economic aspects, the expansion of spiritual tourism helps to increase job opportunities and contributes to the growth of GDP in a country (Rao, 2020). The availability of Muslim needs at destinations, such as Halal food and places to pray, will help to encourage Muslim tourists to travel and fulfil their spiritual needs.

> When a Muslim travel to seek spiritual peace, they need other Muslims to serve them services too. A lot of Muslims now are working in tourism industry. This opportunity will help more Muslim tourists to travel and at the same time can fulfill their well-being like getting Halal foods and suitable places to perform prayers.
>
> (Respondent 11)

> ‘Muslim travellers will look for Halal foods and mosques to perform prayers. These are among of their important spiritual needs in their daily life. In choosing tourism packages, they will look at this services first.
>
> (Respondents 1, 4, 8 and 10)

When being asked about the target market, tour operators were trying to focus on the Millennial group, especially students from schools and higher institutions. This is agreed by Rawal and Sah (2017), who said that the number of spiritual visitors has increased around the world. Spiritual tourism seems striving to engage the Millennial where they are motivated in seeking spiritual experiences to take advantage of their everyday lives in terms of personal growth and development. They realized that this group is becoming increasingly concerned where most of them experience anxiety and stress in their daily life.

> Millennials have begun to embark on spiritual journeys. One of the activities they involve is attending a motivational camp to increase their self-motivation.
>
> (Respondent 11)

> Most of our clients are university students who aim to conduct certain researches and spend their semester break. They want to relax and run away from their stressful daily routine. This type of holiday can help to increase their self-confidence while socializing with friends and locals.
>
> (Respondent 8)

> Our clients also consist of students from foreign countries who further study in Malaysia, basically from the Millennial group. They need to go out and want to enjoy their life.
>
> (Respondent 5)

> Our target customers consist of students of schools and institutes of higher learning.
>
> (Respondent 1)

## 13.5 Research limitations

This study was conducted in peninsular Malaysia. Therefore, it is proposed that the study on spiritual tourism should be extended to other countries, as each country might has different planning on how to develop this niche tourism market. The study on the tourists' side should be done in order to further clarify their specific needs, wants and motivations to involve in spiritual tourism. To meet the needs of present and future Millennial visitors, this will help to provide tourism service providers a guide in preparing and market their tourism packages effectively. In different circumstances, a comparative study on the concept of spiritual tourism between tourist and tour operators can be done, as well as the implementation of the tourism package at the national and international levels.

## 13.6 Research implications

Spiritual tourism is shown to have excellent market potential not only in Malaysia, but also in other countries. Accordingly, this study can benefit tourism service providers in planning and designing specific spiritual tourism promotions. By identifying the issues, threats and prospects from the perspective of tourism suppliers, governments would have ideas about how to enhance the effort of tourism development in the country and its sustenance. This is to ensure that the spiritual tourism products and services offered will meet tourists' needs and wants.

The study shows that the Millennial group is the emerging generation to shape the future global tourism industry. They can be seen as an innovative group, where their decisions will influence how the rest of society views tourism. Due to this fact, targeting Millennial group to niche tourism market, such as spiritual tourism, can help tour operators focus on how to promote their tourism packages effectively. Not only that, but they also need to have a strong networking with other tourism stakeholders, particularly the government, to address the issues and threats that derive from spiritual tourism. This effort will ensure spiritual tourism to grow further, as well as increasingly gaining attention among tourists.

## 13.7 Conclusion

In the marketing of tourism destinations, operators and travel agencies must play an important role, which is to provide information to potential visitors and to develop and promote tourism packages (Velenik & Boskovic, 2006). As Malaysia is a Muslim country, the marketing of spiritual tourism is not widely used. Tour operators tend to promote religious and pilgrimage tourism packages in order to attract tourists. Based on Cheer et al. (2017), the unique motivation for spiritual tourism includes placing an emphasis on religious observance, ritualized practice, identity reaffirmation and cultural performance. Corresponding to this, the branding and marketing of spiritual tourism needs to focus on the purpose of the visit and unique experience that can be gained by tourists when they want to buy this tourism packages. This will help in positioning the development of spiritual tourism worldwide.

Spiritual tourism is not only focus destination-oriented. It can be done anywhere around the world because it revolves an individual's sense of connectivity, attachments and the journey of travelling. Based on the study, it can be concluded that spiritual tourism is also inspired by religious tourism, including visits to holy places and participate in religious events. However, visiting mosques, temples and churches is not intended for worship. Yet the aim is more to appreciate the cultural differences practiced by the local community, where tourists can see the way of life of the outside world in an open-minded or non-discriminatory manner.

The objective of spiritual tourism is to enhance physical, mental, social and spiritual well-being. Thus, the activities provided and the drivers of tourists' motivation to involve in this tourism segment usually associated with leaving positive impacts with regard to a clean environment, involving local communities, increasing economic performance and the preservation of culture. Given that the impacts are not harmful, spiritual tourism has the potential to be one of drivers of sustainable tourism development. Hence, tourists may enjoy those journeys in which the motivation is to get peace while travelling.

## References


Agrawal, M., Choudhary, H., & Tripathi, G. (2010). Enhancing Buddhist tourism in India: an exploratory study. *Worldwide Hospitality and Tourism Themes.*

Ayorekire J. Joseph O., Mugizi F. and Byaruhanga, B. M. (2020). Opportunities and challenges of religious tourism development in Uganda: Policy, planning and institutional perspectives. *International Journal of Religious Tourism and Pilgrimage.* 8(3), 144–158.

Banerjee M. (2015). Spiritual Tourism: A tool for Socio-Cultural and Sustainable Development. *International Journal of Science and Research (IJSR).* 4 (10), 1244–1249.

Bhuiyan A. H., Siwar C., Ismai, S. M., Islam R., and Ehsan D. (2011). Potentials of Islamic Tourism: A Case Study of Malaysia on East Coast Economic Region. *Australian Journal of Basic and Applied Sciences.* 5(6), 1333–1340.

Blackstone A. (2012). Principles of Sociological Inquiry – Qualitative and Quantitative Methods. *Chapter 7: Sampling.* Saylor Foundation.

Cavagnaro E., Staffieri S. and Postma A. (2018). Understanding Millennials' tourism experience: Values and meaning to travel as a key for identifying target clusters for youth (sustainable) tourism. *Journal of Tourism Futures.* DOI 10.1108/JTF-12-2017-0058.

Cheer J. M., Belhassen Y. and Kujawa J. (2017). The Search for Spirituality In Tourism: Toward A Conceptual Framework For Spiritual Tourism. *Tourism Management Perspective.* 24. 252–256.

Choe J. (2020). Wellness and Spiritual Tourism After Covid-19. *Anthropology of Tourism Interest Group of the American Anthropological Association.* Retrieved at http://atig.americananthro.org/wellness-and-spiritual-tourism-after-covid-19/

Cochrane J. (2009). Spirits, Nature and Pilgrimage: The "Other" Dimension in Javanese Domestic Tourism. *Journal of Management, Spirituality and Religion.* 6(2), 107–119.

Cowling C. and Lawson C. (2016). Dipping Qualitative Toes into Qualitative Worldview: Methodological Manoeuvres in a Multicultural in Context. In Harreveld B., Danaher M., Lawson C., Knight B., Busch G. (eds), *Constructing Methodology for Qualitative Research* (pp. 31–44). Palgrave Macmillan, London.

Din, K. H. (1989). Islam and tourism: Patterns, issues, and options. *Annals of Tourism Research,* 16(4), 542–563.

Haq F. and Wong, Y. H. (2010). Is Spiritual Tourism A New Strategy for Marketing Islam?. *Journal of Islamic Marketing.* 11(2), 136–148.

Heidari A., Yazdani H. R., Saghafi F. and Jalilvand M. R. (2018). The Perspective of Religious and Spiritual Tourism Research: A Systematic Mapping Study. *Journal of Islamic Marketing.* 9(4), 747–798.

Jauhari V. and Sanjeev, G. M. (2010). Managing Customer Experience for Spiritual and Cultural Tourism: An Overview. *Worldwide Hospitality and Tourism Themes.* 21(5), 467–476.

Kale S. H. (2004). Spirituality, religion, and globalization. *Journal of Macromarketing.* 24(92), 92–107.

Kuralbayev A. and Abishev N. (2017). The Measuring of the Spiritual Tourism in Regions of South Kazakhstan. *Revista ESPACIOS.* 38(51). ISSN 0798 1015.

Laderlah S. A., Rahman S. A., Awang K. and Man Y. C. (2011). A Study on Islamic Tourism : A Malaysian Experience. In *2nd International Conference of Humanities, Historical and Social Sciences.* 21–23 October. Cairo, 184–189.

Lituanas C. R. A. (2017). An Inquiry at the Millennial Youth Spiritual Identity: Experiences of DLSU Students. *Arts and Culture: Heritage, Practices and Futures*. Manila, Philippines.

Mapjabil J., Razak R. R. A., Marzuki M. and Zainol R. M. (2015). Pelancongan Islam: Suatu Tinjaunan Konseptual dan Relevannya di Malaysia. *Malaysia Journal of Society and Space*. 11(1), 172–182.

Mitroff I. and Denton E. (1999). A Study of Spirituality in the Workplace. *Sloan Management Review*. 40 (4): 83–92.

Mya K. (2020). Why Millennial Travelers Are Seeking More Authentic Experiences. Retrieved on 7 March 2021 at https://www.under30experiences.com/blog/why-millenial-travelers-are-seeking-more-authentic-experiences.

Norman A. (2012). The Varieties of the Spiritual Tourist Experience. *Literature & Aesthetics*. 22 (1): 20–37.

Patra D. (2019). *The Growing Inclination of Millennials Towards Spirituality. What is driving fast-paced millennials towards Spirituality and Meditative Practices?* Retrieved 10 April 2021 at http://bwpeople.businessworld.in/article/The-Growing-Inclination-Of-Millennials-Towards-Spirituality/26-06-2019-172453/

Piewdanga S., Mekkamola P. and Untachai S. (2013). Measuring Spiritual Tourism Management in Community: A Case Study of Sri Chom Phu Ongtu Temple, Thabo District, Nongkhai Province, Thailand. *Proceeding of Social and Behavioral Sciences*. 88: 96–107.

Rao M. S. (2020). *Spiritual Tourism: Tourists First, Tour Operators Second, and Destinations Third. World Financial Review*.

Rawal Y. S. and Sah V. (2017). Emergence of spiritual tourism and its impact on hospitality industry with special reference to Uttarakhand region. *Atithya: Journal of Hospitality*. 3(2), 21–26.

Roof W. C. (1993). *A generation of seekers: The Spiritual Journeys of the Baby Boom Generation*. San Francisco: *Harper Collins*.

Sims A. (1994). "Psyche": Spirit As Well As Mind?. *British Journal of Psychiatry*. 165, 441–446.

Sukaatmadja I. P. G. (2020). The Influence of Spiritual Tourism on Tourist Perceived Value and Its Implication on Tourists' Revisit Intention to Bali. *International Journal of Advanced Science and Technology*. 29(5), 13140–13153.

Suri R. and Rao J. (2014). Impact of spiritual marketing on different segments of tourists and their evaluation of the site. *Journal of Business and Economic Policy*. 1(1), 26–34.

Taylor S. J., Bogdan R., and DeVault M. L. (2016). *Introduction to Qualitative Research Methods: A Guidebook and Resource*, Fourth Edition. *New York: John Wiley and Sons, Inc.*

UNWTO (2013). *First UNWTO International Conference on Spiritual Tourism for Sustainable Development*. Ninh Binh Province, Viet Nam. 21–22 November 2013.

Velenik R. and Boskovic D. (2006). *The Role of Travel Agencies and Tour Operators in the Development of Rural Tourism in Istria Country*. In the *11th International Scientifical Conference Tourism, Regional Development and Education*. Tabor, 85–91.

WARC (2018). Spiritual Tourism Grows Strongly in India. Retrieved on 6th December 2020, at https://www.warc.com/newsandopinion/news/spiritual-tourism-grows-strongly-in-india/40500

Wellness Tourism Worldwide (2020). *Improving Well-being and Economic Growth Through Travel*. Hungary: Xellum Ltd.

Zamani-Farahani, H. and Eid, R. (2016). Muslim world: A study of tourism and pilgrimage among OIC member states. *Tourism Management Perspectives*. 19, 144–149.

# 14 Strategic approach to spiritual tourism destination branding development among millennials

*Ann Suwaree Ashton*

## 14.1 Introduction

Spiritual tourism is a secular approach to wellness tourism (Ashton, 2018; Chandler, Holden, & Kolander, 1992; Gill, Packer, & Ballantyne, 2019; Smith & Puczkó, 2013; Voigt & Pforr, 2014). It is special interest tourism that includes a notion of mind/soul for awakening individual feelings to realize a peaceful and tranquil life (McCartney, 2020; Smith & Diekmann, 2017; Smith & Puczkó, 2015). Spiritual retreat tourism can be related to religion (Smith, 2003; Yoon et al., 2015) or non-religion, and the location of the activity can sometimes can be located in a rich and natural environment, or set in a temple area, for example a spiritual retreat in Thailand (Ashton, 2018; Heintzman, 2013; Tilson, 2005; Voigt & Pforr, 2014). This type of tourism is about the practice of mind/soul recovery, especially for those seeking a work–life balance (Heintzman & Mannell, 2003; Linders & Lancaster, 2013).

Spiritual tourism is not a new topic, but it has caught scholars' attention for a number of years (see, among others, Choe & O'Regan, 2020; Haq & Medhekar, 2020; Haq & Wong, 2013; Mourtazina, 2020; Mukherjee, Bhattacharjee, & Singha, 2020; Piewdang, Mekkamol, & Untachai, 2013; Sharma, 2020; Singleton, 2017; Sirirat, 2019; Suravichai & Namwong, 2020). Past studies have focused on a variety of topics, such as tourist experience (Choe & O'Regan, 2020; Mourtazina, 2020; Norman, 2014); motivation (Bone, 2013; Jiang, Ryan, & Zhang, 2018; Suravichai & Namwong, 2020); and developing and planning (Buzinde, 2020; Mourtazina, 2020; Sharma, 2020; Sharpley & Jepson, 2011). Another study explored on religiousness or spirituality and hedonic well-being related to life satisfaction (Yoon et al., 2015).

This significant gap within the spiritual tourism context has been addressed in this study by developing specific topics for destination branding strategy development. A rare study on wellness branding strategy in the countries of the Balkan region focused on health tourism (Sziva et al., 2017). Another study emphasized the importance of wellness tourism and destination competitiveness development (Prideaux, Berbigier, & Thompson, 2014), while

DOI: 10.4324/9781003170631-18

some paid attention to Asian countries, including India, with the aim to explore on how to use a wellness service for positioning and promoting branding in India (Ravichandran & Suresh, 2010). As a result, it was noticed that the spiritual tourism destination branding strategy topic is still lacking, especially for the Millennial generation; consequently, this study's intention is to bridge this academic gap.

It is important to bridge this gap for a number of reasons, firstly, in reflecting an increasing interest in the work–life balance, the development of destination branding for spiritual tourism can help in responding to the high demand of tourists who are interested in retreat centres (Heintzman, 2013). Spiritual destination branding, if planned properly, will help in marketing destinations, because branding comprises signage, terminology, or symbols designed to attract tourists to visit (Pritchard & Morgan, 1998; Ruiz-Real, Uribe-Toril, & Gázquez-Abad, 2020). Furthermore, spiritual tourism destination branding development can help in building social well-being for tourists, because such destinations set up activity centres and organize tourists into groups, thereby encouraging spending life with new friends and locals during spiritual retreat activity (Bone, 2013; Buzinde, 2020). The numbers of the groups range between 10 and 20 persons at any one time during the activity, hence helping the members in a group to learn about cultural diversity as they will have to stay in tents or camp together (Bone, 2013). As a result, the wellness life can be fulfilling, and if one has been accepted into the society, he/she can learn about and exchange with new cultures, for instance language and local lifestyle during participating in the retreat activity (Srivastava & Kaul, 2014).

The aim of this study is to propose a strategic approach for spiritual tourism destination branding development among Millennial tourists. Three objectives were raised: firstly, to study on the functional attributes of spiritual tourism destination branding; secondly, to study psychology attributes, such as how to create feelings or ideas towards spiritual tourism destination branding; and lastly, to study the unique experience of spiritual tourism.

## 14.2 Spiritual tourism activity as non-religious and religious

Spiritual tourism activity can be part of religion or non-religion or it can be subjective spirituality; however, subjective spirituality is not tied to religious doctrines, but related to beliefs in supernatural spirits (Yoon et al., 2015). Typically, spiritual tourism activity is related to the treatment of mind/soul, leading to a healthy physical appearance (Ashton, 2018). Spiritual retreat activity, such as spa & beauty and meditation, helps with self-awareness and inner peacefulness (Aggarwal, Guglani, & Goel, 2008; Bone, 2013; Buzinde, 2020; Eddy, 2014; McCartney, 2020; Norman & Pokorny, 2017). There are different types of yoga, for example, Hatha-Yoga and Raja Yoga (Sharma, Tomar, Ahola, Tomar, & Ahola, 2008 ). Hatha-Yoga (or kriyas) is a technique for preventing old age symptoms and has been proven as a cure for several ailments, while Raja Yoga focuses on consciousness; it creates the energy that allows the heartbeat and the senses to act properly. In addition, this type also helps with the transcendental state, cultivating abilities such as clairvoyance, telepathy, healing, transferring energies, and more (Sharma et al., 2008 ). Likewise, yoga helps with spiritual composure by relieving stress with equanimity, positive thinking, meditation, non-judgement, patience and non-violence (Buzinde, 2020).

Meditation is one of the most important activities for spiritual retreat, and whether or not people participate based on sacred and secular experience, it is all linked to mind (Astin, 1997; Eddy, 2014; Norman & Pokorny, 2017). For example, the experience of Zen meditation can be separation from daily life; the landscape values of the location; the temple atmosphere; contact and sharing experiences with monks and mentors: all contribute to a sense of personal wellness that participants obtain (Jiang et al., 2018).

In contrast, the spiritual tourism activity related to religion can be sought by participation in different activities. Firstly, organizational religiousness involves attending services and other activities in a place of worship; secondly, private religious practices can be performed with private prayer, reading sacred texts, and meditation; and, finally, daily spiritual experience is performed by day-to-day religious and spiritual experiences and personal interaction with the involvement of God in daily life (Yoon et al., 2015).

Spiritual and holistic holiday typology activities include: therapeutic (spas, thalassotherapy, massage, and aromatherapy); healing (ayurvedic treatments, reiki, reflexology, crystal healing, astrology and tarot); contemplative (retreats, meditation, prayer, visualization); active (yoga, pilates, tai chi); creative (painting, writing, drama, dance); enlightening (pilgrimage, sacred sites, mystical landscapes); interactive (volunteering, community engagement, participation in rituals or festivals); and personal development (counselling, stress management, life strategies) (Smith, 2003).

In summary, it can be concluded that spiritual tourism activity can comprise meditation, yoga and anatomy, all regardless of religion or culture, but also diverse participants' faiths, including Christian, Hindu, Sheikh, Buddhism and non-denominational (Buzinde, 2020). In addition, activities such as yoga should be practiced to improve the quality of life, with a sense of extending strength, optimism and peacefulness (Sharma et al., 2008 ).

## 14.3 Millennial generation personal traits

The Millennial generation is one of the most interesting tourist groups because it is larger in number than other generations, however, there is no census on the exact years they were born between; some say the Millennial generation are persons born between 1980 and 1992 (DeVaney, 2015). Similarly, others say the members of the Millennial generation are those born in or after 1980 (Sweeney, 2006); by contrast, others contend they were born between 1979 and 1995 (Cutler, 2015). However, the Millennial generation is sometimes defined as Generation Y or the Nexus generation (Sweeney, 2006). The personal traits, including mindset and attitudes of this generation, can be such as optimistic, civic-minded, close parental involvement, value work–life balance, impatient, multitasked, and team-oriented (DeVaney, 2015). This generation is seriously seeking for a work–life balance more so than other generations, and for high-service position employers should offer free meals, swimming, spa, and on-site doctors in order to keep them happy at the work place (Sweeney, 2006). In addition, they are digital natives because this generation's has been connected to technology since their first and second school grades (DeVaney, 2015).

In conclusion: in this study, the concepts of spiritual tourism and the internal mindset of the Millennial tourist are related; they can be either religious or non-religious, usually performed through specific activities, especially meditation or yoga, and are for those seeking spiritual well-being, enlightenment and an increase in self-awareness (Ashton, 2018; McCartney, 2020).

## 14.4 Contemporary design branding

The management of a destination branding strategy has become a significant challenge in contemporary tourism, and it is crucial for all stakeholders to identify and adapt to suit with the new changes (Rather, Najar, & Jaziri, 2020; Tasci, 2011). The definition of a brand can be referred to as a name, term, sign, symbol or design, or a combination of them intended to identify the goods and services of one seller or a group of sellers and differentiate them from those of competitors (Keller, 1993, p. 2). Destination branding is a set of marketing activities that support the creation of a name, symbol, logo, word, mark or other graphics that readily identifies and differentiates a destination (Adamus-Matuszyńska, Dzik, Michnik, & Polok, 2021). Therefore, the way of promoting destination branding nowadays has shifted from the old traditional ways, such as brochures, leaflets or advertising on TV, to social media (Goyal, 2021; Pitcher, 1985). In this new era the development of IT has become the best option to promote a destination banding awareness because it is easy, of lower cost and instantly reaches all target market groups once posted on social media, such as YouTube, Facebook or Instagram (Adamus-Matuszyńska et al., 2021; Alegro & Turnšek, 2021). Other examples are visual aids and video that advance the most popular contents: emotional stability and a state of well-being, natural or cultural heritage, tourists' activities (activities in nature), and storytelling, one of the best contents to promote tourism destination branding (Alegro & Turnšek, 2021).

In relation to this study, spiritual tourism can be achieved by promoting tourism destination branding using both the physical and spiritual aspects, and especially by promoting destination icons such as historical, cultural, the local way of life, spiritual beliefs and values, uniqueness of activities: also by taking the advantage and creating destination brand awareness through the development of the new IT and media support (Anjomrouz, Dehyadegari, Hosseini, & Anjomrouz, 2021).

## 14.5 Destination branding strategy and spiritual tourism theory

Destination branding strategy has been proposed in different ways; for instance, García, Gómez, and Molina (2012) proposed the strategy of stakeholder responsibility and consumers' goal achievement. Firstly, the entrepreneur's responsibility refers to the role of entrepreneurs whose duty is to transfer cultural values at a destination to the visitor, and secondly, local people whose role is working within the tourism hospitality industry. Locals play an important part in their society, for instance, they are the people who will provide great service and sometimes may be asked to provide local knowledge to visitors. Thirdly, response to consumer goals: achievement of a visitor's goal can influence intention to revisit (García et al., 2012; Stylos, Vassiliadis, Bellou, & Andronikidis, 2016); as said, to develop tourist destination branding one must know tourist demand (Ryan & Zahra, 2010).

On the other hand, one important strategy for building destination branding is the development of destination brand image (Ashton, 2014; Quintal, Phau, & Polczynski, 2014; Wita & Namwong, 2018). Ashton (2014) proposed that there are three components to consider when developing destination brand image: firstly, sufficient tourism attributes, especially that of infrastructure, and good quality of product and service (Ashton, Scott, & Breakey, 2008). For instance, to develop destination branding, one must create products with good value, especially for price and quality perception (Sziva et al., 2017; Tsai & Wang, 2017).

Secondly, provide a clear picture or idea of a destination's brand image in the tourist's mind, for example, the significance and uniqueness of the place and surrounding atmosphere. More importantly, the brand name of a destination should derive from the uniqueness of the local geography, traditions or culture, including events or festivals (Blain, Levy, & Ritchie, 2005; Crockett & Wood, 1999; Hankinson, 2001; Mueller & Peter, 2008; Ryan & Zahra, 2004). Finally, provide a unique experience for the tourist, such as special events/festivals and unique activity developed from the local culture, traditions and religion principles (Ashton, 2014; Haq & Wong, 2013).

In relation to spiritual tourism, the destination attributes must comprise nature, lots of trees, diversity of native wildlife, plenty of areas for walking tracks, stunning scenery and unique biodiversity (Bone, 2013). Similarly, to provide a spiritual tourism destination one must respond to tourist 'push and pull' motivation: push motivation includes relaxing, self-esteem, novelty, escape, and physical appearance, while pull motivation includes authentic experience, historical significance, natural setting, far from the usual places, and peaceful atmosphere (Ashton, 2018). A study carried out by Bhumiwat and Ashton (2020) proposed that wellness tourism experience must comprise well-being in life, especially for physical, mental, social and emotional wellness.

Thus, as discussed in the literature review, the theory foundation proposal of a strategic approach to spiritual tourism destination branding development study will be based on the three main components of the destination branding concept, especially for destination image development. Firstly, conduct a study on the functional attributes of the spiritual tourism destination branding, including infrastructure, human resources, service quality and stakeholder roles. Secondly, study on how to create feelings or ideas for spiritual tourism destination branding. Lastly, study on how to create unique experiences for spiritual tourism destination branding (Ashton, 2014; Echtner & Ritchie, 1993; Haq & Wong, 2010; Haq & Wong, 2013).

## 14.6 Methodology

A qualitative method was adopted, and the Millennial tourist group was selected as the participant target for this study. The Millennial generation can be defined in different birth years, such as those born between 1979 and 2000, becoming the largest population segment in the United States in 2015 (Garikapati, Pendyala, Morris, Mokhtarian, & McDonald, 2016) (as seen in Table 14.1). The reason for selecting this group for the study was to fill the gap where the Millennial spiritual tourism and branding strategy development research is lacking within this area. Past studies of this generation focused on defining generations, especially where the Millennials end and Generation Z begins (Dimock, 2019); Millennials and Finance: the Amazon generation (Cutler, 2015); activity and travel behaviour of Millennials (Garikapati et al., 2016); attitude (Kowske, Rasch, & Wiley, 2010). The population number in this generation is higher than other groups; for example, a comparison was made in 2015 regarding the numbers in each generation and the years in which they were born; the number of boomers (1946–1962) was between 76–79 million, while for the same year (2015) young adults (aged between 18–34) or the Millennial generation (1979–1995), the number rises to 83.1 million (Cutler, 2015).

A famous meditation institute was selected as the study area for a number of reasons: firstly, this institute has been accepted as one of the best meditation practices in Thailand and it has opened branches worldwide (280 branches in Thailand, and approximately eight branches overseas). For example, the Canadian Education Department has recognised it as a very good meditation institute with potential for helping people to have mind and spiritual recovery.

*Table 14.1* US Census Bureau, "Young Adults Then and Now," December 2014

| | | |
|---|---|---|
| Year of data | 1980 | 2013 |
| Year of birth | 1946–1962 | 1979–1995 |
| No. of population (millions) | 67.1 | 72.8 |
| Bachelor's degree or higher (%) | 15.7 | 22.3 |
| Median annual earning ($) | 35,845 | 33,883 |
| Living in poverty (%) | 14.1 | 19.7 |
| Living with parents (%) | 22.9 | 30.3 |
| Never married (%) | 41.5 | 65.9 |

Source: Cutler (2015)

The researcher enrolled in a meditation course with this institute; the course took six months, and after graduating all students received a certificate enabling them to teach basic meditation practice (Astin, 1997; Eddy, 2014; Schedneck, 2015; Sharma, 2020). Hence, data received for this study was in-depth, based on insightful information of the overall spiritual tourism context and meditation activity designed by the institute.

The data collecting process used face-to-face interviews with the participants who were in the Millennial generation, especially those born during the years 1979–1995. Secondly, the participants were required to have meditation experience of at least six months or had finished the course (to be sure they practiced meditation because they really enjoyed the activity). The participants were employed by the researcher to contact persons who had studied in the same course, including others who had already graduated, by using the snowball technique, where one who the researcher spoke with referred another who had the same experience (Handcock & Gile, n.d.). The face-to-face interviews took at least 45 minutes, and no more than one hour; the interview process was permitted for voice recording and after finishing it was transcribed to perform the data analysis process.

In this study content analysis technique was used to analyse data. The content analysis technique helps performance by searching for an underlying theme in the text, helping in systematically explaining the meaning of the information received from the participants (Denzin & Lincoln, 2005; Elias, 2005; Krippendorff, 2004). In addition, the process of data analysis is also used to quantify qualitative data received from all participants and for sorting, comparing and summarizing in response to the study aim.

## 14.7 Participant profiles

The 15 participants in this study were Thai, and were interviewed on a face-to-face basis. There were 11 females and 4 males. The participants represented the ages between 25 and 39 years, and surprisingly most were working in a university as a lecturer or researcher; there were a few master students, a banker, hospitality staff and those working in a company (as seen in Table 14.2).

## 14.8 Results

The main aim of this study is to propose a strategic approach for spiritual tourism destination branding development. Three objectives were raised: firstly, to study on the functional attributes of spiritual tourism destination branding. Secondly, to study on psychology attributes, such as how to create feelings or ideas towards spiritual retreat tourism destination branding, and, finally, to study spiritual tourist unique experience, as the details reveal below.

*Table 14.2* Participants' Profile

| *No.* | *Title* | *Sex* | *Age* | *Organization* |
|---|---|---|---|---|
| 1 | Piano Teacher, Piano Accompanist | Female | 26 | Private Music School |
| 2 | Master's degree Student | Female | 25 | University/Institute |
| 3 | University Lecturer | Female | 33 | University/Institute |
| 4 | University Lecturer | Female | 35 | University/Institute |
| 5 | Sales Manager | Female | 38 | Assurance Company |
| 6 | University Lecturer | Female | 39 | University/Institute |
| 7 | Banker | Male | 36 | Private Bank Company |
| 8 | Researcher | Female | 37 | University/Institute |
| 9 | University Lecturer | Male | 28 | University/Institute |
| 10 | University Lecturer | Female | 28 | University/Institute |
| 11 | Banker | Female | 29 | Private Bank Company |
| 12 | Officer | Female | 29 | Private Company |
| 13 | University Lecturer | Female | 37 | University/Institute |
| 14 | Hotelier | Male | 34 | Hotel |
| 15 | Master's degree Student | Male | 31 | University/Institute |

### *14.8.1 Research objective one: the functional attributes that can be created toward spiritual tourism destination branding*

There are several components of spiritual tourism attributes creating destination branding; the participants revealed several important specific attributes to create spiritual destination branding.

Firstly, the place must be suitable for spiritual activity, as such, meditation activity should be conducted in a quiet and peaceful place conducive for focusing the mind and gaining spiritual recovery. Moreover, they are in the main conservational tourists concerned with being environmentally-friendly, thus, a spiritual retreat tourism destination should be located in an environment endowed with green scenery, a peaceful atmosphere, as living with nature is most crucial. Participants emphasized their seriousness about the infrastructure and accessibility at a destination. For example, they highlighted: "The destination should have a quiet, peaceful and silent atmosphere because it allows participants to slow down their everyday rhythms and create inner space for a better understanding of thought and emotion [ID 3, 10, 11, 12, and 15]."

Furthermore, for the environmental aspect most participants were concerned about natural setting, local culture, nature-based activities, and peaceful atmosphere. For instance, one said, "The shelter should be quiet, and have a silent environment suitable for doing meditation, resulting in helping with the transformational journey and leading to the acquisition of inner well-being, hence self-happiness [ID 3]." Another participant expressed, "Experience with the trial exceeded my expectation, and especially the environmentally friendly destination influenced me to finally return to the same place again [ID 5]."

Secondly, service quality must cover basic needs, including shelter or simple accommodation, basic food, water, and simple clothing suitable for each activity, such as meditation. For instance, the accommodation place for spiritual tourists will be at the temple area, rehabilitation centre, or hotel/resort. The four basic needs will be provided, but they are not luxury places, because for disciplines of spiritual activity one must be able to leave a comfortable life at home and practice a mindset of acceptance and discover a new reality in life. However, the four basic needs are such as a simple and very minimal comfortable accommodation, with no bed or mattress available, but just the floor to lie down on, basic and simple food/water only (some places will provide only vegetarian food), and a simple shower/toilet, such as no hot water. In addition, to participating in meditation activity one must observe the basic rule of being able to live in nature and survive in any condition, especially in less comfortable living conditions [ID 3, 5, 8, and 13]. In addition, it would be great to provide some equipment if necessary, as one participant recommended: "They may be able to provide equipment to accommodate the elderly if needed, for example a chair for the elderly to sit on when doing meditation because they cannot sit on the floor [ID 1]." Additionally, the provision of basic equipment for spiritual activities such as sitting and walking, meditation and yoga is important. The spiritual activities provider prepares the necessary equipment when practicing yoga. In the words of one participant: "They provide a mat for yoga or when practicing meditation, including meditation costumes [ID2]."

Thirdly, staff performance: overall, participants agreed that to create destination branding for spiritual retreat tourism, the staff working within the spiritual retreat venue must perform well, especially, the skill and knowledge of the job must be professional and expert. For example, participants said, "The professional performance of therapists, life coaches, or people involved in spiritual activities must have knowledge and be expert in what they do [ID 1, 5, and 15]." In addition, the spiritual leaders' knowledge is important; they must know how to perform expertly in spiritual activities, with an ability to guide participants to follow, such as in yoga and meditation. For instance, one participant said:

> The service quality in the rehabilitation centre must be good, especially with friendly staff providing activities and an atmosphere that meets the needs of the tourist, such as to set standards that can give good treatment; hence the tourist can receive good service quality.
>
> [ID 7]

Fourthly, providing spiritual retreat activity programmes for tourists: the spiritual retreat activity is basically adapted from Buddhist practices to respond to the modern western world so that all religions can participate. It was found that overall spiritual retreat activity for this study is mainly the practice of meditation, and there are some activities organized under this programme. The programme can include short walks and doing meditation around the place, sitting on a hard surface and meditating (normal practice is usually around 30 minutes), or hiking up mountains (similar to monks on a pilgrimage). The participants usually follow the eight lay precepts of Buddhism, which are: first, abstaining from killing living beings; second, abstaining from stealing; third, abstaining from sexual misconduct; fourth, abstaining from telling lies; fifth, abstaining from intoxicating drinks and drugs; sixth, abstaining from eating in the afternoon; seventh, abstaining from entertainment and beautifying the body; and eight, abstaining from using luxurious furniture. There is the use of a silent period to experience the process that can give opportunities to be touched by nature, by escaping from social speaking. The retreat course aims to make people understand the nature of life by the teachings of a highly respected spiritual leader, such as a monk or life coach. It was discovered that many participants who take part in a spiritual retreat programme will face up to some harsh crisis that occurred previously in their life, and it is expected that this way might cure some such wound. For example, one participant said:

> I engaged in life coaching while on holidays by reserving the spiritual tourism package. The programme is about 4–5 days and includes accommodation, transportation, activities (life coaching), and food/drink. The retreat programme is organized once per year, and this retreatment centre is not advertised on any media, but people know by word of mouth

She said,

> Especially, the life coach programme helped to address the hard life I am facing, change the way I operate my daily life and work, and handle my mistakes or failures, especially when my married life was broken. They guided me on how to deal with failure, and rather changed my negative attitude to positive thoughts about my married life. It helped me discover my own self-actualisation, understand myself from the spiritual perspective and discover a meaningful life."
>
> [ID 5]

### *14.8.2 Research objective 2: the feelings or ideas that can be created toward spiritual tourism destination branding*

The results revealed that to create feelings and ideas toward tourists' attitudes could help to create effective spiritual tourism destination branding. Overall, participants agreed that there are three symbols that can be created to build a tourist's feelings or ideas of a spiritual tourism destination development. Firstly, they believed that meditation is one activity symbolic of representing spiritual retreat tourism because the way of this activity helps in encouraging people to enhance the quality of their well-being in life [ID 2, 3, 5, 6, 8, and 14].

Secondly, the picture of a heart symbolizes the spirituality of mental, physical and emotional well-being [ID 1, 12, and 15]. Thirdly, a bird symbol is one picture that should be used as a logo for branding the spiritual retreat tourism destination. For instance, one participant said, "I feel we should use the bird as a symbol to brand a spiritual tourism destination because the bird represents emotional freedom" [ID 4].

### *14.8.3 Research issue 3: the unique experiences that can be created towards spiritual tourism destination branding*

The results revealed that there are some unique experiences that can be created to develop spiritual tourism destination branding. It can be stated in summary that use of local culture, unique places to create spiritual activity and organizing spiritual events or festivals are among the features that help to develop spiritual tourism destination branding.

Firstly, exploring the local food, lifestyle and unique traditions of a region, for example experiencing typical local and healthy food, can be advantageous, especially if they want to have fresh ingredients from local farmers. At the same time, this will help to support the farmer and generate income, thereby boosting the local economy. Moreover, they preferred to live like the locals, wearing simple and informal clothing or local costumes.

The participants also remarked that they have experienced a spiritual retreat programme many times in different regions around Thailand. They revealed that many places provide similar activities, including meditation, both walking and sitting, and follow routine activities such as listening to nature. However, the actual spiritual programme experiences differ depending on where they go, such as North, South, West or East regions. One participant stated that the activities are quite similar in each region, but that the unique experiences gained from each destination will lead to an intention to revisit the place [ID 5]. Another participant made the following observation:

> The temple-style spiritual attractions such as Wat Ram Poeng, Wat Phra That Si Chom Thong, and Wat Phra That Doi Suthep are in Chiang Mai province in the north area of Thailand. They provided Dhamma meditation style and a comfortable place with the necessary amenities where I participated in the spiritual retreat programme.
>
> [ID 3]

These unique places also give great experiences; however, many spiritual programmes are organized in the temples, while some are located at resorts. The most unique nature of such a place is when there is a good atmosphere of calmness and charm, surrounded with a large landscape with tall green trees: such is suitable to practice meditation.

In addition, it was found that organizing a spiritual retreat festival helps to create a unique experience, apart from participating in the meditation practice. A festival will help to promote and attract spiritual tourists to visit the destination, as evidenced from those participants who asserted that:

> There is a yoga festival nearby the sea at Hua Hin in Pra Chuanp Khiri khan province, Thailand. There are many yoga instructors, and yoga players joining this event come from all over the world.
>
> [ID 2]

Another said,

> The ISAN SOUL festival (North East region of Thailand) is arranged by the related tourism organization and the governor within that region. Each year the event will be moved to other provinces within the North East of Thailand. The festival presents the local music style and is one of the activities that spiritual tourists can join.
>
> [ID 15]

It can be summarized that spiritual tourism is able to attract tourists through the unique experiences of the atmosphere, spiritual events and local festivals found in each destination. These unique experiences can result in spiritual tourism destination branding development.

## 14.9 Conclusion and discussion

This study investigated the experience of the Millennial generation on spiritual tourism destination branding and interrogated their perception and experience on spiritual retreat activity. The findings of this study enable the use of a strategic approach for spiritual tourism destination branding development. It can be concluded that spiritual tourism for the Millennial generation can be generated through different components: destination attributes; creating a brand, symbol and/or logo of spiritual tourism in the tourist's mind; and by creating unique experiences through both local culture and spiritual tourism events.

Firstly, destination attributes for spiritual tourism comprise place, service quality, spiritual activity leader performance, and the spiritual retreat activity programme package. Likewise, Sziva et al. (2017) contended that to develop a destination branding one must create product brands with good value, especially for price and quality perception. In addition, Ashton (2014) supported that destination brand image development must provide sufficient tourism attributes, especially infrastructure, and good quality of product and service.

Secondly, with regard to initiating a spiritual brand name, logo or symbol, according to an insightful result derived from this study, it can be proposed that three icons should represent spiritual tourism: these include pictures of meditation practice, a heart and a bird. Similarly, a logo, sign, or symbol could represent an activity, product or service, local geography, culture or the attributes of an activity (Blain et al., 2005; Crockett & Wood, 1999; Hankinson, 2001; Mueller & Peter, 2008; Ryan & Zahra, 2004).

Lastly, unique experience must be provided by the inclusion of local culture, unique places for spiritual activity, and by organizing a spiritual event or festival. In turn, these can be features that help with a strategic approach towards spiritual tourism destination branding. Thus, the current findings are embedded with a study carried out by Haq and Wong (2013), agreeing that for destination branding one must provide a unique experience for the tourist, especially with special events/festivals and unique activity developed from the local culture's traditions and religion principles. However, the centre's events created would be based on a spiritual tourism context, including such as yoga tourism or meditation practices. Similarly, Ravichandran and Suresh (2010) confirmed that unique wellness experiences will give well-being on emotional, physical, intellectual and spiritual themes. Likewise, memorable experiences of wellness life can be created to affect the well-being of one's physical, mental, social and emotional status (Bhumiwat & Ashton, 2020).

## 14.10 Implications of the study

Spiritual tourism destination branding strategy theory offers a framework in which future academicians, educators and researchers can explore the constructs of wellness life within the spiritual tourism context. This theory will increasingly become crucial for destination branding strategy as society continues to obsess over wellness living, including work–life balance and quality of life leading to spiritual well-being. Scholars must explore how the three components of spiritual tourism destination branding can contribute and educate for the increasing efficiency of the marketing strategy in branding development for the tourism industry.

The three components of spiritual tourism destination branding strategy comprise firstly, spiritual tourism functional attributes, including four attributes: place; service quality; spiritual leader qualification; and spiritual activity program.

The first attribute, the place for spiritual tourism, should be a peaceful and quiet environment in which to perform the spiritual activity. The second attribute, spiritual service quality, refers to the basic human survivor needs; it is necessary to include those simple things that are most crucial in life, such as accommodation/shelter, food/water, clothing (suitable for meditation or yoga practice), and safety/security (a safe place to stay even when doing meditation alone). For instance, meditation practice clothing should fit, be comfortable, give a relaxed feeling, be made from the right type of fabric (such as organic cotton), be warm (suitable for the weather of each destination), not made from expensive material or over-designed; and the colour should impart a sense of calmness when seeing or wearing it. For the third attribute, spiritual leader qualification, staff who lead all spiritual activity programmes must be well trained, professional and expert in the exercise they instruct, such as meditation practice, yoga, and other related spiritual activity. With regard to the fourth attribute, spiritual activity, programmes must cover what the tourist expects and refer to the spiritual tourism package which should include the spiritual activity and social interaction, as after the activity is finished tourists may gain some additional knowledge, such as an opportunity to listen to a highly respected person, a religious leader or intellectual person within the related field (for example, invite a high-ranking monk to teach about life). In addition, programmes to visit the local community could be organized, including trying local food, learning about the lifestyle, and participating in festivals or traditions when having the opportunity.

Second component: spiritual psychology attributes refer to destination branding image, feelings, ideas or symbols related to the spiritual tourism context. It is important that when creating destination branding one creates a symbol or logo that can attract tourist attitude (García et al., 2012; Perkins, Khoo-Lattimore, & Arcodia, 2020; Ruiz-Real et al., 2020). In this study, creating a brand image into the tourist mind can be achieved by using three important symbols, especially, the images of meditation practice, a heart, or bird representing the spiritual tourism branding image. In turn, tourists will perceive the logo and symbols, and immediately recognize a spiritual tourism destination; it can be good branding to propose within this context.

The third component, spiritual uniqueness experience, comprises two perspectives: firstly, create a spiritual experience to have effectiveness on the wellness of physical, mental, social and emotional states (Bhumiwat & Ashton, 2020). The physical activity can help delay, prevent and manage many of the chronic diseases; especially, adults aged 50 and older are at risk (Watson et al., 2016). Mental well-being in life can be modified through some physical activities, including forms of exercise, starting from walking (Windle, Hughes, Linck, Russell, & Woods, 2010). Social well-being with family, friends and people at a destination can be helpful for the maintenance of good health (Su, Long, Wall, & Jin, 2016). Finally, emotional well-being refers to the emotional quality of an individual's everyday experience: the intensity of experiences of joy, stress, sadness, anger and affection that make one's life either pleasant or unpleasant (Kahneman & Deaton, 2010, p. 16489); it includes positive mood and high self-esteem (Schutte, Malouff, Simunek, McKenley, & Hollander, 2002). Secondly, create a spiritual event or festival for the tourist to explore further spiritual experience, such as a meditation event, or yoga festival; this event can be designed based on religious culture and resources, conducted in an environmentally unique location suitable with the spiritual activity and could include a religious ceremony, providing a unique feature experience for tourists at a destination.

This study provides not merely significant theoretical contribution, but also useful guidelines for managers and stakeholders related to the spiritual tourism industry, especially among the Millennial generation. For managerial contributions, this study provides great benefits for both tourist spiritual/mind retreats and marketing strategy competitive advantage. Creating great destinating branding is a crucial marketing strategic tool because it can add value to many businesses (Ashton, Limisariyapong, & Islam, 2019; Braun, 2005). In addition, the findings help to create great reputation for the destination and nation as a whole, including travel agents, service businesses and hospitality operators (Hall, 2010; Jetter & Chen, 2011; Saraniemi, 2010); reputation is the aggregate of stakeholders' images of a country over time (Passow, Fehlmann, & Rahlow, 2005). The useful study findings can help relevant stakeholders in several ways: firstly, in terms of destination attributes when building a destination branding for spiritual tourism, the place must have a silent landscape with a peaceful environment, because spiritual and mind recovery activity needs a very peaceful atmosphere, and easy access to the site with various modes of transportation.

Secondly, logos or symbols of the spiritual tourism destination should be designed, and clear information provided to all interested tourists around the world. The feelings or ideas that could be represented are such as an icon or logo of the meditation destination; a bird and/or heart well represents spiritual tourism. For the activity instructors, the extent of their professionalism and expertise will help tourist decision in participating at a destination; many of them search for information, mainly online. The tourist will look for a famous trainer if they can, hence creating a weblog and advertising the good qualifications of the instructors or trainers will be an advantage. Currently, spiritual tourism information is still lacking, and it would be of great competitive advantage if spiritual service providers provide clear information about a destination, including spiritual activity, and perhaps not only about religion, but for all believers who want to relax, escape from the daily routine, and have mind/soul retreat (as seen in Figure 14.1).

## 14.11 Limitation of the study and future research

This study conducted face-to-face interviews with participants who take meditation practice courses, some of them joining in meditation programmes at temples, mainly in Thailand. Future researchers should conduct in-depth interviews with those who participate in other spiritual activities, for example yoga, regarded as valuable in spiritual or mind/body recovery; additionally, study on how to develop activities and provide the spiritual retreat experiences expected, for example, how to encourage tourists to participant and gain experience in mental, physical, emotional and social well-being activities. Researchers could also explore in other countries where there are famous spiritual retreats and unique destinations, such as India, Kashmir, and Nepal.

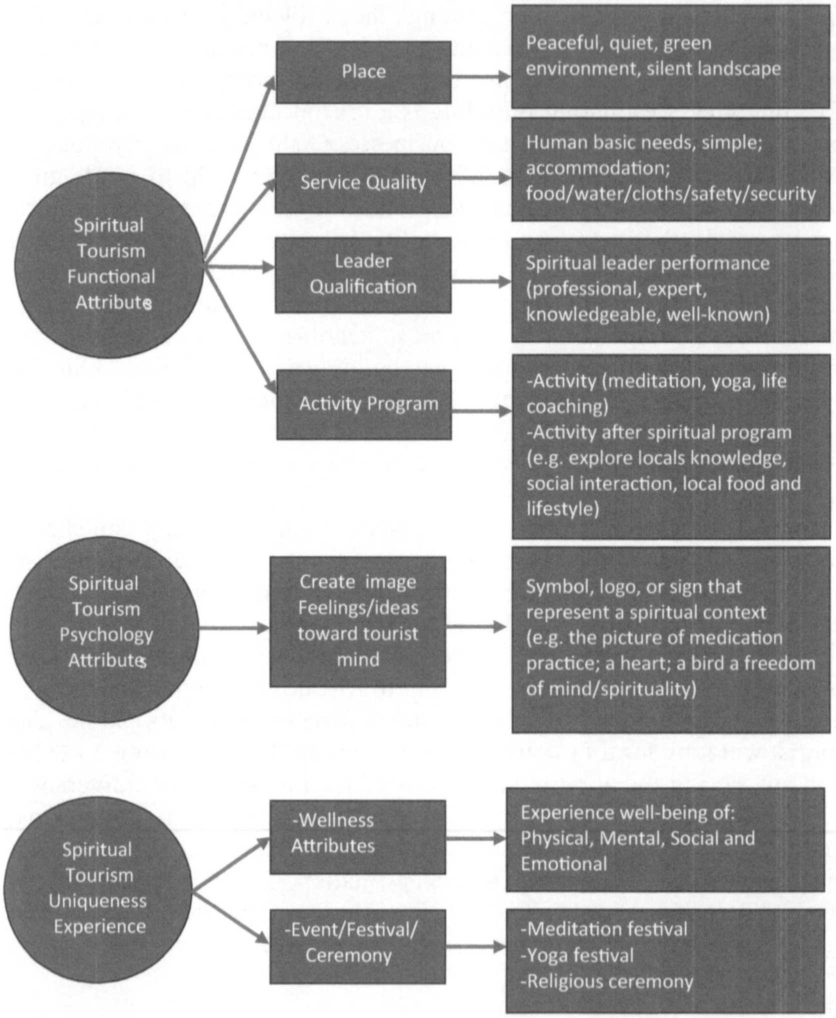


*Figure 14.1* Spiritual Tourism Destination Branding Development for the Millennial Generation.

## References

Adamus-Matuszyńska, A., Dzik, P., Michnik, J., & Polok, G. (2021). Visual Component of Destination Brands as a Tool for Communicating Sustainable Tourism Offers. *Sustainability*, 13(2), 731.

Aggarwal, A. K., Guglani, M., & Goel, R. K. (2008). *Spiritual & Yoga Tourism: a Case Study on Experience of Foreign Tourists Visiting Rishikesh, India. In Conference on Tourism in India –Challenges Ahead.*

Alegro, T., & Turnšek, M. (2021). Striving to Be Different but Becoming the Same: Creativity and Destination Brands' Promotional Videos. *Sustainability*, 13(1), 1–16.

Anjomrouz, E., Dehyadegari, S., Hosseini, K., & Anjomrouz, H. (2021). The Role of Urban Branding to Attract Foreign Tourists. *Independent Journal of Management & Production*, 12(1), 310–328.

Ashton, A. S. (2014). Tourist Destination Brand Image Development an Analysis Based on Stakeholders' Perception A Case Study from Southland, New Zealand. *Journal Vacation Marketing*, 20(3), 279–292.

Ashton, A. S. (2018). Spiritual Retreat Tourism Development in the Asia Pacific Region: Investigating the Impact of Tourism Satisfaction and Intention to Revisit: A Chiangmai, Thailand Case Study. *Asia Pacific Journal of Tourism Research*, 23(11), 1098–1114. doi: 10.1080/10941665.2018.1526198.

Ashton, A. S., Limisariyapong, S., & Islam, R. (2019, 7–8 September). *The Development of Value Perception toward Cultural Tourism Destination: A North East Thailand Case Study*. Paper presented at the *Penang International Conference on Hospitality 2019 (PiTCH 2019) and 2nd National Healthcare Food Service Conference 2019 (NHFSC 2019)*, 7–8 September, 2019, Penang, Malaysia.

Ashton, A. S., Scott, N., & Breakey, N. M. (2008, 10–14 February). *Hotel Restaurant Co-branding: The Impact of Consumer Evaluation on Perceived Risk, Perceived Value and Intention to Purchase*. Paper presented at *the Council for Australian University Tourism and Hospitality Education (CAUTHE)*, Gold Coast, Australia.

Astin, J. A. (1997). Stress Reduction through Mindfulness Meditation. *Psychotherapy and Psychosomatics*, 66(2), 97–106.

Bhumiwat, A., & Ashton, A. S. (2020). Emotional Tourist Experience Model: Exploring the Health-wellbeing for the Active 50-plus Tourist *Journal of Tourism, Hospitality & Culinary Arts*, 12(3), 84–105.

Blain, C., Levy, S. E., & Ritchie, J. R. B. (2005). Destination Branding: Insights and Practices for Destination Management Organizations. *Journal of Travel Research*, 43(4), 328–338.

Bone, K. (2013). Spiritual Retreat Tourism in New Zealand. *Tourism Recreation Research*, 38(3), 295–309.

Braun, P. (2005, 6–7 September ). *Creating Value to Tourism Products through Tourism Networks and Clusters: Uncovering Destination Value Chains*. Paper presented at the *Global Tourism Growth: A Challenge for SMEs Korea Conference*.

Buzinde, C. N. (2020). Theoretical Linkages between Well-being and Tourism: The Case of Self-determination Theory and Spiritual Tourism. *Annals of Tourism Research*, 83(July), 1–11.

Chandler, C. K., Holden, J. M., & Kolander, C. A. (1992). Counseling for Spiritual Wellness: Theory and Practice. *Journal of Counseling & Development*, 71(2), 168–175.

Choe, J., & O'Regan, M. (2020). Faith Manifest: Spiritual and Mindfulness Tourism in Chiang Mai, Thailand. *Religions*, 11(4), 177.

Crockett, S. R., & Wood, L. J. (1999). Brand Western Australia: a Totally Integrated Approach to Destination Branding. *Journal of Vacation Marketing* 5(3), 276–289.

Cutler, N. E. (2015). Millennials and Finance: The" Amazon Generation". *Journal of Financial Service Professionals*, 69(6), 33–39.

Denzin, N. K., & Lincoln, Y. S. (2005). Introduction: The Discipline and Practice of Qualitative Research In N. K. Denzin & Y. S. Lincoln (Eds.), *The Sage Handbook of Qualitative Research* (3rd ed.). New Delhi, India Sage Publications, Inc.

DeVaney, S. A. (2015). Understanding the Millennial Generation. *Journal of Financial Service Professionals*, 69(6), 1–5.

Dimock, M. (2019). Defining Generations: Where Millennials end and Generation Z begins. *Pew Research Center*, 17, 1–7.

Echtner, C., & Ritchie, J. R. B. (1993). The Measurement of Destination Image. An Empirical Assessment. *Journal of Travel Research*, 31(4), 3–13.

Eddy, G. (2014). The Vipassana Retreat Experience: A Consideration of the Meditation Retreat as a Religious Paradigm of Travel. *Tourism Management Perspectives*, 22(1), 38–59.

Elias, P. (2005). Methodological, Statistical and Practical Issues Arising from the Collecting and Analysis of Work History Information by Survey Techniques In D. De Vaus (Ed.), *Research Design* (Vol. 3, p. 141). Great Britain: Sage Publications Ltd.

García, J. A., Gómez, M., & Molina, A. (2012). A destination-branding model: An empirical analysis based on stakeholders. *Tourism Management*, 33(3), 646–661.

Garikapati, V. M., Pendyala, R. M., Morris, E. A., Mokhtarian, P. L., & McDonald, N. (2016). Activity Patterns, Time use, and Travel of Millennials: A Generation in Transition? *Transport Reviews*, 36(5), 558–584.

Gill, C., Packer, J., & Ballantyne, R. (2019). Spiritual Retreats as a Restorative Destination: Design Factors Facilitating Restorative Outcomes. *Annals of Tourism Research*, 79, 102761.

Goyal, A. (2021). Communicating and Building Destination Brands with New Media *Impact of New Media in Tourism* (pp. 1–20): IGI Global.

Hall, C. M. (2010). Tourism Destination Branding and its Effects on National Branding Strategies: Brand New Zealand, Clean and Green but is it Smart? *European Journal of Tourism, Hospitality and Recreation*, 1(1), 68–89.

Handcock, M. S., & Gile, K. J. 2011 Comment: on the Concept of Snowball Sampling. *Sociological Methodology*, 41(1),367–371.

Hankinson, G. (2001). Locating Branding: A Study of the Branding Practices of 12 English Cities. *Journal of Brand Management*, 9(2), 127–142.

Haq, F., & Wong, H. Y. (2010). Is spiritual Tourism a New Strategy for Marketing Islam? *Journal of Islamic Marketing*, 1(2), 136–148.

Haq, F., & Wong, H. Y. (2013). Branding Islamic spiritual tourism: an exploratory study in Australia and Pakistan. *European Journal of Business and Management*, 5 (11), 154–162.

Haq, F. M., & Medhekar, A. (2020). The Rise of Spiritual Tourism in South Asia as Business Internationalization. *Journal of Management and Research*, 7(1), 52–78.

Heintzman, P. (2013). 6 Retreat Tourism as Transformational Tourism. In Y. Reisinger (Ed.), *Transformational tourism: Tourist perspectives* (Vol. 68). Boston, MA: CABI.

Heintzman, P., & Mannell, R. C. (2003). Spiritual Functions of Leisure and Spiritual Well-being: Coping with Time Pressure *Leisure Sciences*, 25(2–3), 207–230.

Jetter, L., & Chen, R. (2011). Destination Branding and Images: Perception and Practices from Tourism Industry Professionals. *International Journal of Hospitality and Tourism Administration*, 12(2), 174–187.

Jiang, T., Ryan, C., & Zhang, C. (2018). The spiritual or secular tourist? The experience of Zen meditation in Chinese temples. *Tourism Management*, 65, 187–199.

Kahneman, D., & Deaton, A. (2010). High Income Improves Evaluation of Life but not Emotional Well-being. *Proceedings of the National Academy of Sciences*, 107(38), 16489–16493.

Keller, K. L. (1993). Conceptualizing, Measuring, and Managing Customer Based Brand Equity. *Journal of Marketing*, 57(January), 1–22.

Kowske, B. J., Rasch, R., & Wiley, J. (2010). Millennials' (lack of) Attitude Problem: An Empirical Examination of Generational Effects on Work Attitudes. *Journal of Business and Psychology*, 25(2), 265–279.

Krippendorff, K. (2004). *Content Analysis an Introduction to its Methodology*. USA: Sage Publications, Inc.

Linders, E. H., & Lancaster, B. L. (2013). Sacred Illness: Exploring Transpersonal Aspects in Physical Affliction and the Role of the Body in Spiritual Development *Mental Health. Religion & Culture* 16(10), 991–1008.

McCartney, P. (2020). Yoga-Scapes, Embodiment and Imagined Spiritual Tourism. In C. Palmer & H. Andrew (Eds.), *Tourism and Embodiment* (pp. 86–106). NY: Routledge.

Mourtazina, E. (2020). Beyond the Horizon of Words: Silent Landscape Experience within Spiritual Retreat Tourism. *International Journal of Culture, Tourism and Hospitality Research*, 14(3), 349–360.

Mueller, S., & Peter, M. (2008). The Personality of Freestyle Snowboarders: Implications for Product Development. *Original Scientific Paper*, 56(4), 339–354.

Mukherjee, S., Bhattacharjee, S., & Singha, S. (2020). Religious to Spiritual Tourism-An Era of Paradigm Shift in India. *Available at SSRN*, doi: 10.2139/ssrn.3546903.

Norman, A. (2014). The Varieties of the Spiritual Tourist Experience. *Literature & Aesthetics*, 22(1), 20–37.

Norman, A., & Pokorny, J. J. (2017). Meditation Retreats: Spiritual Tourism Well-being Interventions. *Tourism Management Perspectives*, 24 (October), 201–207.

Passow, T., Fehlmann, R., & Rahlow, H. (2005). Country Reputation-from Measurement to Management: The Case of Liechtenstein. *Corporate Reputation Review*, 7(4), 309–326.

Perkins, R., Khoo-Lattimore, C., & Arcodia, C. (2020). Understanding the Contribution of Stakeholder Collaboration towards Regional Destination Branding: A Systematic Narrative Literature Review. *Journal of Hospitality and Tourism Management*, 43(June), 250–258.

Piewdang, S., Mekkamol, P., & Untachai, S. (2013). Measuring Spiritual Tourism Management in Community: A Case Study of Sri Chom Phu Ongtu Temple, Thabo District, Nongkhai Province, Thailand. *Procedia-Social and Behavioral Sciences*, 88, 96–107.

Pitcher, A. E. (1985). Advertising voices: The role of branding in international advertising. *International Journal of Advertising*, 4(3), 241–246.

Prideaux, B., Berbigier, D., & Thompson, M. (2014). Wellness Tourism and Destination Competitiveness. *Wellness Tourism: A Destination Perspective*, 45–58.

Pritchard, A., & Morgan, N. (1998). Mood Marketing: The New Destination Branding Strategy: A Case Study of "Wales" The Brand. *Journal of Vacation Marketing*, 4(3), 215–229.

Quintal, V., Phau, I., & Polczynski, A. (2014). Destination Brand Image of Western Australia's South-West Region: Perceptions of Local versus International Tourists. *Journal Vacation Marketing*, 20(1), 41–54.

Rather, R. A., Najar, A. H., & Jaziri, D. (2020). Destination Branding in Tourism: Insights from Social Identification, Attachment and Experience Theories. *Anatolia*, 31(2), 229–243.

Ravichandran, S., & Suresh, S. (2010). Using Wellness Services to Position and Promote Brand India. *International Journal of Hospitality & Tourism Administration*, 11(2), 200–217.

Ruiz-Real, J. L., Uribe-Toril, J., & Gázquez-Abad, J. C. (2020). Destination Branding: Opportunities and New Challenges. *Journal of Destination Marketing & Management*, 17, 1–13. doi: 100453.

Ryan, C., & Zahra, A. (2004). The Political Challenge: The Case of New Zealand's Tourism Organizations. In N. Morgan, A. Pritchard & R. Pride (Eds.), *Destination Branding Creating the Unique Destination Proposition* (2nd ed., pp. 79–110). Burlington, MA: Butterworth-Heinemann, Elsevier Ltd.

Ryan, C., & Zahra, A. (2010). The Political Challenge: The Case of New Zealand's Tourism Organizations. In N. Morgan, A. Pritchard & R. Pride (Eds.), *Destination Branding Creating the Unique Destination Proposition* (2nd ed., pp. 79–110). Great Britain: Butterworth-Heinemann, Elsevier Ltd.

Saraniemi, S. (2010). Destination Brand Identity Development and Value System. *Tourism Review*, 65(2), 52–60.

Schedneck, B. (2015). *Thailand's International Meditation Centers: Tourism and the Global Commodification of Religious Practices*. London, New York: Routledge.

Schutte, N. S., Malouff, J. M., Simunek, M., McKenley, J., & Hollander, S. (2002). Characteristic Emotional Intelligence and Emotional Well-being. *Cognition & Emotion*, 16(6), 769–785.

Sharma, N. P. (2020). Meditation tourism, Buddhist philosophy in practice: A case study of Lumbini, Nepal. *The Gaze: Journal of Tourism and Hospitality*, 11(1), 37–59.

Sharma, V., Tomar, R. S., Ahola, N. N., Tomar, R. S., & Ahola, N. N. (2008 15–17 May). *Branding India for Health and Spiritual Tourism*. Paper presented at the *Conference on Tourism in India- Challenges Ahead India*.

Sharpley, R., & Jepson, D. (2011). Rural Tourism: A Spiritual Experience? *Annals of Tourism Research*, 38(1), 52–71.

Singleton, A. (2017). The Summer of the Spirits: Spiritual Tourism to America's Foremost Village of Spirit Mediums. *Annals of Tourism Research*, 67 (November), 48–57.

Sirirat, P. (2019). Spiritual tourism as a tool for sustainability: a case study of Nakhon Phanom province, Thailand. *International Journal of Religious Tourism and Pilgrimage*, 7(3), 97–111.

Smith, M. (2003). Holistic holidays: Tourism and the Reconciliation of Body, Mind and Spirit. *Tourism Recreation Research*, 28(1), 103–108.

Smith, M., & Puczkó, L. (2013). *Health and Wellness Tourism*. London, New York: Routledge.

Smith, M., & Puczkó, L. (2015). More than a Special Interest: Defining and Determining the Demand for Health Tourism. *Tourism Recreation Research*, 40(2), 205–219.

Smith, M. K., & Diekmann, A. (2017). Tourism and Wellbeing. *Annals of Tourism Research*, 66, 1–13.

Srivastava, M., & Kaul, D. (2014). Social Interaction, Convenience and Customer Satisfaction: The Mediating Effect of Customer Experience. *Journal of Retailing and Consumer Services*, 21(6), 1028–1037.

Stylos, N., Vassiliadis, C. A., Bellou, V., & Andronikidis, A. (2016). Destination Images, Holistic Images and Personal Normative Beliefs: Predictors of Intention to Revisit a Destination. *Tourism Management*, 53(April), 40–60.

Su, M. M., Long, Y., Wall, G., & Jin, M. (2016). Tourist–community interactions in ethnic tourism: Tuva villages, Kanas Scenic Area, China. *Journal of Tourism and Cultural Change*, 14(1), 1–26.

Suravichai, S., & Namwong, S. (2020). Destination Development for Spiritual Tourism and Tourist Behavior in Chiang Mai Province. *Dusit Thani College Journal*, 14(1), 145–164.

Sweeney, R. (2006). Millennial Behaviors and Demographics. *Newark: New Jersey Institute of Technology*, 12(3), 1–10.

Sziva, I., Balázs, O., Michalkó, G., Kiss, K., Puczkó, L., Smith, M., & Apró, E. (2017). Branding Strategy of the Countries in the Balkan Region-focusing on Health Tourism. *GeoJournal of Tourism and Geosites*, 19(1), 61–69.

Tasci, A. D. (2011). Destination Branding and Positioning. *Destination Marketing and Management: Theories and Applications*, 113–129, doi: 10.1079/9781845937621.0113.

Tilson, D. J. (2005). Religious-spiritual Tourism and Promotional Campaigning: A Church–state Partnership for St. James and Spain. *Journal of Hospitality & Leisure Marketing* 12(1–2), 9–40.

Tsai, C.-T. S., & Wang, Y.-C. (2017). Experiential Value in Branding Food Tourism. *Journal of Destination Marketing & Management*, 6(1), 56–65.

Voigt, C., & Pforr, C. (2014). *Wellness Tourism A Destination Perspective*. London, New York: Routledge.

Watson, K. B., Arlson, S. A., Gunn, J. P., Galuska, D. A., O'Connor, A., Greenlund, K. J., & Fulton, J. E. (2016). Physical Inactivity Among Adults Aged 50 Years and Older-United States, 2014. *Morbidity and Mortality Weekly Report (MMWR)*, 65(36), 954–958.

Windle, G., Hughes, D., Linck, P., Russell, I., & Woods, B. (2010). Is Exercise Effective in Promoting Mental Well-being in Older Age? A systematic Review. *Aging & Mental Health*, 14(6), 652–669.

Wita, N., & Namwong, S. (2018). Tourism Perception toward Destination Brand Image Sustainability: Mae Kam Pong Community Case Study. *Journal of International and Thai Tourism*, 14(July–December ), 94–123.

Yoon, E., Chang, C. C.-T., Clawson, A., Knoll, M., Aydin, F., Barsigian, L., & Hughes, K. (2015). Religiousness, Spirituality, and Eudaimonic and Hedonic well-being. *Counselling Psychology Quarterly*, 28(2), 132–149.

# 15 Understanding the Millennial tourist

## Proclivities and penchants of indigenous Rastafarian cultural experiences among Generation Y

*Johnnel Smith*

### 15.1 Introduction

There is a natural mystic floating in the air when it comes to the isles of the Caribbean. From steel pans and reggae bands with pulsating rhythms to positive vibrations filling the atmosphere, the Caribbean is a melting pot of cultures and sub-cultures. From the tip of the Bahamas to the south of Guyana, the isles of the Caribbean amount to over 27 island states and archipelagos. Each of them is unique, possessing vibrant and captivating cultures that will leave any traveler wanting to experience more.

Jamaica – the third-largest island in the Caribbean – offers a magnificent palette of experiences. It is a land of unique culture, engaging activities, and warm, welcoming people with a most diverse range of attractions and culture (Jamaica Tourist Board, 2019). Jamaica is renowned around the world for its picturesque topology and cultural delights, decorated by its athletic prowess with world record holders such as nine-time Olympic gold medal champion Usain Bolt and the Jamaican bobsleigh team. Of greater note and equal global impact, Jamaica is the birthplace of reggae music and the home of legendary reggae artist Bob Marley, along with Rastafarianism and the Rastafari faith. Religion and spirituality has been a key component of Jamaican culture, attracting visitors to the island who seek to explore and experience the home of 'Rastas'.

Religion has often been considered as a key motivator to travel, accentuating tourism-related activities (Mosammam et al., 2016). Religious tourism can be defined as the visit of sacred places, to participate or follow-up in religious ceremonies and the pilgrimage in the form of visits or activities to fulfill religious duties in the evaluation of tourism understanding. Religious tourism incorporates visits to religious settings that are considered relevant to either one's own faith or to the faith of others (Raj and Morpeth, 2007). Religious sites such as cathedrals, temples and mosques are attracting an increasing number of visitors worldwide, not only for their spiritual value but also for recreational, educational and cultural purposes (Woodward, 2004). Rinschede (1992) explained that travel behaviour, including travel patterns, transportation choices, seasonal demand and socialization processes, are all affected by beliefs, especially in the context of religious tourism, with this

DOI: 10.4324/9781003170631-19

information it is also important for us to differentiate between religious tourism and spiritual tourism.

Heidari et al. (2018) further cite that spiritual tourism aims to explore the elements of life which lie beyond the self and contribute to a balance between body, mind and spirit. These may or may not have an affiliation to religion (Hill and Smith, 2010). Spirituality is a set of ideas that one can find in religious traditions, but which also exist independently of them, including an emphasis and culture of the self, wholeness, holism and (inter)connectedness, meaning- search/quest and experience-orientation, non-theistic cosmology, peacefulness/tolerance and similar positive value commitments (Heidari et al., 2018). Aggarwal et al. (2008) asserted that spirituality means having understanding with deep, often religious, feelings and beliefs, including a person's sense of peace, purpose, connection to others, and beliefs about the meaning of life. Therefore, spiritual holidays seek to address the interdependence of physical, emotional, mental and spiritual aspects. It was established prior that Jamaica is the birthplace of Rastafarianism and this is has been a great point of attracting visitors to the island, so let's take a closer look at the evolution of Rastafarianism and its appeal to travellers.

### *15.1.1 Rastafarianism in Jamaica*

Rastas/Rastafari and Rastafarianism are almost synonymous with 'brand Jamaica' and arguably the best-known Rastafari is the late Bob Marley. Rastafari, also known as Rastafarianism and the Rastafari movement, is a religious sect that developed in Jamaica during the 1930s among impoverished and socially disenfranchised Afro-Jamaican communities (Gjerset, 1994). It is classified as both a new religious movement and a social movement by scholars of religion. There is no central authority in control of the movement and much diversity exists among practitioners, who are known by a variety of labels: Rastafari, Rastafarians, or Rastas. Rasta beliefs are based on a specific interpretation of the Bible (Chawane, 2014). Central to these is a monotheistic belief in a single God, referred to as 'Jah', who is deemed to partially reside within each individual. Rastas accord Haile Selassie, the emperor of Ethiopia between 1930 and 1974, key importance; many regard him as the Second Coming of Jesus and Jah incarnate, while others see him as a human prophet who fully recognized the inner divinity in every individual (Clarke, 1986). Rastafari is Afrocentric and focuses attention on the African diaspora, which, it believes, is oppressed within Western society, or what the Rastas call "Babylon". Its Afrocentric ideology was largely a reaction against Jamaica's then-dominant British colonial culture. Rastas refer to their practices as "livity". Communal meetings are known as "groundations", and are typified by music, chanting, discussions, and the smoking of cannabis, the latter regarded as a sacrament with beneficial properties. Rastas emphasize what they regard as living "naturally", adhering to 'ital' (natural) dietary requirements, twisting their hair into dreadlocks, and following patriarchal gender roles (Barrett, 1997).

In the 1960s and 1970s, it gained increased respectability within Jamaica and greater visibility abroad through the popularity of Rasta-inspired reggae musicians, most notably Bob Marley. Enthusiasm for Rastafari declined in the 1980s, following the deaths of Haile Selassie and Marley, but the movement survived and still retains a presence in many parts of the world. One recent study suggested that there are an estimated 700,000 to 1,000,000 Rastas across the world (Hansing, 2001). The largest population is in Jamaica, but small communities can be found in most of the world's major population centres. Rastas come from various ethnic groups, although the majority are of black African descent.

### *15.1.2 The Millennials – a vibrant and promising market for Spiritual Tourism*

Over the past decade, Millennial travel has emerged as an increasingly important segment of the international tourism sector. In 2010, Millennials (youth born in the 1980s and early 1990s) generated US $165 billion in tourism receipts and 187 million international visits around the world, accounting for 20% of total global travel (WYSE Travel Confederation and UNWTO, 2011). Furthermore, Millennial travel is forecast to grow in the future. The number of international Millennial trips is expected to nearly double to reach about 300 million trips per year by 2023 (Canada Tourism Commission, 2015).

The motives of tourists is particularly important and has been studied by many authors (Beard and Ragheb 1983; Crandall 1980; Crompton 1979; Dann 1981; Manfredo et al., 1996; Gitelson and Kerstetter 1990; Hollender 1977; Krippendorf, 1987; Lee & Crompton 1992; Loker and Perdue 1992; Pearce 1988; 1991; Shoemaker 1989). Slivar et al. (2019) cite that researching leisure travel behaviour and travel trends of new generations is the first step for tourism providers towards modifying their offers in order to match target markets' needs. Millennials or Generation Y and Post-Millennials or Generation Z, influence the characteristsics of contemporary tourism.

Sofronov (2018) cites that Millennials travel more than any other demographic. On average, they take 35 days of vacation each year, and they're also bucking the trend by increasing their annual travel spend as other generations tend to be more conservative with their cash. The Millennial travel segment places a high value on travel as a life experience and considers travel to be one of the most important aspects of their lives, higher than any other age segment. For this generation, travel is more than just vacation; it is often an essential component of their personal growth and a learning process. As a result, Millennials tend to plan more trips than individuals in other age categories (Deloitte and the Tourism Industry Association of Canada, 2010).

Cavagnaro and Staffieri (2015) share that understanding youngsters' tourism experience, values and motivations to travel are key to identifying target clusters for sustainable tourism. Young tourists are key for tourism's future at least for three reasons: the sheer amount on young tourists travelling in the present (Richards, 2006); the fact that their original choices may lead to new attitudes towards tourism by the wider society (Crocetti et al., 2011; Leask, 2016); and the anticipation that young tourists will continue to travel in the future (Barton et al., 2013). Sofronov (2018) further shares that Millennial travellers are interested in authenticity, fulfillment and sustainability. Having grown up with near-ubiquitous Internet access and beginning their professional careers in the wake of a financial crisis, Millennials are using technology to make savvy travel-buying decisions, largely based on the perceived authenticity of the experience they are offered.

Jamaica is renowned for offering authentic cultural experiences especially in indigenous communities that offer a strong spiritual component, such as the Rastafarian indigenous villages (Jamaica Tourist Board, 2019). Therefore, in fulfilling the overarching objectives of this book – millennials and spiritual tourism – the following aims helped to guide the study:

1. To analyze the spiritual tourism product offerings for the Millennial traveller in Jamaica.
2. To investigate the motivations, tourist behaviours and travel adoption among Millennial travellers within indigenous spiritual communities in Jamaica.
3. To deduce the demand/preferences of Millennial travellers for indigenous spiritual products in Jamaica.

The chapter first takes us through literature on the Experience Economy and Experiential Marketing theories that underpin product design and delivery. It goes on further to frame the characteristics, motivations and values of the Millennial traveller, and their demand for immersive, authentic cultural experiences. It then examines the experiences within the current realities of our world by particularly referencing the impact of COVID-19 on these tourism activities and lastly summarizes the value of the chapter by identifying the social and practical implications of the study.

## 15.2 Literature review

### *15.2.1 The Experience Economy and Experiential Marketing*

The contextual framework for this study examines the authentic spiritual experiences of the Millennial travel market. Therefore, the researcher sought to establish the theoretical lens within which this study would be framed and draws upon the work of Pine and Gilmore's Experience Economy (1999) as the main theoretical underpinning for this study. More specifically, the branch of theory which examines Experiential Marketing (McCole, 2004). Experiential Marketing is a component of the Experience Economy and is a growing trend which involves marketing a product or a service through experiences that engage the customers and create emotional attachment to the product/service (McCole, 2004). The economic pattern of a society varies with peoples' lifestyles. It has evolved from the earliest agrarian economy and industrial economy into the most recent service economy. The most prominent characteristic of the service economy is building consumer rapport by selling products that meet consumers' needs for personalization. However, the widespread development of the service economy has curbed peoples' desires for personalized consumption, which prompted a shift in consumption to providing new commercial services, subsequently leading to the emergence of the next economy pattern—the Experience Economy. The Experience Economy focuses on creating touching, surprising, and unforgettable sensations for consumers, and, therefore, experience becomes the most essential element of this economic pattern. The Experience Economy is formed when an industry uses experiences to satisfy consumers' needs and create economic value (Pine II and Gilmour, 1999).

> Millennials particularly seek social and experiential travel activities and experiences that will lead to personal growth. They value authenticity, new and unique things, self-confidence and technology. Young travelers also place significant importance on their community and friends. This segment is more environmentally and socially aware than any other age segment. These values and cultural aspects have major implications on how young travelers develop their travel plans and where they decide to travel.
>
> (Canadian Tourism Commission, 2015 p. 7)

Beyond the sheer size of the youth travel segment, the unique character of this market offers many other benefits to the tourism sector: Millennials often spend more in destinations than other tourists because they travel for longer periods; Millennials travel more frequently and further than other age groups and are likely to return to the destination over their lifetime; Millennials are pioneer travellers who discover and promote destinations not yet visited by traditional tourists (Canada Tourism Commission, 2015). In addition, Millennials are "time-richer" than other age segments—the greater amount of discretionary time they enjoy allows them to travel more often and spend more time in their chosen destination (Canada Tourism Commission, 2015).

Barton et al. (2013) cite that Millennials in the United States value diversity, embrace a global perspective, and are open to new experiences. These characteristics—which suggest a generation of natural travellers—are good news for the travel industry. Millennials are already engaging with brands, forming perceptions, and developing travel habits and preferences. These nascent tendencies will solidify as members of this generation age. Airlines, hotels and destinations that do not reach out to Millennials now, seeking to understand and address their unique needs, may miss the boat entirely. In addition to focusing on long-term planning and development, an effective strategy must consider how branding and marketing, loyalty programmes, distribution channels and amenities must change to meet the needs and preferences of Millennials (Barton et al., 2013).

The Millennial travel segment tends to embrace social values, innovation and environmental consciousness. They are characterized as optimistic persons, social, open to changes and have high expectations from themselves and others (Princeton One, 2017), are more entrepreneurial, realistic, have a shorter attention span and higher expectations (Lundin, 2018). Their characteristics influence their travel habits. The shift towards placing the value on *experiences*, rather than material things is evident (Lundin, 2018). Any businesses in pursuit of sustainable development must commit to organizing special events that are worth remembering or cherishing for customers. These memories or feelings form the core product characteristic in the minds of customers, thus leading to the rise of the Experience Economy (Manfredo et al., 1996). The rise of the Experience Economy has not only prompted businesses to produce products but also provided a platform on which consumers can gain unforgettable experiences.

### *15.2.2 Experiential marketing and hand-touch product design*

The concept of Experiential Marketing is derived to facilitate the further development of the Experience Economy. Schmitt (1999) proposed the concept of Experiential Marketing, defining it as a process in which individual consumers observe or take part in an event where they are motivated to identify with an idea or engage in consumption behaviour, and realize that a complete process of development is inevitably required to strengthen product value and create experiences for consumers.

The primary characteristic of a hand-made product is creating memorable experiences for consumers by allowing consumers to touch and feel the product. The experiences brought by hand-touch products lead to the formation of an Experience Economy for hand-touch products, which facilitates the promotion and sale of these products. Du et al. (2019) suggests that hand-touch products are products that emphasize tactile sensations. These products can generate sensory stimulation and create memorable experiences for consumers, thereby driving purchase decisions. The economic value of hand-touch products is undoubtedly associated with the experience economy. Consumers generally focus on how much experience value is created by the hand-touch product itself. An Experiential Marketer purports to create an environment that engages customers in visual, stimulated, interactive, and tactile experiences, thereby encouraging consumers to adopt consumption (Beard and Ragheb, 1983).

With thorough understanding of the literature on the value of experiences and preference of Millennials for experiences, we are now able to take a closer look at experiential marketing and investigate the development and integration of the tangible (hand-touch products) within the intangible (an experience, i.e spiritual tourism).

## 15.3 Tangible and experiential elements of Rastafarianism

### *15.3.1 Smoking Cannabis*

Bob Marley introduced much of the world to the tenets of Rastafari, including the smoking of ganja (marijuana). The principal ritual of Rastafari is the smoking of marijuana, also known as 'ganja', 'cannabis', 'callie', 'Iley', 'the herb', 'the holy herb', 'the grass', and 'weed'. In Rastafari, smoking cannabis is considered a spiritual act and a sacrament; they regard it as having healing properties (Barnett, 2005) and eulogize it for inducing feelings of "peace and love" and claim that it cultivates a form of personal introspection that allows the smokers to discover their inner divinity (Edmonds, 2012). Some Rastas believe that cannabis smoke serves as an incense that counteracts immoral practices in society (Sibanda, 2016). Cannabis is usually smoked during groundings (a communal ceremony for grounding the spirit), although some practitioners also smoke it informally in other contexts (Clarke, 1986). Some Rastas smoke it almost all of the time, something other practitioners regard as excessive, and many practitioners also ingest cannabis in a tea, as a spice in cooking, and as an ingredient in medicine (Edmonds, 2012). However, not all Rastas use ganja. Some abstainers explain that they have already achieved a higher level of consciousness and thus do not require it (Clarke, 1986).

Rastas typically smoke cannabis in the form of a large, hand-rolled cigarette known as a 'spliff' (Barrett, 1997/1988). This is often rolled together while a prayer is offered to 'Jah' (the divine God); the spliff is lit and smoked only when the prayer is completed. At other times, cannabis is smoked in a water pipe referred to as a 'chalice'. The pipe is passed in a counter-clockwise direction around the assembled circle of Rastas (Edmonds, 2012). In many countries – including Jamaica – cannabis is illegal. Rastas have always advocated for the legalization of cannabis and in 2015, Jamaica decriminalized personal possession of marijuana up to two ounces and legalized it for medicinal and scientific purposes (Margolin, 2018).

### *15.3.2 Reggae Music*

As Rastafari developed, music became its chief communicative medium (King, 2002). Rasta music is performed to praise and commune with Jah, and to reaffirm the rejection of Babylon (Clarke, 1986). Rastafari music is developed at reasoning sessions, and Rastas believe that their music has healing properties, with the ability to cure colds, fevers and headaches; many of these songs are sung to the tune of older Christian hymns, but others are original Rasta creations (Clarke, 1986). The different components of the music are regarded as displaying different symbolism; the bassline symbolizes blows against Babylon, while the lighter beats, denoting hope for the future, the drumming is accompanied with chanting, and dancing (Edmonds, 2012).

The year 1968 saw the development of reggae in Jamaica, a musical style typified by slower, heavier rhythms and an increased use of Jamaican creole (language) (King, 2002). Reggae was a medium for social commentary

(Barrett, 1997/1988) and also adopted Rasta chants, language, motifs and social critiques (Edmonds, 2012). Reggae gained widespread international popularity during the mid-1970s, coming to be viewed by black people in many different countries as the music of the oppressed (King, 2002).

The rest of the chapter will see a weaving of these experiential elements within spiritual communities in Jamaica and how they are consumed by the Millennial traveller.

## 15.4 Methodology

The methodology takes the form of qualitative semi-structured interviews coupled with a qualitative content analysis conducted online by reviewing websites of indigenous Rastafari villages in Jamaica, blogspots, YouTube channels and review boards on TripAdvisor.com. The interviews were conducted in Jamaica. The location for the research was chosen based on the background and the professional network of the researcher which provided greater ease and access to the information needed to conduct the study. The close proximity of the island to travelers from the US and Canada was also an advantage as it was able to justify the findings of the research on the Millennial travel market in these two countries as shared in the Literature Review.

The sample was drawn from communities in Jamaica where indigenous villages were located, including St. James, Westmoreland, Kingston and St. Andrew and St. Ann. Ten interviews were conducted with Jamaican travel agents (two respondents), destination managers (three respondents) from destination management companies (DMCs), and tribal leaders from Rastafari indigenous communities (five respondents) who have been welcoming tourists to their communities for years. The sample was very insightful about the needs of this particular consumer market as they all recorded over ten years of professional experience with the Millennials and spiritual tourism (establishing this parameter for the sample helped to narrow the number of respondents and to give more detailed viewpoints on the growth, evolution and trends in this growing market). The interview questions were prepared and premised on the research questions. The interviews were conducted in two groups via email, phone, Zoom meetings and social networking sites such as LinkedIn, Facebook and WhatsApp. The first group of interviews were conducted in September 2020 and the second group of interviews in January 2021. The responses were analyzed via thematic analysis.

## 15.5 Findings – the components of spiritual tourism in indigenous Rastafari villages

The spiritual tourism product from Indigenous Rastafari villages in Jamaica were found to be that of mostly daily excursions to the villages or homestays within the villages. The offerings boast a diverse array, all of which espouse the Rastafarian values of equality, healthy living and connectedness with nature.

### *15.5.1 Excursions to indigenous Rastafari villages*

#### *15.5.1.1 The Bob Marley Museum and other indigenous villages*

Our first respondent (an Indigenous Rastafari Host) provides this description of what an excursion to the village entails:

> A typical tour can take 2–5 hours to complete and are located in all parts of the island from St. James in the West to Kingston in the East. For example, a visit to the Bob Marley Museum will have you enter his house, his recording studio, walking through his bedroom and backyard. The vibrations of his music fill the air and so does the smell of marijuana. This all leaves travelers wanting more.

The tour of the Bob Marley Museum usually lasts around two hours and ends with a 20-minute documentary of the life of Bob Marley. At the end of the tour, you also meet and interact with former members of his band and partake in the sacramental smoking of cannabis and playing of drums and other instruments.

> There is also a tour of the birthplace and mausoleum of Bob Marley, which is surrounded by green hills of Nine Miles in Ocho Rios, Jamaica. In order to get there tourists are taken along a winding road through villages and farmland, which is an experience in itself.
>
> (Respondent #3, Indigenous Rastafari Host)

> Other tours create a space to connect you to Rastafari, to make universal human connections, and to preserve and protect the Rasta culture of humanity. They (tour guides) take or 'trud' (walk) travelers together across a river and through the bamboo glade, where they become attuned to the energy of water, air, living things.
>
> (Respondent #2, Destination Manager)

Upon arrival at a rasta village, guests are greeted and get a chance to savour coconut water served in a calabash bowl (an eco-friendly cup made by rastas).

> After learning about Rasta philosophy and liberty in the Big Hut (home of the chief), they are invited to meet and talk with community members in their different cottage enterprises and see their hand-made drums, soaps, jewelry and art while tourists also get to nurture the organic permaculture medicinal herb farm. Other activities include swimming in crystal clear mineral springs, discovering a Ganja Field, see and taste exotic fruits and plants. The experience comes to a close with an interactive drumming session with community members. Leaving patrons feeling refreshed and renewed.
>
> (Respondent #8, Indigenous Village Operator)

### *15.5.2 Homestays to indigenous Rastafari villages*

The respondents further shared that travellers can also visit the Rastafari Indigenous Village as part of a multiple days stay and activity-filled experience that offers a more comprehensive, deeper experience of Jamaica and its indigenous spiritual offerings.

> A visit to Nine Mile and the Bob Marley Mausoleum will facilitate sightseeing around the village, an overnight or weekend in a homestay, a healing music session, followed by hot rock massage at Bath Fountain (a nearby mineral pool) and a one-on-one 'counselling session' with a Rasta Healer. While others will have a visit to a Rastaman in the countryside with a short-term stay at his farm. Travelers are taught about fruits and herbal plants, cooking and also learn Rastafari Nyabinghi Drumming lessons with a healing drumming session.
>
> (Respondent #9, Travel Agent)

Through warm interaction and musical performances, travellers learn about the Rastafarians' spiritual belief system, traditional skills, organic farming, and the rich cultural heritage they contribute to Jamaica. Rastafarian villagers invite travellers into the daily life of their community. Travellers learn the techniques they use to build traditional drums, grow organic vegetables, make herbal remedies and medicines, and spice their traditional cooking. After a leisurely wander around the village's meditative labyrinth, they are given an herbal tea and fresh fruits while listening to the villagers play music and chant. Longer visits include treks into the countryside where travellers can take a dip in a natural pool.

### *15.5.3 Tourist behaviour, travel adoption and trends among Millennials*

The respondents were asked to provide their opinions on travel behaviour, adoption, and trends of the spiritual experiences that are to be had in the indigenous villages, especially with the interactive and hand-touch elements of the product. Their responses are documented as follows:

1) Of all the varying demographic groups that frequent these attractions, it is most popular with young people (Millennials). These Millennials admire the creativity and business acumen of these communities. They are also fond of: the respect that indigenous communities demonstrate for people and the environment, their understanding of the medicinal benefits of various herbs and spices, and of course their amazing drumming. (Respondent # 1 Destination Manager)
2) We love young people! They are open-minded and love new adventures. When they come (Millennial tourists), they are introduced to the village leaders. Some villages have over 500 residents, village residents and visitors are encouraged to interact with one another. They love talking with

the elders in the villages. It's a sight to see. As a real community of people, travellers are asked to be respectful, especially when taking photos, and millennials usually are. The ritual of smoking marijuana ('wisdom weed' or 'holy herb') has spiritual significance, and is used during communal meetings to enhance feelings of community, heighten meditation, and produce visions of a religious nature. The young people really like this part the most and are avid partakers. (Respondent #4, Indigenous Rastafari Village Host)

3) At the Rastafari Indigenous Village you learn about Rastafari customs, heritage and rich culture. We see that younger people are more interested in hearing about this way of life. Rastas live their lives in nature with knowledge of history, herbal plants and a loving heart. River/waterfall bathing and outdoor cooking are options, so is camping out and sleeping in hammocks, these offer particular intrigue to those (tourists) who have never done this before. Live chanting and storytelling in these tours allow you to mix and mingle with community members and to see some of the Rastafarians' creations and inventions. Through storytelling, we take them back in time when being a rasta was a crime and they can learn from the Rastas themselves about how this very influential group became known and impactful worldwide. We receive the most repeat business from them (millennials). They keep coming back and they bring their friends. They love it! (Respondent #9, Travel Agent)

The respondents continued by sharing that "They (the young travellers) love this as these experiences help them break away from their lives back home and just escape to a different world. It's more psychological than anything. They come to experience something new, something exotic!" (Respondent #5, Indigenous Village Host).

### *15.5.4 Impacts of COVID-19 on indigenous Rastafari villages and their interactions with travellers*

All respondents agree that COVID-19 has had a detrimental impact on tourism activities but they all agree that keeping customers safe is a top priority during this pandemic. "Face masks are required for travellers in public areas and social distancing is enforced throughout the experience. Regularly high-trafficked are sanitized using natural sanitizers like tea tree oil and other natural antibacterial plants" (Respondent #1, Destination Manager).

> Smoking is still offered but participants are asked to roll their own 'cigars' and do not share or pass it on to others. Drumming still takes place. However, these instruments are sanitized before and after use and the capacity of visitors to the villages have been reduced in accordance with the guidelines from the Ministry of Health in Jamaica. Social distancing has also had an impact of the experience.
>
> (Respondent #5, Indigenous Village Host)

## 15.6 Discussion

Spiritual tourism is a combination of spirituality and tourism. The results of this study showed that people are seeking psychological separation from their everyday life through spiritual travel practices. Through such practices people can learn, reflect and exercise spirituality in a way that may be unavailable in the home country. Spiritual tourism offers knowledge to tourists as they have the opportunity to experience a kind of engagement and illumination that they may not yet have experienced. Millennials are inclined towards these travels as unexpected solo trips or small group excursions that allow them to interact with the nature, foster an intercultural exchange, and participate in meditation, wellness and spiritual healing. They can be with themselves or have active engagement and immersion with local, indigenous spiritual practices and lifestyle as evidenced in the study conducted.

It was also evident that there continues to be a great demand among Millennials for these experiences. Tour operators of these indigenous cultural and spiritual experiences found that out of all the range of demographics that visited the cities, Millennials were the most open-minded with limited inhibitions and therefore they wanted to delve into these experiences and seemed the most satisfied with it. The research also showed that the hosts believed Millennials saw these spiritual/cultural experiences as bringing out a new side of them; at times, a connection with lost heritage and a way of escape for mundane of modern lifestyles. This form of spiritual tourism in this instance is also seen as a means of self-actualization and expansion. Another significant find was the level of repeat business from previous visitors and that most Millennials heard about these experiences through reading blogsites and word-of mouth (W-O-M) marketing. Millennials talk about their experiences with friends and share experiences with their online community via social media.

The study also produced the following significant findings:

(1) The proclivity for choosing indigenous spiritual experiences is much higher than other attractions among Generation Y (Millennials);
(2) Millennial consumers valued the hand-touch products experiences in their travel and preferred travel that allow for a more immersive experience instead of simply being an observer;
(3) Spiritual tourism may enable sustainability as the penchant for hand-touch/immersive experiences by Millennial consumers can improve the longevity of the product through the continuous infusion of live cultural elements.

## 15.7 Limitations of the study

This study is not without its limitations. The limitations of the study includes the generalizability in assessing the needs and preferences of the Millennial traveller for indigenous spiritual tourism products. While this study captured a vibrant market of millennial travellers who were interested in these experiences, the same may not be the case for all Millennial travelers. It is also noted that the Millennials discussed in the study were primarily from the United States and Canada. Differing Millennial markets may yield differing results.

Another limitation of note is that this study was conducted during the COVID-19 pandemic where usual and anticipated demand was affected by global travel restrictions. The reflections of the respondents were made based on the travel market at the usual level of demand, before the pandemic.

## 15.8 Conclusion

This chapter further establishes the fact that Millennials have a preference for travel with more immersive experiences. It illuminates that Millennials seek and opt for having an individualized experience that provides a greater sense of novelty. Through novelty and wanting new challenges to expand themselves, learn, grow and evolve, Millennial tourists frequently visit and participate in spiritual experiences and see these experiences as a means of breaking away from mundane lifestyles. The chapter helps to advance knowledge about the demand for Experiential Marketing in tourism and makes a unique contribution to literature by integrating the theory of hand-touch product design within the context of spiritual tourism.

From the information provided herein, the social and practical implications of this work are that a new narrative can be created and developed around the tourism product and experiences of Jamaica and other small island developing states and their continuous engagement with varying demographics of tourists. It is also hoped that these valuable spiritual cultural products will receive the recognition and support from governments and communities so that further development and growth can be fostered. This work contributes to advancing theories around this evolving niche and the compendium of literature for this book, aimed at filling the lacuna of research on Millennials and spirituality.

## References

Barnett, Michael (2005). "The Many Faces of Rasta: Doctrinal Diversity within the Rastafari Movement". *Caribbean Quarterly* 51(2): 67–78.

Barrett, Leonard E. (1997/1988). *The Rastafarians*. Boston: Beacon Press. ISBN 978-0-8070-1039-6.

Barton, C., Haywood, J., Jhunjhunwala, P., Bhatia, V. (2013) Traveling with Millennials. https://www.bcg.com/en-au/publications/2013/transportation-tourism-marketing-sales-traveling-millennials.

Beard, J. G., Ragheb, M. G. (1983). Measuring leisure motivation. *Journal of Leisure Research* 15(3): 219–228.

Canada Tourism Commission. (2015). Canada Millennial domestic travel summary report. http://www.destinationcanada.com/sites/default/files/201611/Programs_MillennialTravel_DomesticReport_EN.pdf.

Cavagnaro, E., Staffieri, S. (2015). Understanding millennials' tourism experience: values and meaning to travel as a key for identifying target clusters for youth (sustainable) tourism. *Journal of Tourism Futures*. doi: 10.1108/JTF-12-2017-0058.

Chawane, Midas H. (2014). The Rastafarian Movement in South Africa: A Religion or Way of Life?. *Journal for the Study of Religion* 27(2): 214–237.

Clarke, Peter B. (1986). *Black Paradise: The Rastafarian Movement*. New Religious Movements Series. Wellingborough: The Aquarian Press. ISBN 978-0-85030-428-2.

Crandall, R. (1980). Motivations for leisure. *Journal of Leisure Research* 12(1): 45–54.7.

Crocetti, E., Fermani, A., Pojaghi, B., & Meeus, W. (2011, February). Identity formation in adolescents from Italian, mixed, and migrant families. In *Child & Youth Care Forum* (Vol. 40, No. 1, pp. 7–23). Springer US.

Crompton, J. L. (1979). Motivations for pleasure vacation. *Annals of Tourism Research* 6(1): 408–424.

Dann, G. M. S. (1981). Tourist motivation: An appraisal. *Annals of Tourism Research* 8(2): 187–219.

Du, S., Zhang, H., Xu, H., Yang, J., & Tu, O. (2019). To make the travel healthier: a new tourism personalized route recommendation algorithm. *Journal of Ambient Intelligence and Humanized Computing* 10(9): 3551–3562.

Edmonds, Ennis B. (2012). *Rastafari: A Very Short Introduction*. Oxford: Oxford University Press. ISBN 978-0-19-958452-9.

Gitelson, R. J., Kerstetter, D. L. (1990). The relationship between sociodemographic variables, benefits sought and subsequent vacation behavior: A case study. *Journal of Travel Research* 28(3): 24–29.

Hill, P. C., & Smith, G. S. (2010). Coming to terms with spirituality and religion in the workplace. In *Handbook of workplace spirituality and organizational performance* (pp. 187–200). Routledge.

Hollender, J. W. (1977). Motivational dimensions of the camping experience. *Journal of Leisure Research* 9(2): 111–141. https://www.amazon.com/Understanding-Millennial-Mind-Menace-Amazing/.../09771.

http://resonanceco.com/reports/2018-future-millennial-travel-report/

https://www.aarp.org/content/dam/aarp/research/surveys_statistics/life-leisure/2017/2018-travel-

https://www.amazon.com/Understanding-Millennial-Mind-Menace-Amazing/.../09771.

https://www.forbes.com/sites/jefffromm/2018/09/05/transitioning-travel-to-the-millennial-

https://www.huffingtonpost.com/george-beall/8-key-differences-between_b_12814200.html?

http://www.jtbonline.org

https://www.tripadvisor.com/TripAdvisorInsights/tripbarometer

https://www.tripadvisor.com/TripAdvisorInsights/w620

https://www.viator.com/tours/Montego-Bay/Rastafari-Indigenous-Village-Tour/d432138355P24?mcid=56757

Ipsos Reid poll (2010). Commissioned by Deloitte and the Tourism Industry Association of Canada (TIAC).

Jamaica Tourist Board. (2019) . *Annual Travel Statistics 2018*. Kingston Jamaica.

King, Stephen A. (2002). *Reggae, Rastafari, and the Rhetoric of Social Control*. Jackson: University Press of Mississippi. ISBN 978-1-60473-003-6.

Krippendorf, J. (1987). *The Holiday Makers: Understanding the Impact of Leisure and Travel*. London: William Heinemann.

Lee, T., Crompton, J. (1992). Measuring novelty seeking in tourism. *Annals of Tourism Research* 19 (4): 732–752.

Leask, A. (2016). Visitor attraction management: A critical review of research 2009–2014. *Tourism Management* 57: 334–361.

Loker, L. E., Perdue, R.R. (1992). A benefit-based segmentation of a non-resident summer travel market. *Journal of Travel Research* 31: 30–35.

Lundin, K (2018): Post-Millennial Marketing: How to Market Effectively to Generation Z, available at: https://www.crowdspring.com/blog/marketing-generation-z/

Manfredo, M. J., Driver, B. L., & Tarrant, M. A. (1996). Measuring leisure motivation: A meta-analysis of the recreation experience preference scales. *Journal of Leisure Research* 28(3): 188–213.

Margolin, Madison (August 20, 2018). Now Decriminalized, Could Jamaica Become Destination for Legal Weed? *Rolling Stone*. Retrieved October 9, 2020 https://www.rollingstone.com/culture/culture-features/jamaica-legal-weed-rastafari-criminalized-711745/.

McCole, P. (2004). *Refocusing marketing to reflect practice: The changing role of marketing for business. Marketing Intelligence & Planning* 22(5): 531–539.

Mosammam, H. M., Sarrafi, M., Nia, J. T., & Heidari, S. (2016). Typology of the ecotourism development approach and an evaluation from the sustainability view: The case of Mazandaran Province, Iran. *Tourism Management Perspectives* 18: 168–178.

Pearce, P. L. (1988). *The Ulysses Factor: Evaluating Visitors in Tourist Settings*. New York: Springer-Verlag.

Pearce, P. L. (1991). Travel stories: An analysis of self-disclosure in terms of story structure, valence, and audience characteristics. *Australian Psychologist* 26(3): 172–174.36.

Pine II, B. J., Gilmour, J. (1999). *The Experience Economy*. Boston, MA: Harvard Business School Press.39.

Princeton One (2017): Understanding Generation Y, What you need to know about the Millennials. www.princetonone.com/news/PrincetonOne%20White%20Paper2.pdf.

Raj, R., & Morpeth, N. D. (2007). Introduction: establishing linkages between religious travel and tourism. *Religious Tourism and Pilgrimage Festivals Management: An International Perspective*, 1–14.

Richards, R. J. (2006). Ernst Haeckel and the struggles over evolution and religion. *Annals of the History and Philosophy of Biology* 10: 89–115.

Rinschede, G. (1992). Forms of religious tourism. *Annals of Tourism Research* 19(1): 51–67.

Schmitt, B.H. 1999 *Experiential Marketing: How to Get Consumer to Sense, Feel, Think, Act, and Relate to Your Company and Brands*; New York, NY: Free Press.

Shoemaker, S. (1989). Segmenting the US travel market according to benefits realized. *Journal of Travel Research* 28: 8–21.

Sibanda, Fortune (2016). One Love, or Chanting Down Same-Sex Relations? Queering Rastafari Perspectives on Homosexuality. In Adriaan van Klinken, Ezra Chitando (eds.). *Public Religion and the Politics of Homosexuality in Africa*. Abingdon and New York: Routledge. pp. 180–196. ISBN 978-1-317-07342-0.

Slivar, Iva, Aleric, Dražen, Dolenec, Sanja. (2019). Leisure travel behavior of generation Y & Z at the destination and post-purchase. *E-Journal of Tourism*. 147. doi: 10.24922/eot.v6i2.53470.

Sofronov, Bogdan, (2018). Millennials: A New Trend for the Tourism Industry Annals of Spiru Haret University. *Economic Series* 18(3): 109-122, doi: 10.26458/1838.

Troeger, C., Blacker, B., Khalil, I. A., Rao, P. C., Cao, J., Zimsen, S. R., ... & Reiner Jr, R. C. (2018). Estimates of the global, regional, and national morbidity, mortality, and aetiologies of lower respiratory infections in 195 countries, 1990–2016: a systematic analysis for the Global Burden of Disease Study 2016. *The Lancet Infectious Diseases* 18(11): 1191–1210.

Woodward, S. C. (2004). Faith and tourism: Planning tourism in relation to places of worship. *Tourism and Hospitality Planning & Development* 1(2): 173–186.

WYSE Travel Confederation/UNWTO. The Power of Youth Travel, 2011. *The UNWTO defines youth travel to include all independent trips for periods of less than one year by people aged 16-29*. Youth travel forecast prepared by the UNWTO.

# 16 Imminent future of Millennial generations

## A tourism management perspective

*Sandeep Kumar Walia and Aruditya Jasrotia*

### 16.1 Introduction

According to the Urban Dictionary, "millennial is an identity given to a broadly and vaguely defined group of people." We can divide this group into two parts: Generation Y and Generation Z, which are often at odds with each other. Generation Y "grew-up on personal computers, cell phones and video game systems", while Generation Z "has grown up on tablets, smartphones, and apps". Even though, it is difficult to identify a strong differentiation between these two age groups, it can be said that the Millennial generation is somewhere born between 1981 and 2001. Both Generation Y and Generation Z have been transmuting communication and individuality on an international scale. The Millennials are showcasing the trend of delaying the completion of their studies, starting careers, getting married and starting a family. The Millennials desire a purposeful life and intend to make a difference in the world, which is completely opposite of the 'Baby Boomers' who prioritize education, getting a job and getting married. In other words, Millennials prefer to get involved in unique experiences, like going to far-flung areas or interacting with local communities (Veríssimo & Costa, 2018).

The Gen Y or Generation Y or the Millennial generation, is a section of customers who are born and raised in the fast-changing millennium era (KPMG, 2017). Moreover, travel and tourism are extremely crucial for the Millennial generation and their top priority is travelling to new places (Cavagnaro, Staffieri, & Postma, 2018; OECD, 2018). Millennial generation travels more frequently than any other generation, and there is more probability of them travelling as they have more money and disposable income to spend (Nielsen, 2017). The major source market for Millennials travelling outbound is countries like the USA, China, the UK and Germany (ITB World Travel Trends, 2019). The Millennials from Europe take at least 4–5 trips every year, which makes them a substantial economic generator in the tourism industry (Cavagnaro et al., 2018). Additionally, according to an estimate, Millennials travellers are 40 percent of the Europe's total outbound travel, which is the largest age group for international travel (ITB World Travel Trends, 2019). Hence, it's clear that today's Millennials are a key sector of the tourism industry and their travel behaviours bring forward remarkable

DOI: 10.4324/9781003170631-20

fluctuations in the tourism market (Ketter, 2019; OECD, 2018). There are somewhat diverse views concerning the assortment of years in which Millennials were born. One viewpoint is given by Cavagnaro et al. (2018), which refers to the year 1980 to the mid-1990s. Likewise, a report by the OECD proposed that Millennials are born in the initial 1980s to the mid-1990s, and another report by KPMG (2017) defined the years of birth of Millennials as between 1980 and 1995. Other standpoints defined this range as between 1980 and 1999 (Garikapati, Pendyala, Morris, Mokhtarian, & McDonald, 2016). According to Skinner, Sarpong, and White (2018), where many definitions refer to the birth year of the Millennial era as 1980, many researchers point out it as 1977. Consequently, it is clear that there is some kind of overlapping between younger Millennials and the older Generation Z, which is born in the late 1990s. As there is an increase in the number of Millennials, this present study aims to shed light on their characteristics and travel behaviours, and how their travel trends are shaping the future of the tourism industry.

It is apparent from the studies that young tourists are crucial for tourism's future because of the following reasons; the high number of Millennial tourists travelling throughout the world in this era (Richards, 2006), the fact that their behaviour and travel choices can create new attitudes and trends in the tourism industry (Crocetti et al., 2011; Leask, Fyall, & Barron, 2013) and, last but not least, the projection that young tourists will keep travelling in the future (Barton, Haywood, Jhunjhunwala, & Bhatia, 2013). The last reason is of the utmost importance here from the future perspective, as the young generation will take the place occupied by older generations in future in the tourism industry. These days, the young generation (Gen Y, Gen X and the Millennial generation) is showcasing diverse wants and needs which are quite different than older generations in general (Howe & Strauss, 2000) and in tourism in particular (Glover, 2010). In order to be ready for the future challenges and opportunities, the tourism industry needs to understand and accommodate these changing needs. Consequently, this study aims to shed light on the features and travel behaviours of Millennials, and how their travel trends are going to shape the future of the tourism industry across the globe. Moreover, this study also attempts to outline young tourists that can be approached by destination management organizations in the future.

## 16.2 Literature review

The researchers in the tourism domain have started to give attention to on young travellers in general and especially on those travellers who are born between 1980 and 2000, the so-called "Millennial" Generation (Glover, 2010; Pendergast, 2010; Richards and Wilson, 2006). This has led to the concept of Youth Tourism, which refers to the sovereign trips taken by people of age group of 16–29 years which are driven by the urge to understand and indulge in other cultures/religions, gain unique life experience and take advantage from different learning prospects available in the tourism destinations, for a time less than one year (UNWTO and WYSE Travel Confederation, 2008). The age group which falls under this definition matches the Millennial generation. As per most of the definitions, the Millennial generation is born between 1980 and 2000. The Millennials can all be called as youth travellers and their travel involvement can therefore be understood with the help of literature on youth tourism. The UNWTO and World Youth Student & Educational Travel Confederation mentioned that Generation Z constitutes around 30 percent of the global population, represent 29 million international travellers from all over the world. The Generation Z is considered the generation with the Internet in their pockets, this characteristic differentiates them from other generations such as 'Baby Boomers' and even than the older Generation Y born before the invention of the Internet (WYSE, 2016).

Even though there is a rise in the attention on the Millennial generation, prevailing research on youth tourism is comparatively underdeveloped (Richards and Wilson, 2006; Staffieri, 2016). This is astonishing for the reason that young tourists have a considerable material and tremendous influence on the present and future of the tourism industry. Primarily, they epitomize an increasingly noteworthy economic strength, as in the year 2015, almost one 23 percent of global tourists were aged between the range of 16–29; one in every three people staying in the hotels were Millennials and the total worth of international youth tourism was expected to reach US$400 billion by the year 2020, which is almost twice the value of year 2009 (UNWTO and WYSE Travel Confederation, 2016). Secondly, in comparison with the elder generations, young people are more robust as they have a tendency to visit destinations that are under the influence of environmental and socio-political pressure (UNWTO and WYSE Travel Confederation, 2016). Young tourists tend to focus more on the experiences and activities on the destinations and adjust on the transportation and accommodation part. Richards (2011) mentioned that on a significant trip young people employ an average of US$2,600, which is nearly three times what an average tourist spends on a trip. Hence, Millennials epitomize a significant economic opportunity for the tourism industry in general and for economically and politically delicate destinations in particular.

From the socio-cultural perspective, it has been noted that youngsters are a ground-breaking power who, with their travel selections, might lead to new tourism trends and styles to be followed by the much broader society (Martinengo & Savoja, 1993; Fermani et al., 2011; Cavagnaro & Staffieri, 2015). Consequently, variations and progresses in tourist behaviour in future can be justified by looking at the current travel behaviour of millennials (Leask et al., 2013). Millennials are not just pertinent for their existing influence on tourism industry, but they are undoubtedly the tourists of the future (Pendergast, 2010; Richards and Wilson, 2006; UNWTO and WYSE Travel Confederation, 2008). Millennials are also portrayed as natural travellers as more than 60 per cent of millennials look at travel as a significant fragment of their lives. They go on four to five trips annually and are anticipated to seek even more extensive tourism experiences in the old age (Barton et al., 2013; Ovolo Hotels, 2013). Slowly, the Millennial generation is moving upwards and replacing the older generations on the population pyramids. In order to prepare for the future, the tourism industry needs to start with manufacturing such products and services which are future-proof, while keeping in mind the choices of the Millennial generation. As with the young tourists of today, the middle-aged tourist in the 2020s and 2030s will have entirely diverse needs and wants. Considering all of these facts it is clearly necessary to investigate this target group and to identify their tourism behaviour, such as the values and beliefs that matter to them.

## 16.3 Social media as an influence

The biggest question which the tourism industry faces today is: what motivates Millennials to travel and how will they make their place in the tourism industry in future? The key reason is that the social media has attracted young generations from across the world. Moreover, social media also provides opportunities for accessing numerous applications which makes it easy to travel and search information online. Millennials are also very eager to see the world and cannot until their retirement age. Millennials have become doubtful about traditional retirement plans and they do not want to be dependent on an uncertain future. Furthermore, travelling also provides access to numerous prospects for progress and growth in a specific job or in studies. Students and job-seekers travel to reflect on and assess if their education, job, and life decisions were right, and what can they do to be happy in life (Machado, 2014). Pleasure and happiness are the principal factors which distinguish Millennials and Generation X. It seems that Millennials want to follow a deeper understanding of their life and feel happy with their jobs. Because of advances in technology, not only can Millennials explore the world, but they can also work while travelling.

> Digital nomads, also known as location independent workers, are still small in numbers, but the travel industry has taken note. […] Their impact on some destinations is noticeable: more-co-working spaces are springing up and digital nomads have an influence on other young travellers through blogging travel and lifestyle advice.
>
> (Mohn, 2017)

Making new connections online and staying connected with friends and family are one of the most significant priorities for most of today's young travellers. As mentioned by Shallcross (2015), Millennials have access to a big chunk of applications and they use information technology to get in touch with their global audience which is an important fragment of their complete travel experience. Recognized as extremely active social media users, Millennials consider mobile applications and the Internet to be an essential portion of their travel experience. They desire to travel solo or with friends, as an alternative for family or a partner (Veríssimo & Costa, 2018). Today's travellers upload pictures and videos on social media websites such as Facebook, YouTube and Instagram. They also like to write and record vlogs about their travel experiences, tricks and tips. By looking at these websites, blogs, social media groups, and pages Millennials become enthused and knowledgeable about tourism destinations. Millennials like to share tips, guidance and the best places to visit. There are also several websites, for example, 'Itineree' and 'Triphackr', which provide travel hacks (Harrison, 2017). These days, travellers depend on peer-to-peer photographic user-generated data to enjoy their vacations and get the best insider tips from travel influencers of their own age.

## 16.4 Experiencing local culture

Millennials are the actual explorers and adventure-oriented souls. They are continuously and regularly looking for travel content on the Internet which allows them to find out about the destinations to which they would like to travel in future. They are also interested in collecting material and information on the local ethnicities, local traditions, local cuisines and intend to discover each and every part of them. Millennials gather all of the necessary information with the help of the Internet and social media like YouTube and Facebook where numerous travel blogs are uploaded every day, by travel influencers and enthusiasts. A lot of travellers, upload numerous videos and photographs that showcase the cultural values followed in a particular area or city and a diverse set of gastronomies that are offered and served by the local people. Moreover, locals also offer various adventure tour packages to the Millennials, who are seeking something new and adventurous during their accommodation bookings. Millennials these days are motivated to 'Gain a feeling of belongingness', desire to 'interact with local people' and to 'meet fellow travellers'. This allowed us to distinguish between friendships built with people living in the destination and those forged with fellow travellers. This provides a measure of the extent to which the traveller wants to escape the 'tourist bubble' or enclave (Wilson & Richards, 2008) and make contact with locals. Further, Millennials are looking for the opportunities to improve and develop their creativity and are looking for innovative and creative styles of travel (Richards, 2011). It is now evident that the Millennial travellers are no longer going to Southeast Asia, Europe and South America for leisure and for enjoyment. On the other hand, they are looking for authentic experiences, uniqueness and cultural immersion in order to achieve their personal as well as professional goals. Several studies have showcased that the primary purpose of the Millennials is to travel to experience new cultures and traditions.

Millennials are inclined towards trying out the local cuisines and drinks and they consider this experience as an imperative portion of their travels, which is debatably a part of an areas or city's culture. There is also a new trend in the tourism industry, known as the "experiential tourism", in which the connections and communication between the local residents and the tourists are a vital part of the overall tourism experience and the customer satisfaction. New-fangled technologies and information technology-based applications are enabling tourists to plan and design their own itineraries. There are several companies like TripAdvisor, Booking.com and Airbnb, which are dominating the tourism industry. This has led to the explosion of new trends in cultural tourism, representing how and why culture has emerged as a dominant factor in overall tourism experiences.

## 16.5 Discussion and conclusion

In future, Millennial travellers are expected to have an immense influence on the travel and tourism industry as compared to any other group. Their prominence in the tourism industry will rise continuously and it is predicted that their children are also likely to follow their steps and indulge in the same travel trends. Furthermore, adventure tourism activities, visits to rural destinations, and online trip designing/planning are anticipated to increase and dominate the kingdom of tourism. Similarly, there will be an increasing inclination of travelling to foreign destinations, solo travelling, as well as more indulgence of social media in the travel and tourism industry. Additionally, Millennial travellers are also contributing towards rising trends in the use of the Internet of Things enabled travel solutions for the hotel industry. The prime reason behind this is that Millennial travellers want faster and improved services. Regarding this, hotels are now moving towards mobile applications, artificial intelligence, Internet-enabled products and services in hotels in order to provide an improved service to the customers. Continuous updates in technology and new tourism trends can enable the hotel industry to prepare for the future and to recognize budding business opportunities. Millennials are also relying on technology for planning the entire trips staring from the checking reviews online, booking the hotels, booking flight tickets, web check-in, to staying connected to friends and family. Hence, if the entire business travel ecosystem can be completely unified and technology-supported, the Millennial traveller experience will be enhanced significantly. It is now becoming apparent that numerous Millennials are moving out of their parents' house at an early stage, buying vehicles at a young age, postponing marriage and family plans, and prioritizing travel over work. Millennials are at present shaping the world's changing demographics, which will affect future travel trends, tourist behaviours and their consequences on the tourism industry. The exceptional habits and desires of the increasingly significant millennial generation will impact and shape the travel industry in the future. It is going to be exhilarating to witness how the tourism industry will incorporate technology and innovativeness to attend Millennials in the future.

## References


Barton, C., Haywood, J., Jhunjhunwala, P., & Bhatia, V. (2013). *Traveling with Millennials*. Boston, MA: The Boston Consulting Group (BCG).

Cavagnaro, E., & Staffieri, S. (2015). *A study of students' travellers values and needs in order to establish futures patterns and insights*. Journal of Tourism Futures.

Cavagnaro, E., Staffieri, S., & Postma, A. (2018). Understanding millennials' tourism experience: values and meaning to travel as a key for identifying target clusters for youth (sustainable) tourism. *Journal of Tourism Futures*, Vol. 4 No. 1, pp. 31–42.

Crocetti, E., Fermani, A., Pojaghi, B., & Meeus, W. (2011, February). Identity formation in adolescents from Italian, mixed, and migrant families. *Child & Youth Care Forum*, Vol. 40, No. 1, pp. 7–23. Springer US.

Garikapati, V.M., Pendyala, R.M., Morris, E.A., Mokhtarian, P.L., & McDonald, N. (2016), Activity patterns, time use, and travel of millennials: a generation in transition?, *Transport Reviews*, Vol. 36 No. 5, pp. 558–84.

Glover, P. (2010), Generation Y's future tourism demand: some opportunities and challenges, in Benckendorff, P., Moscardo, G. and Pendergast, D. (Eds), *Tourism and Generation Y*, Cambridge, MA: CAB International, pp. 155–63.

Harrison, D. (2017). Tourists, mobilities and paradigms. *Tourism Management*, Vol. 63, pp. 329–337.

Howe, N., & Strauss, W. (2000). *Millennials rising: The next great generation*. Vintage.

ITB World Travel Trends 2018–2019 (2019), "What are the trends to look out for? ITB and IPK travel trends report", March 5, available at: www.itb-berlin.de/media/itbk/itbk_dl_all/ITB_2019_WTTR_Factsheets_ALLE_Web_4.pdf.

Ketter, E. (2019), Eating with EatWith: analysing tourism-sharing economy consumers, *Current Issues in Tourism*, Vol. 22 No. 9, pp. 1062–75.

KPMG (2017), "Meet the Millennials", June 1, available at: https://home.kpmg/content/dam/kpmg/uk/pdf/2017/04/Meet-the-Millennials-Secured.pdf.

Leask, A., Fyall, A., & Barron, P. (2013). Generation Y: opportunity or challenge–strategies to engage Generation Y in the UK attractions' sector. *Current Issues in Tourism*, Vol. 16 No. 1, pp. 17–46.

Machado, A. (2014). How Millennials are changing travel. *The Atlantic*, 18.

Martinengo, M. C., & Savoja, L. (1993). *Giovani e turismo*.

Mohn, T. (2017). The digital nomad life: Combining work and travel. *The New York Times*, Vol. 3, available at: https://mobile.nytimes.com/2017/04/03/business/digital-nomads-work-tourism.html..

Nielsen (2017), "Young and ready to travel: a look at millennial travelers", January 1, availableat: www.nielsen.com/wp-content/uploads/sites/3/2019/04/nielsen-millennial-traveler-study-jan-2017.pdf.

OECD (2018), *OECD Tourism Trends and Policies 2018*, Paris: OECD Publishing, available at: https://doi.org/10.1787/tour-2018-en.

Ovolo Hotels (2013), "*Millennial marketing in hospitality*". Presentation given at *UNWTOs 5th global summit on city tourism*", Luxor, 1–2 November 2016, available at: http://cf.cdn.unwto.org/sites/all/files/pdf/3_3_amadou_doumbia-ovolo_hotels-millennial_marketing_in_hospitality.pdf (accessed 11 February 2018).

Pendergast, D. (2010). Getting to know the Y generation. *Tourism and generation Y*, Vol. 1, pp. 1–15.

Rauch, J. (2018). *The happiness curve: Why life gets better after midlife*. Bloomsbury Publishing.

Richards, G. (2011). Creativity and tourism: The state of the art. *Annals of Tourism Research*, Vol. 38, No. 4, pp. 1225–1253.

Richards, G., & Wilson, J. (2006). Youth and adventure tourism. *Tourism Business Frontiers: Consumers, Products and Industry*, 40–47.

Richards, K. (2006). *Language and professional identity: Aspects of collaborative interaction*. Springer.

Shallcross, J. (2015). *How Millennials are traveling better than you.* Conde Nast Traveler.

Skinner, H., Sarpong, D. and White, G.R. (2018), "Meeting the needs of the Millennials and Generation Z: gamification in tourism through geocaching", *Journal of Tourism Futures*, Vol. 4 No. 1, pp. 93–104.

Staffieri, S. (2016). "L'esperienza turistica dei giovani italiani". Retrieved from http://www.editricesapienza.it/sites/default/files/5411_Staffieri_EsperienzaTuristicaGiovaniItaliani.pdf.

United Nations World Tourism Organisation and World Youth Student & Educational Travel Confederation (UNWTO and WYSE Travel Confederation) (2008), *Youth travel matters: understanding the global phenomenon of youth travel*, Madrid: UNWTO.

United Nations World Tourism Organisation and World Youth Student & Educational Travel Confederation (UNWTO and WYSE Travel Confederation) (2016), *Affiliate members global reports, volume thirteen – the power of youth travel*, Madrid: UNWTO.

Veríssimo, M., & Costa, C. (2018). Do hostels play a role in pleasing Millennial travellers? *The Portuguese case. Journal of Tourism Futures*, Vol. 4 No. 1, pp. 57–68.

Wilson, J., & Richards, G. (2008). Suspending reality: An exploration of enclaves and the backpacker experience. *Current Issues in Tourism*, Vol. 11 No. 2, pp. 187–202.

WYSE (2016), "*Student and youth travel in cities*", Presentation given at *UNWTOs 5th Global Summit on City Tourism*, Luxor, 1-2 November.

# Index

Note: Page numbers in **bold** indicate tables, page numbers in *italic* indicate figures.